Dynamics of Internal Gravity Waves in the Ocean

ATMOSPHERIC AND OCEANOGRAPHIC SCIENCES LIBRARY

VOLUME 24

Editor-in-Chief

Robert Sadourny, *Laboratoire de Météorologie Dynamique du CNRS,*
École Normale Supérieure, Paris, France

Editorial Advisory Board

Dynamics of Internal Gravity Waves in the Ocean

by

Yu. Z. Miropol'sky †

Formerly of P.P. Shirshov Institute of Oceanology,
Russian Academy of Sciences,
Moscow, Russia

translated and edited by

O.D. Shishkina

Institute of Applied Physics,
Russian Academy of Sciences,
Nizhny Novgorod, Russia

KLUWER ACADEMIC PUBLISHERS
DORDRECHT / BOSTON / LONDON

Library of Congress Cataloging-in-Publication Data

Miropol'skii, IU. Z. (IUrii Zakharovich), 1943-
 [Dinamika vnutrennikh gravitatsionnykh voln v okeane. English]
 Dynamics of internal gravity waves in the ocean / by Yu. Z. Miropolsky ; edited and
translated by O.D. Shishkina.
 p. cm. -- (Atmospheric and oceanographic sciences library ; v. 24)
 Includes bibliographical references and index.
 ISBN 0-7923-6935-1 (alk. paper)
 1. Internal waves. 2. Gravity waves. 3. Fluid dynamics. I. Shishkina, Ol'ga Vasil'evna.
II. Title. III. Series.

GC211.2 .M5713 2001
551.47'02--dc21
                                                                    2001029313
ISBN 0-7923-6935-1

Published by Kluwer Academic Publishers,
P.O. Box 17, 3300 AA Dordrecht, The Netherlands.

Sold and distributed in North, Central and South America
by Kluwer Academic Publishers,
101 Philip Drive, Norwell, MA 02061, U.S.A.

In all other countries, sold and distributed
by Kluwer Academic Publishers,
P.O. Box 322, 3300 AH Dordrecht, The Netherlands.

This work is an updated translation of
Dinamika Vnutrennikh Gravitacionnykh voln V Okeane.
Leningrad: Gidrometeoizdat, 1981

This book is dedicated
to the great worldwide
efforts of the numerous
army of theoretical and
experimental
researchers in the field
of physical
oceanography

Contents

Foreword

The author of this book, Yuri Zakharovich Miropol'sky, will never see it. An aggressive disease cut the life of this young and talented person. Almost up to the last hour of his short life he continued working on the manuscript.

Yuri Zakharovich Miropol'sky was born in June 1943. In 1966 he graduated from the Moscow Physical–Technical Institute. All of his scientific activity was connected with the Shirshov Institute of Oceanology of the Russian Academy of Sciences. Being already sick, he defended his professorial thesis in 1977.

The scientific interests of Prof. Miropol'sky were connected basically with two of the most important problems of oceanology—a theory of the upper quasi-homogeneous ocean layer and a theory of internal waves. One of the first nonstationary models of the upper quasi-homogeneous layer belongs to Prof. Miropol'sky and his colleagues. The authors of this model pointed out for the first time an important property of self-similarity of temperature distribution in the upper thermocline. A study of these problems was the last point in Prof. Miropol'sky's scientific life: on the basis of the revealed self-similarity he developed an elegant theory describing the seasonal evolution of the upper thermocline in the ocean.

Studies by Prof. Miropol'sky reflect almost all of the aspects of internal waves problems. He investigated the problems of effectiveness of the generation of internal waves by atmospheric pressure and by a wind field; questions of propagation of small amplitude internal waves from local initial disturbances (the Cauchy–Poisson generalized problem) and of small amplitude waves in the presence of horizontal inhomogeneity of the density field in the ocean, as well as the influence of the fine vertical structure on the internal waves' propagation. A number of works contain a nonlinear theory of oceanic internal waves. There are the problems of the propagation of nonlinear stationary internal waves in shallow water and in thin stratified layers of the deep ocean. A nonstationary theory of weakly nonlinear internal waves built by Prof. Miropol'sky and his colleagues appeared to be extremely important.

Yu. Miropol'sky was very interested in experimental investigations, natural measurements, and data processing, and worked on numerous expeditions. He took part in the creation of interpretation methodology of measurements of temperature and velocity fluctuations in the ocean, which allowed one to obtain both dispersion relations and the most important statistical characteristics of the internal waves field. He studied in detail the fine structure of the temperature field and pointed out a number of its universal properties. These and other results are presented in the book.

Colleagues of Prof. Miropol'sky esteemed any possibility of scientific cooperation with him and loved him very much. As are many gifted persons, he was talented in several fields: he was highly educated, had an outstanding sense of humor, and a real charm. In the memory of all who knew him he will always be a bright, generous, and merry person.

Andrey S. Monin
Corresponding Member of the Russian Academy of Sciences
Shirshov Institute of Oceanology,
Russian Academy of Sciences,
Moscow

Preface

The book *Dynamics of the Internal Gravity Waves in the Ocean* by Yu. Miropol'sky considers modern theoretical methods of investigating wave motions, results of their application to internal waves problems, as well as ocean observational data. Nonlinear effects and their connection with the formation of oceanic fine structure of the density vertical distribution are analyzed. It interprets oceanic observational data and supplies methods of distinguishing internal waves and turbulence.

The book is meant for specialists in physics of the ocean, oceanography, geophysics, and hydroacoustics.

In writing the monograph the author wanted to create a systematic interpretation of the theoretical and the most practical experimental aspects of the dynamics of internal waves in the Ocean. The author, drawing the reader's attention first of all to the physical effects important from an oceanological standpoint, nevertheless was inspired by the beauty of a mathematical description. He distinguished ideas popular among oceanographers but unproved by theory or by experiment, from quantitative theoretical conclusions and observational data.

A secondary target set by the author, giving some detailed theoretical results, was the fast introduction of an inexperienced reader to the range of modern ideas and methods in the study of wave processes in dispersive media.

A number of books in the Western literature have considered internal gravity waves. Some fragmentary data on internal waves are summarized in corresponding chapters of the books by Phillips *The Dynamics of the Upper Ocean* [299], by Turner *Buoyancy Effects in Fluids* [367] and by Monin, Kamenkovich and Kort *The Variability of the Ocean* [262], which present only qualitative pictures of some particular aspects of the problem. Despite the appearance of the detailed book by Whitham *Linear and Nonlinear Waves* [391], internal waves—due to their anisotropy and nonuniformity—present the non-trivial object of applying of these methods.

Specialists in this scientific field have to rely basically on what has been written on this in parts of more general text books by Tolsoy *Wave Propagation* [361], by LeBlond and Mysak *Waves in the Ocean* [201], by Yih *Stratified Flows* [403], some passages in text books by Lighthill *Waves in Fluids* [212], which focuses on the theory of internal wave generation and propagation, *Atmosphere–Ocean Dynamics* by Gill [100], which focuses on the role of internal waves in atmospheric and oceanic processes, and by Craik *Wave Interactions and Fluid Flows* [55] discussing the topics of wave interaction phenomena in fluids at rest and shear flows.

Few books dealing entirely with internal waves are known: *Interne Wellen* by Krauss (in German) [179], *Internal Gravity Waves* by Olbers [277] and Munk *Internal Waves* [268], the latter is more like an extensive literature review, as well as the papers [83], [95], [210].

However, all books published in this topic in the Western literature view internal waves only as special cases within a more general framework, so that their different aspects are covered only partially.

Since the point at which a sufficiently complete state of the physical theory of internal waves appeared in journals and comprehensive experimental data were accumulated, there has been a need for writing this monograph, in which a wide knowledge on the internal ocean waves is generalized and systematized.

The book starts with a systematic statement of principles of linear (Part 2) and nonlinear (Part 3) internal waves theory after a brief description of principles of the Ocean's thermohydrodynamics (Part 1). It gives a clear notion of the internal waves amongst other oceanic wave motions and of the degree of approximations permissible for the theoretical description of the internal waves. Unfortunately the limited content of the monograph has not allowed us to give as detailed an account of the available experimental data (Part 4) as the author desired.

Nonlinear effects in internal waves propagation (Part 3) are studied by the Hamiltonian formalism, though many of the results in the nonlinear theory of internal waves are obtained by different asymptotic decompositions directly from the initial hydrodynamic equations. Application of the Hamiltonian formalism method gives the theory of nonlinear internal waves a required generality and clarifies the connection between the theory of internal waves and the general physical theory of wave propagation in dispersive media. From the methodical viewpoint, using the Hamiltonian formalism in the present monograph is described thoroughly in Chapter 7 in the terms of some simple mathematical procedures.

Of course, the book cannot capture all of the numerous works dealing with internal waves. Therefore the author has tried to give the most detailed information only on those aspects of the physics of internal waves which were based on his personal experience or which he considered especially important. The other problems are supplemented by a quite detailed bibliography.

Some of the results presented were obtained by the author in collaboration with A.I. Leonov, A.G. Voronovich, B.N. Filyushkin, R.E. Tamsalu and Yu.D. Borisenko. Chapter 7, containing the method of the Hamiltonian formalism in the internal waves theory, and sections 9.1, 9.4, 10.1–10.3 were written by A.G. Voronovich. Experimental data and the references on the main topics were updated by the interpreter.

Acknowledgments

The translator of this book is grateful to Dr. Leo Maas (Netherlands Institute for Sea Research, The Netherlands) and to Dr. Bruno Voisin (Laboratoire des Écoulements Géophysiques et Industriels, CNRS–UJF–INPG, France) for the initiation of this translation and their permanent further assistance.

Technical preparation of the copy ready version was supported by the Russian Foundation for Basic Research under grant no. 99-05-64394.

Introduction

Investigations in the field of internal gravity waves has occupied one of the central places in oceanology in recent years. The topicality of the problems is caused first of all by propagation of internal wave motions in the entire depth of the World Ocean, and, in this connection, by their important role in all oceanic dynamic phenomena.

The practical importance of internal waves is evident for hydroacoustics, submerged navigation, hydrobiology, hydro-optics and marine mud formation. Particular recent interest in internal waves study has been caused by their great, and perhaps decisive, role in the processes forming horizontal and vertical exchange in the Ocean. In this connection investigations of internal gravitational waves, their appearance, propagation, and especially problems of unsteadiness and breaking, become decisive for understanding ocean dynamics.

The main factor forcing internal waves in the Ocean is a stable stratification according to wave density increase towards the gravity direction, i.e., from the upper surface to the ocean's bottom. Such stratification stability takes place everywhere in the Ocean and is connected first of all with the free water surface heated by the solar radiation. Only polar zones and temperate latitudes in winter time, where cooling from above is possible, may be considered as exceptions. But evenin these cases neutral and stable stratification is observed. The sea water density ρ depends not only on the temperature T but also on the salinity s, which is why, besides thermodynamic processes, evaporation and atmospheric

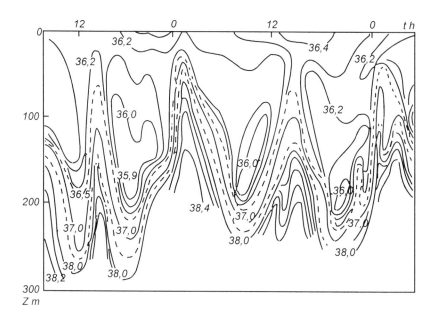

Figure 0.1.   Oscillations of the salinity ($^o/_{oo}$) in the Straits of Gibraltar, 16–18 May 1961 [26].

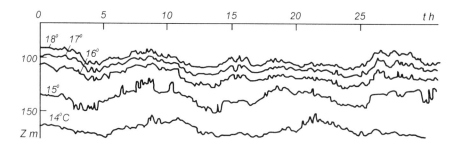

Figure 0.2.   Depth variation of isotherms in the Atlantic Ocean [249].

precipitation processes also take part in the stratification formation. Nevertheless, thermal processes are evidently dominant.

Let us consider that some disturbance has appeared in a fixed stable stratified medium and has displaced liquid particles there from their mechanical equilibrium state. Then, under influence of the gravity and of the buoyancy forces aimed at the equilibrium state restoration, os-

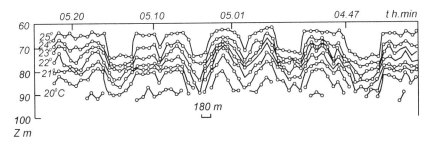

Figure 0.3. Depth variations of isotherms in the Atlantic Ocean caused by short period internal waves [276].

cillations in the fluid volume will start. Those oscillations present a phenomenon of internal waves propagating everywhere in the Ocean.

The parameters of ocean waves vary within a rather wide range. Low frequency internal waves may have lengths of tens and hundreds of kilometers, and propagation velocities up to several meters per second. Short period internal waves with periods from 5–10 min to 2–5 hrs. have lengths from several hundred meters to kilometers and propagation velocities of the order of several tens of centimeters per second. Amplitudes of low frequency internal waves may reach 100 m (Fig. 1). Short period internal waves evidently have amplitudes lower than 10–20 m.

Internal waves observed in the Ocean have different shapes. Long enough internal waves may be almost sinusoidal (Fig. 2). Shorter internal waves frequently have a shape different from sinusoidal and preferably propagate as wave trains (Fig. 3). Sometimes, especially in shallow water regions or in the coastal zone, solitary internal waves are observed (Fig. 4).

As a rule internal waves, having maximum amplitudes at some depth, cause only insignificant displacements at the free surface. But though these displacements are small they are not zero: seismographic study at the Arctic ice revealed oscillations with periods typical for internal waves (Fig. 5). Internal waves passing inside the ocean water are also confirmed by horizontal lines and highlights seen from satellites and airplanes (see, e.g., [6]).

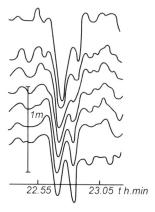

Figure 0.4. A solitary internal wave observed in the Caspian Sea at the distance of 40 km from the coast at the total depth of 56 m [36].

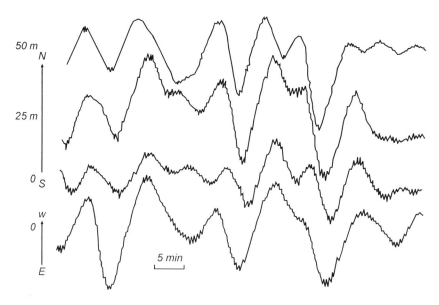

Figure 0.5. The oscillogram of ice oscillations caused by the passage of internal waves under the ice cover (the drifting station SP-20 [332]).

I
PRINCIPLES OF THERMOHYDRODYNAMIC DESCRIPTION OF INTERNAL GRAVITY WAVES IN THE OCEAN

Chapter 1

BRIEF INFORMATION ABOUT OCEANIC THERMOHYDRODYNAMICS

In this chapter the principles of oceanic thermohydrodynamics are presented briefly. For the relatively short internal waves with length scales less than 100 km, considered in this book, it is obviously possible to neglect the radius of the Earth restricting the local description.

This chapter contains basic information about oceanic thermodynamics and the notion of stratification. Adiabatic approximation is discussed quite thoroughly in the present chapter, since it is the basis for the study of various wave ocean motions. In conclusion, a brief classification of ocean waves is presented which allows us to reveal the place of internal waves amongst other wave motions.

1. SEA WATER THERMODYNAMICS. STRATIFICATION

Before the consideration of dynamic and non-equilibrium effects a brief discussion of the basis of the equilibrium thermodynamics of sea water will be useful.

Sea water presents a compound multi-component medium which may be considered as the two-component system 'water + salt'— only as an assumption (quite enough for this book's goals).

The sea water's density ρ depends not only on the pressure p and the temperature T, but also on the salinity s, defined as a salt concentration in the water. The dependence $\rho = \rho(T, p, s)$, called the equation of state, plays an important role in the formulation of hydrodynamic equations. For the Ocean the ranges of variation of state variables is as follows: for the temperature, from 4 to $40°C$; for pressure, from 0 to 1000 bar; for salinity[1], from 0 to $40\,^0/_{00}$; here the sea water density varies from 1.00 to 1.04 g $\times$ cm^{-3} accordingly.

At present, a rather reliable empirical equation of state for sea water in the mentioned range of variables T, p, s [48], [79], [246] is proposed and will be used further.

The differential form of the equation of state is as follows:

$$d\rho = -\alpha\rho dT + \beta\rho dp + \gamma\rho ds \qquad (1.1)$$

The coefficients α, β, γ in (1.1) are functions of T, p, s. They are known from an equation of state and have a simple physical content: α is expansion coefficient; β and γ are barometric and salinity compressibility coefficients respectively.

Direct, but nevertheless quite accurate, measurements of the adiabatic velocity c of sound propagation in sea water allow us to describe the dependence $c\,(p,T,s) = \left[\left(\dfrac{\partial p}{\partial \rho}\right)_{\eta,\,s}\right]^{1/2}$ well, where $\eta = \eta\,(T,p,s)$ is the entropy[2]. A heat capacity c_p and c_v for constant pressure and specific volume respectively and also an adiabatic temperature gradient $\Gamma = \left(\frac{\partial T}{\partial p}\right)_{\eta,\,s}$ may be defined with the help of the parameters α, β, γ, c:

$$c_p = \frac{\ae}{\ae - 1}\frac{\alpha^2 T}{\beta\rho}; \quad \ae = \frac{c_p}{c_v} = \beta\rho c^2; \quad \Gamma = \frac{\alpha T}{\rho\, c_p} \qquad (1.2)$$

To simplify the statement of various problems, an equation of state linearized relative to some standard set of variables ρ_0, T_0, p_0, s_0 is used in some cases in the following form (1.1),

$$\frac{\rho - \rho_0}{\rho_0} = -\alpha_0(T - T_0) + \beta_0(p - p_0) + \gamma_0(S - S_0). \qquad (1.3)$$

Values of α_0, β_0, γ_0, ρ_0 may be obtained from tables [48] versus T_0, p_0 for $S_0 = 35\,^0/_{00}$. Let us note that variation of orders of magnitude of these parameters, as well as parameters defined by (1.2), in the range of ρ, T, p, s mentioned above are as follows: α from 1×10^4 to 4×10^4 $°C^{-1}$; β from 3.3×10^{-5} to 4.5×10^{-5} bar^{-1}; $\gamma \approx 0.8 \times 10^3$ $g \times cm^{-3}$; c_p from 4.0 to 4.2 $j \times g^{-1} \times °C^{-1}$; Γ from 0.35×10^{-3} to 1.8×10^{-3} $°C \times bar^{-1}$. Variation range of Γ is taken from [265]. Variation range of sound velocity c from 1.45×10^3 to 1.60×10^{-3} $m \times s^{-1}$. As was pointed out in the Introduction, internal waves exist only in a stratified fluid. Therefore let us briefly dwell on a notion of 'ocean stratification'.

Ocean stratification means density distribution of the fluid in the gravity field. This notion is the main one for the dynamics of internal waves because internal waves can not exist without stratification.

Let us consider the density distribution in a motionless fluid $\rho_0(z)$ to be known. In this case, from the examination of the balance of the stratified fluid in the gravity field we will obtain the simple equation

$$\frac{d\rho_0}{dz} = -\rho_0 \, g. \tag{1.4}$$

Here g is acceleration due to gravity, the axis z has the opposite direction to the action of gravity, i.e., straight upwards. Owing to the dependence $\rho_0 = \rho_0(z)$ we will suppose all thermodynamic parameters and functions depend on the z coordinate also.

Let us consider the problem of the stability of stratification. If we neglect the density's dependence on pressure, the solution of this problem is obvious: the stratification will be stable if more heavy fluid particles settle under the lighter ones, i.e., $\frac{d\rho_0}{dz} < 0$. In the opposite case $(\frac{d\rho_0}{dz} > 0)$ the stratification is unstable. But if we take into account the influence of pressure on the density, the problem of the stratification's stability is more complicated.

Let a fluid particle which is in thermodynamic balance in a gravity field at a level z be adiabatically[3] displaced by a small distance δz. The density of this particle at the level $z + \delta z$ will be equal to $\rho_0(z) + (\frac{d\rho_0}{dz})_{\eta, s} \delta z$, and the density of the surrounding particles at the same level is $\rho_0(z) + \frac{d\rho_0}{dz} \delta z$. Then the buoyancy force A acting on a unit volume of the adiabatically displaced particle,

$$A \, \delta z = g \left[\left(\frac{d\rho_0}{dz} \right)_{\eta, s} - \frac{d\rho_0}{dz} \right] \delta z \,,$$

will strive to return the particle to the initial level z only if $A > 0$. So the condition of the stability of stratification is

$$\frac{d\rho_0}{dz} < \left(\frac{d\rho_0}{dz}\right)_{\eta,\,s}. \tag{1.5}$$

If an inequality sign in (1.5) is changed to the opposite one, the stratification is unstable and manifests convective movements. In the case of equality in (1.5) the stratification is neutral.

For the stable stratification a so called 'Brunt–Väisälä frequency' [152], [371] may be introduced, characterizing the frequency of small free oscillations of salt water particles in the vicinity of the level z:

$$N^2(z) = \frac{g}{\rho_0}\left[\left(\frac{d\rho_0}{dz}\right)_{\eta,\,s} - \frac{d\rho_0}{dz}\right]. \tag{1.6}$$

Taking into account that $(d\rho_0)_{\eta,\,s} = c^{-2}\,dp_0$, where c is the velocity of sound, and using (1.4) one can rewrite (1.6) in the form

$$N^2(z) = -\left(\frac{g}{\rho_0}\frac{d\rho_0}{dz} + \frac{g^2}{c^2}\right). \tag{1.7}$$

Another equation for $N^2(z)$ in which thermodynamic laws are taken into account may be found in [152].

The characteristic oceanic distributions of the Brunt–Väisälä frequency are caused by specific features of the $\rho_0(z)$ field formation in the ocean. Usually, while measuring oceanic density field parameters, the term $\rho_0(z)$ means the stationary distributions averaged over considerable time ranges and some times over space also. Such an averaging is quite natural, as in spite of temperature, slighter density and negligible pressure variations, movements of different scales also influence the $\rho_0(z)$ field formation (see Fig. 1.1). The density $\rho_0(z)$ in the upper heated ocean layer is almost constant owing to intensive small scale turbulent movements caused by wind induced mixture. This layer is called quasi-homogeneous and its depth h is 20–100 m.

Below the quasi-homogeneous layer $(-h > z > -H)$ a layer of a seasonal thermocline or a density jump layer is disposed, in which density

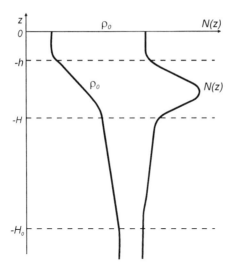

Figure 1.1. Typical density and Brunt–Väisälä frequency distributions in the ocean.

grows as a result of decrease in temperature. The lower border of this layer $H \approx 50$–200 m. Inside the seasonal thermocline a sharp fading of small scale turbulent motions occurs, taking place in the upper quasi-homogeneous layer, so that the intensity of pulsations is quite small at the lower border of the layer.

Below the seasonal thermocline a layer of mean thermocline (the lower border $H_0 \sim 1$ km) appears, where density increases continuously owing to the fall in temperature. The temperature field distribution may be supported here by weak large scale flows.

Below the mean thermocline layer $z < -H_0$ the sea water temperature is almost constant and is at 4–$5\,°$C, which causes almost constant density ρ_0 also.

In accordance with the density distribution described a variation of the Väisälä frequency $N(z)$ occurs (see Fig. 1.1). The largest variations of $N(z)$ are observed in the density jump layer, where a pronounced maximum of $N(z)$ usually takes place. An average value of this maximum in the ocean is $\sim 2 \times 10^{-2}$ s^{-1}. The quasi-homogeneous layer as well as the density jump layer, called an active layer of the ocean, undergo considerable seasonal oscillations; density oscillations with a much greater period may occur in the mean thermocline layer. The theoretical notions about the active ocean layer and about the mean thermocline are being developed intensively at the present time [152], [262].

Besides the low frequency variations mentioned, the density distribution in the ocean has one more feature, a pronounced depth indentation with scales from several centimeters to tens of meters. Such a small scale depth unsteadiness of the density (called a 'fine structure' of the distribution of the density, the temperature and other characteristics and the 'microstructure' at the scales 1 cm to 1 m) vary slowly, with characteristic periods of several hours to one month. This phenomenon is usually caused by nonlinear processes when internal waves propagate (see Chapter 9). In some cases small horizontal variations of the ocean density field have to be taken into account (see Chapter 5).

Thus, because of unsteadiness of the ρ_0 field a coordination of time and space averaging scales of this parameter is necessary. If horizontal inhomogeneity of the density field is neglected in the first approximation, $\rho_0(z)$ will mean the density field smoothed over fine structure features and averaged over a time period of about one month. Such an averaging allows us to consider the density $\rho_0(z)$ to be independent of time when studying propagation of internal waves (with the periods from minutes to several hours) and to investigate the fine structure of the density field separately.

2.   EQUATIONS OF MOTION AND ENTROPY BALANCE FOR SEA WATER

As sea water is, in some approximation, a two-component medium (water + salt), the main equations of ocean thermodynamics are conservation equations for water mass, salt mass, momentum and energy.

Further on all the parameters will be considered in the Cartesian coordinates x_i ($x_1 = x$, $x_2 = y$, $x_3 = z$) with the z axis' direction normal to the earth's surface, fixed at the 'plane' of the Earth, rotating with velocity Ω. Such an approximation is quite appropriate for the examination of internal waves the length of which is small relative to the Earth's radius. A more general statement of the problem is given in the monograph [152].

Conservation laws for salt water mass, salt mass, momentum balance and energy are as follows:

$$\frac{\partial \rho}{\partial t} + \text{div}\,(\rho \mathbf{v}) = 0; \qquad (1.8)$$

$$\frac{\partial \rho}{\partial t} + \text{div}\,(\rho\, s\mathbf{v} + \mathbf{J}_s) = 0; \qquad (1.9)$$

$$\frac{\partial(\rho\, v_i)}{\partial t} + \frac{\partial}{\partial x_k}(\rho\, v_i v_k - \sigma_{ik} + p\,\delta_{ik}) = -\rho\, g\delta_{i3} + 2\rho\,\delta_{ikm} v_k \Omega_m; \qquad (1.10)$$

$$\frac{\partial}{\partial t}(\rho E) + \frac{\partial}{\partial x_k}\left[\rho E v_k + p v_k - \sigma_{ki} v_i + q_k + \frac{\partial \chi}{\partial s} J_k\right] = 0; \qquad (1.11)$$

$$E = \frac{1}{2}\mathbf{v}^2 + \varepsilon + gz.$$

Here ρ is the density, s is the salinity, $\mathbf{v}$ is the velocity with orthogonal components $v_1 = u$, $v_2 = v$, $v_3 = w$, $\mathbf{J}_s$ is the vector of the salt diffusion flow, $\mathbf{q}$ is the heat flow vector, p is the pressure, σ_{ik} is the symmetric tensor of viscous tension (the case without internal rotations is considered), χ is the enthalpy, ε is the internal energy, E is the total energy, $\mathbf{v}^2/2$ is the specific kinetic energy, $\delta_{ikm} v_k \Omega_m = -(\Omega \times \mathbf{v})_i$. The variables δ_{ik} is the Kroneker symbol, δ_{ikm} is the unit tensor antisymmetric for all index pairs. Equation (1.9) is, in fact, a definition of the diffusion flow $\mathbf{J}_s$ and (1.11) is the definition of the heat flow $\mathbf{q}$. A divergent form of (1.8), (1.9), (1.11) means conservation of the values ρ, ρ_s and ρE. Divergencelessness of Eq. (1.10) is caused by rotation only.

As a result of the scalar product of the equation of momentum balance (1.10) with $\mathbf{v}$, an equation for the kinetic energy balance will be obtained:

$$\frac{\partial}{\partial t}\left(\frac{\rho \mathbf{v}^2}{2}\right) + \frac{\partial}{\partial x_k}\left(\frac{\rho \mathbf{v}^2}{2} v_k + p\, v_k - \sigma_{km} v_m\right) = -\rho\, gw + p\,\text{div}\mathbf{v} - \sigma_{ij} e_{ji}. \qquad (1.12)$$

Here $2e_{ij} = \frac{\partial v_i}{\partial x_j} + \frac{\partial v_j}{\partial x_i}$ is the tensor of deformation velocities. Then from the difference of Eqs. (1.11) and (1.12) one can obtain the equation for

internal energy balance from which, with the help of known thermodynamic equations [61], the equation for an entropy balance is obtained

$$\frac{\partial}{\partial t}(\rho\eta) + \operatorname{div}(\rho\eta\mathbf{v} + \mathbf{J}_\eta) = \mathcal{P}_s, \qquad (1.13)$$

$$\mathbf{J}_\eta = T^{-1}\mathbf{q} + \left(\frac{\partial\eta}{\partial s}\right)_{T,p}\mathbf{J}_s, \qquad (1.14)$$

$$T\mathcal{P}_s = -q_i T'\frac{\partial T}{\partial x_i} - J_{s,i}\left(\frac{\partial\mu}{\partial x_i}\right)_T + \sigma_{ij}e_{ji} \geq 0, \qquad (1.14a)$$

where μ is the chemical potential of sea water. Formulae (1.14) show
that entropy does not remain in the nonequilibrium condition owing to
the presence of a source of entropy $\mathcal{P}_s$ in the r.h.s. of equation (1.14a)
connected with dissipative processes (viscous friction, diffusion, heat
transfer). The vector $\mathbf{J}_\eta$ is an entropy flow, generally speaking, independent of the medium's nonequilibrium. Let us stress that the local
expression of the second law of the thermodynamics law is that $\mathcal{P}_s \geq 0$;
the case $\mathcal{P}_s = 0$ corresponds to the condition of thermodynamic equilibrium.

Using ordinary thermodynamic equations one can present the equation of entropy balance (1.13) in equivalent forms written as

$$\rho\, c_p\left(\frac{\partial T}{\partial t} - \Gamma\frac{\partial p}{\partial t}\right) = -\operatorname{div}\mathbf{q} + \sigma_{ij}e_{ji} + J_k\frac{\partial}{\partial x_k}\left(\frac{\partial\chi}{\partial s}\right), \qquad (1.15)$$

$$\frac{c_p}{\alpha}\left(-\frac{d\rho}{dt} + \frac{1}{c^2}\frac{dp}{dt} + \frac{\partial\rho}{\partial s}\frac{ds}{dt}\right) = -\operatorname{div}\mathbf{q} + \sigma_{ij}e_{ji} + J_k\frac{\partial}{\partial x_k}\left(\frac{\partial\chi}{\partial s}\right). \qquad (1.15a)$$

Equations (1.8)–(1.10), (1.13)–(1.15) together with the equation of state
$\rho = \rho\,(T,p,s)$ yield closed combined equations for the definition of all
unknown functions if the expressions for the flows $\mathbf{q}$, $\mathbf{J}_s$ and the 'viscous' tension tensor σ_{ij} are known. As was shown in [192], [193] those
parameters may be presented in the forms

$$\sigma'_{ij} = 2\nu\rho\,e_{ij}, \qquad (1.16)$$

$$q_i = -k\frac{\partial T}{\partial x_i} + \nu\left(\frac{\partial\mu}{\partial x_i}\right)_T, \qquad J_{s,i} = -\rho D\left(\frac{\partial s}{\partial x_i} + \frac{k_p}{p}\frac{\partial p}{\partial x_i}\right), \qquad (1.17)$$

where k is the heat transfer coefficient for sea water, ν is the kinematic viscosity coefficient, and D is the salt diffusion coefficient. The term $(k_p/p)(\partial p/\partial x_i)$ in (1.17) characterizes the biodiffusion phenomenon, here the biodiffusion coefficient k_p for the ocean is apparently insignificant. The coefficients ν, D and the heat transfer coefficient $k/\rho c_p$ for sea water differ slightly from the corresponding coefficients for fresh water. Thus, for $T = 20°C$ and $S = 35\,^0/_{00}$ we have $\nu = 1.049 \times 10^{-2}$ cm$^2 \times$s^{-1}, $\frac{k}{\rho c_p} = 1.49 \times 10^{-3}$ cm$^2 \times$s^{-1}, $D = 1.29 \times 10^{-5}$ cm$^2 \times$s^{-1}. Finally, the combined equations describing sea water thermodynamics consist of Eqs. (1.8)–(1.10), (1.13), (1.14) (or 1.15), the equation of state $\rho = \rho\,(T, p, s)$, and the relations (1.16) and (1.17).

3.   ADIABATIC APPROXIMATION

While studying thermohydrodynamic processes in the ocean, dissipative phenomena (viscosity, heat transfer and diffusion) may be neglected in most cases. A nondissipative approximation presented will be valid for rather large-scale movements in a restricted time range only. Such an approximation will be called 'equilibrium'; in this case, in (1.14) $\mathcal{P}_s = 0$, $\sigma_{ij} = 0$.

Equilibrium processes when additionally $\mathbf{I}_\eta = 0$ will be called 'adiabatic'. So for adiabatic processes

$$\mathbf{q} = 0, \quad \mathbf{J}_s = 0, \quad \sigma_{ij} = 0. \qquad (1.18)$$

Formulae (1.18) correspond to going to zero of the coefficients k, D, ν in the expressions (1.16) and (1.17). In this case, from (1.9), (1.14), and taking into account (1.8), we have:

$$\frac{d\nu}{dt} = 0; \quad \frac{ds}{dt} = 0. \qquad (1.19)$$

Equations (1.19) show that for a fixed water particle the adiabatic approximation is isentropic and isohalinic. Let us remark that the ocean as a whole may be considered in the nonadiabatic but equilibrium approximation while studying heat, salt, and moisture flows from the ocean's surface. Here the transfer processes will be realized as a result of convection (it is natural that they will be effective in a framework of turbulent motions only). But whilst considering wave motions of relatively small scales, such as internal waves, the ocean may considered to be adiabatic.

From (1.1), (1.2), and the thermodynamic identity

$$d\eta = \frac{c_p}{T}dt - \frac{\alpha}{S}dp + \delta\, dS,$$

with the help of formulae (1.19) defining the adiabatic approximation it is easy to obtain

$$\frac{dp}{dt} = c^2\frac{d\rho}{dt}. \tag{1.20}$$

This formula may be obtained from the equation of state $\rho = \rho\,(T, p, s)$ directly, taking into consideration (1.19) and the expression for the velocity of sound c. Then from (1.1) and from the formula for $d\eta$ given above taking into account (1.19), or directly from (1.15), yields

$$\frac{dT}{dt} = \Gamma\frac{dp}{dt}. \tag{1.21}$$

Using (1.20), (1.21) one may introduce the notions of a potential temperature θ and potential density ρ_n with the following formulae

$$\left.\begin{aligned}
\theta &= T - \int_{p_0}^{p}\Gamma\, dp = T - \int_{\rho_0}^{\rho}\Gamma c^2\, dp; \\[2ex]
\rho_n &= \rho - \int_{p_0}^{p}c^{-2}\, dp.
\end{aligned}\right\} \tag{1.22}$$

The parameters θ and ρ_n have the physical meanings of temperature and density of liquid particles brought isentropically and isohalinically to the standard pressure, which is usually equal to 1 bar. In the upper ocean layers those parameters differ insignificantly from the natural temperature and density.

From (1.22) taking into account (1.20) and (1.21) it follows

$$\frac{d\theta}{dt} = 0; \qquad \frac{d\rho_n}{dt} = 0. \tag{1.23}$$

Formulae (1.19) and (1.23) show that in the adiabatic approximation an entropy η, salinity s, potential temperature θ, and potential density ρ_n are adiabatic Lagrange invariants, because they retain their values for a certain particle in salt water.

Now let us write the balance equations of motion in the adiabatic approximation considered. Using (1.18) we will offer for (1.10) the following form

$$\rho \left[\frac{d\mathbf{v}}{dt} + (\mathbf{v} \times \nabla) \, \mathbf{v} \right] + 2\rho \, \Omega \times \mathbf{v} + \nabla p + \rho g \mathbf{k} = 0. \tag{1.24}$$

In Eq. (1.24) $\mathbf{k}$ is the unit vector collinear to the z axis. In the Cartesian coordinates, when the z axis is directed against gravity and starts from the undisturbed water level, the x axis to the East, and the y axis to the North, Ω vector has projections $(0, \Omega \cos \varphi, \Omega \sin \varphi)$, where Ω is the angular velocity of the Earth. Coriolis force $\mathbf{K} = 2\rho \, \Omega \times \mathbf{v}$ has the following projections:

$$(2\rho)^{-1} K_x = - \Omega v \, \sin \varphi + \Omega w \, \cos \varphi;$$

$$(2\rho)^{-1} K_y = \Omega u \, \sin \varphi; \quad (2\rho)^{-1} K_z = - \Omega u \, \cos \varphi. \tag{1.25}$$

Here φ is the latitude $\left(-\frac{\pi}{2} \leq \varphi \leq \frac{\pi}{2}\right)$ depending in our local Cartesian coordinates on the variable y, and u, v, w are the components of the velocity vector $\mathbf{v}$. The combined equations (1.24) together with the mass balance equation (1.18), the equation of state $\rho = \rho \, (T, p, s)$, and the adiabatic conditions (1.19) are seven closed simultaneous equations for seven unknown functions ρ, T, p, s, u, v, w. Initial and boundary conditions should be added to these combined equations.

First let us formulate the boundary conditions. We will consider that the motion of the ocean water takes place in a multiply connected region restricted by the surface Σ with vertical boundaries. Then in the plane $\mathbf{x} = \{x,y\}$ the 'side' ocean boundaries are a finite totality of closed curves Σ_k^0 $(k = 1, 2, \ldots, n)$. At the side boundaries a 'no leakage condition' of fluid should be valid, i.e.,

$$\mathbf{v}_n \, |_{\mathbf{x} \in \Sigma^0} = 0. \tag{1.26}$$

The boundary conditions should be set at the ocean's bottom also. Taking into consideration the bottom's relief $z = -H(x)$, the no leakage condition at the bottom

$$\left[w - u\frac{\partial H}{\partial x} - \frac{\partial H}{\partial y} \right]\bigg|_{z=-H(\mathbf{x})} = 0 \qquad (1.27)$$

means that the velocity vector at a point in the vicinity of the bottom lies in a plane tangential to the bottom surface at that point. If the bottom is horizontal, i.e., $H = \text{const}$, it follows from (1.27) that the following condition should be valid

$$w = 0 \quad \text{at} \quad z = -H. \qquad (1.28)$$

At the free ocean surface $z = \zeta(\mathbf{x},t)$ the kinematic and dynamic boundary conditions should be satisfied. The kinematic boundary condition corresponds to the case of equality of the projections of velocity vectors onto the surface normal at any point on the free surface and for a fluid particle attached to this point (the no leakage condition):

$$w\,|_{z=\zeta} = \frac{d\zeta}{dt} = \left[\frac{\partial \zeta}{\partial t} + u\frac{\partial \zeta}{\partial x} + v\frac{\partial \zeta}{\partial y} \right]\bigg|_{z=\zeta}. \qquad (1.29)$$

The dynamic boundary condition consists in the identity of the plane $\zeta = \zeta(\mathbf{x},t)$ and the plane of a given atmospheric pressure[4] $p_a(\mathbf{x},t)$, i.e.,

$$p(\mathbf{x}, \zeta, t) = p_a(\mathbf{x}, t). \qquad (1.30)$$

Equation (1.30) is frequently used in the differential form $(dp/dt)|_{z=\zeta} = dp_a/dt$, which, taking into account (1.29), may be written as

$$\left[\frac{\partial \Delta p}{\partial t} + \mathbf{v}\nabla(\nabla p) \right]\bigg|_{z=\zeta} = 0, \qquad \Delta p = p - p_a. \qquad (1.31)$$

The initial conditions consist of an assignment of a state of motion at an initial time $t = 0$, i.e., the free surface displacements and all thermohydrodynamic fields versus coordinates should be known for $t = 0$:

$$\mathbf{v}(\mathbf{x}, z, 0) = \mathbf{v}^0(\mathbf{x}, z); \qquad \rho(\mathbf{x}, z, 0) = \rho^0(\mathbf{x}, z);$$

$$T\left(\mathbf{x}, z, 0\right) = T^0\left(\mathbf{x}, z\right); \quad s\left(\mathbf{x}, z, 0\right) = s^0\left(\mathbf{x}, z\right); \quad \zeta\left(\mathbf{x}, 0\right) = \zeta^0\left(\mathbf{x}\right). \quad (1.32)$$

Here an initial pressure distribution $p^0\left(\mathbf{x}, z\right)$ is obtained according to the initial fields ρ^0, T^0, s^0 from the equation of state.

While considering simplified models of the ocean (unlimited in the horizontal or in both the vertical and horizontal directions), the side surfaces or the bottom boundary conditions are replaced by the conditions of periodicity or damping of motions in the fluid for $|\mathbf{x}| \to \infty$ or $|z| \to \infty$. Some other simplifications in the statement of the boundary and initial conditions, which will be discussed in the following chapters, are used.

The balance equation for the total energy E is simplified considerably in the adiabatic approximation. Taking into consideration (1.18) one can write (1.11) in the following form

$$\frac{\partial}{\partial t}\left[\rho\left(\frac{\mathbf{v}^2}{2} + \varepsilon + gz\right)\right] + \operatorname{div}\left[\rho\mathbf{v}\left(\frac{\mathbf{v}^2}{2} + \chi + gz\right)\right] = 0. \quad (1.33)$$

Here $\chi = \varepsilon + \rho^{-1}p$ is the enthalpy. While integrating (1.33) over the total ocean volume, taking into account the kinematic boundary conditions (1.26), (1.27) and (1.30)[5] and the dynamic condition (1.30), we will obtain

$$\frac{\partial}{\partial t}\int d\mathbf{x} \int_{-H(\mathbf{x})}^{\zeta(\mathbf{x},t)} \rho\left(\frac{\mathbf{v}^2}{2} + \varepsilon + gz\right) dz + \int p_{\mathrm{a}}(\mathbf{x}, t)\frac{\partial\zeta}{\partial t} d\mathbf{x} = 0. \quad (1.34)$$

Here the z axis is directed upward and starts from the undisturbed free surface. In the case

$$\int p_{\mathrm{a}}\frac{\partial\zeta}{\partial t} d\mathbf{x} = 0, \quad (1.35)$$

i.e., in the motion of the free surface the average of the work of the pressure forces at the ocean surface goes to zero (the latter is valid, in particular, when $p_{\mathrm{a}} = \mathrm{const}$), the total ocean energy remains in the adiabatic approximation:

$$\overline{E} = \int d\mathbf{x} \int_{-H(\mathbf{x})}^{\zeta(\mathbf{x},t)} \rho\left(\frac{\mathbf{v}^2}{2} + \varepsilon + gz\right) dz = \mathrm{const}. \quad (1.36)$$

Furthermore, the balance equation for the sum of the kinetic and potential energy simplifies considerably in the conditions of the adiabatic approximation.

From (1.33), using (1.9), it is easy to obtain

$$\frac{\partial}{\partial t}\left[\rho\left(\frac{\mathbf{v}^2}{2}+gz\right)\right]+\operatorname{div}\left[\rho\mathbf{v}\left(\frac{\mathbf{v}^2}{2}+\rho^{-1}p+gz\right)\right]+\rho\frac{\partial\varepsilon}{\partial t}=0. \quad (1.37)$$

But under the adiabatic approximation conditions, when relations (1.19) are valid, from the thermodynamic correlation

$$T\frac{d\eta}{dt}=\frac{d\varepsilon}{dt}+p\frac{d\rho^{-1}}{dt}-\mu\frac{ds}{dt}$$

we have

$$\left(\frac{d\varepsilon}{dt}\right)_{\eta,s}=p\rho^{-1}\frac{\partial\rho}{\partial t}$$

Substituting this expression into (1.37) we obtain

$$\frac{\partial}{\partial t}\left[\rho\left(\frac{\mathbf{v}^2}{2}+gz\right)\right]+\operatorname{div}\left[\rho\mathbf{v}\left(\frac{\mathbf{v}^2}{2}+\rho^{-1}p+gz\right)\right]+\rho^{-1}p\frac{\partial\rho}{\partial t}=0. \quad (1.38)$$

Equation (1.38) is used in the next chapter in an application to the internal waves theory.

It is then quite obvious that for any Lagrange adiabatic invariant σ, i.e., for the function $\sigma(\mathbf{x},z,t)$ satisfying the equation $d\sigma/dt=0$, there is a conservation law like (1.36), i.e., the exact equality

$$\int d\mathbf{x}\int_{-H(\mathbf{x})}^{\zeta(\mathbf{x},t)}\rho\sigma dz=\mathrm{const} \quad (1.39)$$

holds. In fact, multiplying the equation $d\sigma/dt=0$ by ρ and using the mass balance equation (1.9) it is possible to write this equation in the divergent form

$$\frac{\partial(\rho\sigma)}{\partial t}+\operatorname{div}(\rho\sigma\mathbf{v})=0.$$

Integrating this relation over the total ocean volume, taking into consideration the kinematic conditions (1.26), (1.27) and (1.29), and then also integrating over time, we will obtain the conservation law (1.39). Thus for the variables η, s, θ, ρ_n we will have conservation laws like (1.39).

For the conclusion of this section let us write the relations (1.9), (1.20), (1.24) in slightly varied forms useful for further analysis. Let us consider the undisturbed density distribution $\rho_0(z)$, i.e., motionless fluid stratification, to be given. Then assuming that

$$\rho = \rho_0(z) + \hat{\rho}(\mathbf{x}, z, t), \quad p = g \int_z^0 \rho_0 dz + \hat{p}, \qquad (1.40)$$

and substituting (1.40) into (1.24), we will obtain

$$(\rho_0 + \hat{\rho}) \frac{d\mathbf{v}}{dt} + 2(\rho_0 + \hat{\rho})\,\Omega \times \mathbf{v} + \nabla\hat{p} + \hat{\rho}g\mathbf{k} = 0. \qquad (1.41)$$

Using equations obvious from (1.40)

$$\frac{d\rho}{dt} = \frac{d\hat{\rho}}{dt} + w\frac{d\rho_0}{dz}, \quad \frac{dp}{dt} = \frac{d\hat{p}}{dt} - gw\rho_0 \qquad (1.42)$$

and substituting (1.42) to (1.20), we will have

$$\frac{d\hat{\rho}}{dt} = \frac{1}{c^2}\frac{d\hat{p}}{dt} + \frac{1}{g}w\rho_0 N^2(z), \qquad (1.43)$$

$$N^2(z) = -\left(\frac{g}{\rho_0}\frac{d\rho_0}{dz} + \frac{g^2}{c^2}\right).$$

Here $N^2(z)$ is the Brunt–Väisälä frequency taken from (1.7), w is the vertical component of velocity vector. At last, from (1.7), (1.20) and (1.42) we will obtain

$$\operatorname{div}\mathbf{v} = \frac{gw}{c^2}\frac{\rho_0}{\rho_0 + \hat{\rho}} - \frac{1}{(\rho_0 + \hat{\rho})c^2}\frac{d\hat{p}}{dt}. \qquad (1.44)$$

The exact (in the adiabatic approximation) equations (1.40)–(1.44) will serve as a basis for different approximations used in geophysical hydrodynamics (see Section 2.1).

4.   CLASSIFICATION OF OCEANIC WAVE MOTIONS

To define the place of the internal waves amongst other oceanic wave motions a brief classification list of free waves in the ocean is presented. Such a classification on the basis of linear analysis and in the absence of flows (in the adiabatic approximation) for the model of wave motion of a layer of finite thickness on a rotating sphere is described in detail in [152]. In the present summary we will give quantitative notions only for physical sources of any type of wave motions and of its basic characteristics observed in the ocean.

All oceanic wave motions are caused by the following physical factors: fluid compressibility; the presence of a free surface; the presence of stratification; the Earth's rotation; the Earth's sphericity; the gravity field. In accordance with the reasons mentioned the following wave motions take place in the ocean.

1. *Acoustic waves.* The physical source causing this wave type is the elastic compressibility of the fluid. All of the other physical factors mentioned give some additional distortions (in the presence of stratification this is sufficient) in propagation of this wave type. Acoustic waves have the highest frequencies in the ocean $\omega > \omega_{\min} \sim c_{\min}/H \sim 1$ rad. (H is the ocean depth of about 3 km) with their maximum length $\lambda \approx H \approx 3$ km.

2. *Surface gravity waves.* This wave type is the result of free surface displacements in the gravity field. Water inhomogeneity and compressibility are almost insignificant for surface waves. The surface waves damp very quickly (in the linear case they do so exponentially) while moving away from the free surface. The shortest surface waves of 1–10 cm length are called 'ripples'. These waves' dynamics is considerably dependent on the surface tension and they are frequently called capillary waves also. The longest surface waves with length of tens of meters are caused by wind. Very long surface waves (up to hundreds kilometers) of a seismic nature ('tsunami') exist too. The surface waves may have, generally speaking, any frequency.

3. *Internal waves.* As all the following chapters of the present monograph are dedicated to this wave type we will characterize them quite
   briefly. The main source of internal waves is fluid stratification in the
   gravity field.

4. *Rossby waves.* a) *barotropic.* The main physical source for its appearance is the joint action of the Earth's rotation and sphericity.
   Compressibility and inhomogeneity of sea water and gravity are almost negligible for these waves. The barotropic Rossby waves have
   lengths of hundreds kilometers.

   b) *baroclinic.* Their existence is caused by the joint action of the inhomogeneity of sea water in the gravity field as well as the Earth's rotation and sphericity. A possible frequency range $\omega < \min\left(\Omega, N_{\max}\right)$,
   where N is the Brunt–Väisälä frequency, Ω is the Earth's angular velocity. Unlike all other wave types the Rossby waves propagate from
   the East to the West only.

5. *Inertial gyro waves.* They can exist under the condition $\Omega > \frac{1}{2}N_{\min}$
   only. Water compressibility and the presence of a free surface have
   no influence on these waves.

As was mentioned above, a classification of the free wave motions
considered relates to the linear approximation only (for small amplitude
oscillations of the free surface). But as the physical sources causing
different types of wave motions differ from each other considerably, it is
possible to spread the presented classification to the nonlinear case as
well.

Notes

1 Sea water salinity is usually measured pro mille, in this case it is written as S, and $s = 10^{-3}S$.

2 Here and below only specific, i.e., related to a unity mass, thermodynamic functions are considered.

3 In the thermodynamics of sea water adiabatic processes mean isentropic and isohalinic processes (see Section 1.2).

4 Here and below the influence of surface straining is omitted.

5 If the ocean is considered to be unlimited in the horizontal direction, condition (1.26) is replaced by the condition of velocities going to zero for $|\mathbf{x}| \to \infty$.

Chapter 2

EQUATIONS OF THE THEORY
OF INTERNAL WAVES

This chapter discusses equations describing oceanic internal waves. These equations are obtained by different simplifications of the basic equations of oceanic thermodynamics derived in Chapter 1. A list of different forms of these equations for special cases — plane (two-dimensional) stationary and nonstationary problems, equations linearized relative to the state of rest and to background flow — is given here for convenience.

1. BASIC APPROXIMATIONS FOR OCEANIC INTERNAL WAVES THEORY

In the following chapters both linear and nonlinear theories of internal gravity waves will be developed. Some simplifications of the combined equations (1.8)–(1.11), together with the equation of state $\rho = \rho(T, p, s)$ and a thermodynamic potential $\varphi(T, p, s)$, are necessary to separate the subject of our study from other types of wave motions and insignificant phenomena. We will consider these simplifications successively.

1.1 EQUILIBRIUM AND ADIABATIC APPROXIMATION

While studying internal waves we shall neglect dissipative effects: viscosity; heat transfer; and diffusion. As explained in Section 1.3, such an approximation limits the ranges of both length and time for which it will be valid.

Let λ be a characteristic length scale of internal waves, namely, for convenience, the length of a wave. Characteristic time scales connected with waves damping owing to viscosity t_ν, of heat transfer t_k, and of diffusion t_D are

$$t_\nu = \lambda^2/\nu, \quad t_k = \rho\, c_p \lambda^2/k, \quad t_D = \lambda^2/D. \qquad (2.1)$$

A characteristic period of internal waves is $T \sim N_0^{-1}$, where N_0 is the maximum of the Brunt–Väisälä frequency. Obviously dissipative effects may be neglected when

$$t_d = \min\{t_\nu,\, t_k,\, t_D\} \gg T \sim N_0^{-1}. \qquad (2.2)$$

Only in the case of the validity of inequality (2.2) dissipative phenomena will be negligible for propagation of internal waves during many wave periods.

From the list of parameters for kinetic coefficients ν, $k/\rho\, c_p$, D given in Section 1.2, it follows that $t_d = t_\nu$. Then from (2.1) and (2.2) we can evaluate a scale of internal wave motions

$$\lambda \gg (\nu/N_0)^{1/2} = \lambda_d. \qquad (2.3)$$

Here λ_d is the maximum scale (defined by viscosity) for which dissipative phenomena are important. Taking into account that $\nu \sim 10^{-2}$ cm$^2 \times$ s^{-1}, $N_0 \sim 10^{-2}$–10^{-4} s^{-1} we obtain $\lambda_d \sim 1$–10 cm. When taking into consideration (see Chapter 3.1) that decreased internal wave frequency means considerably increased length, and assuming (in a conservative estimate) that $\lambda_{min} \approx 10^3 \lambda_d$, the mentioned nondissipative approximation is valid for minimum internal waves length $\lambda_{min} \sim 10$ m. These are quite short internal waves for the ocean. Thus dissipation is almost negligible for the propagation of internal waves. The 'life time' of these waves is determined by other processes generally connected with nonlinear effects examined in Part 3 of this book.

The theory of internal waves discussed here in the adiabatic approximation will neglect mass and energy exchange during the time of waves' generation and propagation. Then all the equations and conservation laws of Chapters 1.3 and 1.4 will be valid for the internal waves.

1.2   APPROXIMATION FOR CORIOLIS FORCE

Omitting very long oceanic internal waves with tidal and inertial periods (which are not discussed in this book) it is possible to neglect the meridional variation of the Earth's rotational velocity by considering local Cartesian coordinates x, y, z tangential to the Earth's surface. Such an approximation ($\Omega = $ const) filters the Rossby waves.

Furthermore, we will consider the 'traditional' approximation for the Coriolis force (1.25) based on the assumption $\Omega_y = \Omega \cos \varphi = 0$. Then from (1.25) we derive ($\mathbf{k}$ is the unit vector oriented along z axis)

$$\left. \begin{array}{l} \mathbf{K} = f\mathbf{k} \times \mathbf{v} = \{-fv,\ fu,\ 0\}\,; \\[2ex] f = 2\Omega \sin \varphi = \text{const.} \end{array} \right\} \tag{2.4}$$

For further calculations we will take $f \sim 10^{-4}$ c^{-1}.

Usually the ground for such an approximation is the assumption of vertical velocities vanishing relative to horizontal components. But this approximation is valid for quite long waves only, and its basic assumption $\cos \varphi = 0$ is not applicable for equatorial regions where terms omitted in (1.25) dominate relative to the remaining terms in (2.4). Nevertheless, a further analysis of internal waves will be performed in the traditional approximation (2.4), as approximation (2.4) is apparently valid for long internal waves in temperate latitudes and the Coriolis force is almost negligible for short internal waves ($\lambda \leq 10$–20 km). The reader can find quite detailed discussions of the 'traditional' approximation in [152], [153], [262]. The non-traditional approximation is also discussed in Chapter 3.7.

1.3   INCOMPRESSIBILITY APPROXIMATION

A characteristic velocity of propagation of internal waves in the ocean is $c_f \sim \lambda N_0$. So a maximum value of c_f for long internal waves with length $\lambda \sim 10$ km is $c_f \sim 1$ m $\times$ s^{-1}. This velocity is 1500 times less than the velocity of sound in sea water. This gives an opportunity to filter fast sound waves from slow internal waves.

Considering Eqs. (1.9) and (1.20)

$$\frac{1}{\rho}\frac{d\rho}{dt} = \text{div } \mathbf{v}, \qquad \frac{d\rho}{dt} = \frac{1}{c^2}\frac{dp}{dt}$$

and assuming formally $c \to \infty$, simple expressions of internal waves theory are derived:

$$\text{div } \mathbf{v} = 0, \qquad \frac{d\rho}{dt} = 0. \tag{2.5}$$

However, such a 'derivation' of Eqs. (2.5) is quite rough with respect to the velocity of sound, which is high but finite ($c \approx 1.5 \text{ km} \times \text{s}^{-1}$). And the assumption $c \to \infty$ remains vague: what value of c may be considered to be infinitely high? To analyze the approximation of water's incompressibility in detail, we turn to Eqs. (1.43), (1.44)

$$\left.\begin{aligned} &\frac{d\hat{\rho}}{dt} = \frac{1}{c^2}\frac{d\hat{p}}{dt} + \frac{w}{g}\rho_0 N^2(z); \\[2mm] &N^2(z) = -\left(\frac{g}{\rho_0}\frac{d\rho_0}{dz} + \frac{g^2}{c^2}\right); \\[2mm] &\text{div } \mathbf{v} = \frac{gw}{c^2}\frac{\rho_0}{\rho} - \frac{1}{\rho c^2}\frac{d\hat{p}}{dt}. \end{aligned}\right\} \tag{2.6}$$

Here $\hat{p}$ and $\hat{\rho}$ are unsteady additions to steady (given) distributions of pressure $p_0(z)$ and density $\rho_0(z)$ [see Eq. (1.40)], and $\rho = \rho_0 + \hat{\rho}$ is the total density. Order of magnitudes are $\hat{\rho} \sim 10^{-3} \text{ g} \times \text{cm}^{-3}$, $\hat{p} \sim 1$ bar.

Now consider the third equation of (2.6). We assume that the second term on the r.h.s. is of the same order as the first one, i.e.,

$$|\rho_0 g w| \geq \left|\frac{d\hat{p}}{dt}\right|. \tag{2.7}$$

The inequality (2.7) is valid only if fast acoustic processes are filtered by scale average $T_a \ll T_w$, where $T_a \sim \omega_a^{-1}$ is the period of acoustic oscillations, $T_w \sim N_0^{-1}$ is the period of the internal waves. So the order of magnitude of $|d\hat{p}/dt| \leq p_a N_0 \sim 10^3$ bar $\times \text{s}^{-1}$ ($p_a \sim 1$ bar is the pressure at the free surface), which is comparable to the order of

$\rho_0 gw \sim 10^3$ bar $\times$ s^{-1} observed in the ocean. Comparing now the first term $\rho_0 gw/\rho c^2$ on the r.h.s. of the third equation in (2.6) with the characteristic term $\partial w/\partial z \sim w/l_z$, where l_z is the characteristic vertical scale of the vertical velocity distribution $w(\mathbf{x}, z, t)$, we have $\partial w/\partial z \gg \rho_0 gw/\rho c^2$ because $H_* = g l_z/c^2 \ll 1$. H_* is the dimensionless ratio of the vertical velocity scale l_z to the characteristic local ocean scale $H_c = c^2/g \approx 200$ km. So $H_* = l_z/H_c < 5 \times 10^{-2}$, assuming that l_z does not exceed the ocean's depth $H \approx 5$ km.

With (2.7) it follows that the r.h.s. of the last equation in (2.6) is negligibly small in comparison with the characteristic terms of the l.h.s., i.e., with an accuracy of 5% we can assume div $\mathbf{v} = 0$. For the first two equations in (2.6), in the presence of a weak stratification the first term in the expression for $N^2(z)$ is compared with the second term $g^2/c^2 \approx 5 \times 10^{-5}$ s^{-2}. With (2.7) and the third equation in (2.6), filtration of acoustic waves then generally leads to

$$\text{div } \mathbf{v} = 0, \qquad \frac{d\rho}{dt} = \frac{1}{c^2}\frac{dp}{dt}. \tag{2.7a}$$

Let now consider the strong stratification case when in (2.6) we will have for $N^2(z)$

$$\left|\frac{g}{\rho_0}\frac{d\rho_0}{dz}\right| \gg \frac{g^2}{c^2} \approx 5 \times 10^{-5} \text{ s}^{-2}. \tag{2.8}$$

This case is characteristic for the most frequent state of oceanic conditions, a pronounced pycnocline (a jump layer) in the upper and the most active ocean layer, as well as for shallow seas. In the main thermocline layer both terms in the inequality (2.8) have the same order, and in the lower homogeneous ocean layer (see Section 1.2) the sign of the strong inequality in (2.8) changes. For the case described by (2.8),

$$N^2(z) \approx -\frac{g}{\rho_0}\frac{d\rho_0}{dz}. \tag{2.9}$$

Assuming the inequality (2.7) to be valid, the term $c^{-2}d\hat{p}/dt = 0$ may be omitted in the r.h.s. of (2.6) compared to the remaining term $-w\rho_0$. Then taking into consideration the first relation in (1.42), we have $d\rho/dt = 0$ asymptotically obtaining Eqs. (2.5). We call this approximation 'isopycnic'.

For the common case the incompressibility condition with respect to the theory of internal waves looks like (2.7). Only for strong stratification typical for the upper ocean layers and shallow seas the 'traditional' combined equations (2.5) filtering acoustic waves are valid.

In the weakly nonlinear theory of internal waves, considered in Part 3 of the book, we have to operate with a series expansion in a small parameter with the higher expansion terms of the disturbance theory. In this case all the approximations of the current subsection, such as (2.5) and (2.7), should be used prudently because the necessity of the complete relations (1.9), (1.20)[1] may appear for higher orders of the disturbance theory.

1.4   'RIGID LID'APPROXIMATION AT THE FREE SURFACE

As mentioned in our classification of oceanic waves in Chapter 1.6, surface gravity waves are caused by the free surface distortion and are rapidly damped in the ocean depth. At the same time, internal gravity waves generally have quite small amplitudes at the free ocean surface. Maximum amplitudes are usually observed near the maximum Brunt–Väisälä frequency. That is why the 'rigid lid' approximation is frequently used to filter internal gravity waves from surface waves. Its meaning is that the free surface disturbances are neglected, i.e., $\zeta = 0$ in conditions (1.29)–(1.31), yielding:

$$w|_{z=0} = 0, \quad p|_{z=0} = p_{\mathrm{a}}(\mathbf{x}, t).$$

1.5   BOUSSINESQ APPROXIMATION

This approximation has no special physical meaning unlike those mentioned in Subchapters 2.1.1–2.1.4 and is used quite frequently to simplify calculations [217]. Consider the momentum balance equation, see Eq. (1.41), in the form

$$\left. \begin{aligned} (\rho_0 + \hat{\rho})\frac{d\mathbf{v}}{dt} + 2\left(\rho_0 + \hat{\rho}\right)\Omega \times \mathbf{v} + \nabla\hat{p} + \hat{\rho}g\mathbf{k} = 0, \\ \hat{p} = p - g \int_z^0 \rho_0\, dz, \\ \hat{\rho} = \rho - \rho_0. \end{aligned} \right\} \qquad (2.10)$$

As mentioned before, the density variations in the ocean do not exceed 3–4%. Therefore $\hat{\rho}$ can be neglected compared to $\rho_0(z)$ in (2.10) for the inertial terms and for the Coriolis force:

$$\rho_0(z) \left[\frac{d\mathbf{v}}{dt} + 2\,\Omega \times \mathbf{v} \right] + \nabla\hat{p} + \hat{\rho}g\mathbf{k} = 0. \tag{2.11}$$

This approximation is called the free convection approximation[2]. As $\rho_0(z)$ differs with the same accuracy (3–4 %) from the constant value ρ_a at the free surface, we can change $\rho_0(z)$ in (2.11) to ρ_a:

$$\rho_a \left[\frac{d\mathbf{v}}{dt} + 2\,\Omega \times \mathbf{v} \right] + \nabla\hat{p} + \hat{\rho}g\mathbf{k} = 0; \qquad \rho_a = \text{const.} \tag{2.12}$$

The approximation (2.12) is called the Boussinesq approximation.

The approximations mentioned are used to a different degree in this book. The adiabatic approximation is used later in the whole book. The Coriolis force is taken into account in the 'traditional' approximation (2.4) only (except of Chapter 3.7). The incompressibility approximation is used almost everywhere in the form (2.5), which is a shortcoming of this book. There are some exceptions for the 'rigid lid' approximation, the free convection approximation in the form (2.11), and the Boussinesq approximation in the form (2.12) in this book.

Thus, along the whole book we will use the following equations for the description of internal waves:

the momentum balance equation

$$\rho \left(\frac{d\mathbf{v}}{dt} + f\mathbf{k} \times \mathbf{v} \right) = -\nabla p - \rho g\mathbf{k}; \tag{2.13}$$

the continuity equation

$$\text{div}\ \mathbf{v} = 0; \tag{2.14}$$

and the adiabaticity equation in the forms

$$\frac{d\rho}{dt} = \frac{1}{c^2}\frac{dp}{dt}, \tag{2.15}$$

$$\frac{d\rho}{dt} = 0. \tag{2.15a}$$

The basic equations (2.13)–(2.15) are simplified in some particular cases which will be considered below.

2. EQUATIONS FOR PLANE FINITE AMPLITUDE INTERNAL WAVES

For relatively short or high frequency waves the Earth's rotation may be neglected. Then considering only short waves propagating in the x, z-plane, i.e., assuming $\mathbf{v}(\mathbf{x}, z, t) = \mathbf{v}(x, z, t)$, where $\mathbf{v} = \{u, 0, w\}$ and $\rho = \rho(x, z, t)$, $p = p(x, z, t)$, we can derive a stream function $\psi(x, z, t)$ from:

$$u = \frac{\partial \psi}{\partial z}; \qquad w = -\frac{\partial \psi}{\partial x}. \tag{2.16}$$

Using the stream function ψ, the combined equations (2.13), (2.14), (2.15a) (i.e., the isopycnic condition) may be transformed by a cross derivation to two equations for ψ and ρ :

$$\rho \left[\frac{\partial}{\partial t} \Delta \psi + J(\Delta \psi, \psi) \right] = g \frac{\partial \rho}{\partial x} - \left\{ \frac{\partial \rho}{\partial z} \left[\frac{\partial^2 \psi}{\partial t \, \partial z} + J\left(\frac{\partial \psi}{\partial z}, \psi \right) \right] \right.$$
$$\left. + \frac{\partial \rho}{\partial x} \left[\frac{\partial^2 \psi}{\partial t \, \partial x} + J\left(\frac{\partial \psi}{\partial x}, \psi \right) \right] \right\}; \tag{2.17}$$

$$\frac{\partial \rho}{\partial t} + J(\rho, \psi) = 0, \tag{2.18}$$

where $\Delta = \left\{ \frac{\partial^2}{\partial x^2} + \frac{\partial^2}{\partial z^2} \right\}$ and $J(\alpha, \beta) = \partial(\alpha, \beta)/\partial(x, z)$ is the Jacobian function.

For the Boussinesq approximation the term in curly brackets in (2.17) must vanish, and ρ in the first term of the l.h.s. must be changed to $\rho_a = $ const. For the free convection approximation, ρ on the l.h.s. of the equation must be assumed as $\rho_0(z)$, $\partial \rho / \partial z$ in the first term in curly brackets must be changed to $d\rho_0/dz$, and the second term must be neglected. If ψ and ρ are known the pressure may then be derived from

$$\Delta p - \frac{1}{\rho} \left[\frac{\partial \rho}{\partial x} \frac{\partial p}{\partial x} + \frac{\partial \rho}{\partial z} \frac{\partial p}{\partial z} \right] = 2 \left(\frac{\partial^2 \psi}{\partial z \, \partial x} \right)^2 - 2 \frac{\partial^2 \psi}{\partial x^2} \frac{\partial^2 \psi}{\partial z^2}, \tag{2.19}$$

where terms in the rectangular brackets vanish in the Boussinesq approximation. The boundary conditions (2.27), (2.29), (2.30) for combined equations (2.17), (2.18) are

$$\frac{\partial \psi}{\partial x} = -\frac{\partial \psi}{\partial z}\frac{\partial H}{\partial x} \quad \text{when} \quad z = -H(x); \tag{2.20}$$

$$\frac{\partial \psi}{\partial x} = -\frac{\partial \zeta}{\partial t} - \frac{\partial \psi}{\partial z}\frac{\partial \zeta}{\partial x} \quad \text{when} \quad z = \zeta(x,t); \tag{2.20a}$$

$$p(x,z,t)\,|_{z=\zeta} = p_a(x,t). \tag{2.20\,b}$$

For a flat bottom $H = \text{const}$ and neglecting surface waves, i.e., for $w = 0$ and $z = 0$, the boundary conditions for the combined equations (2.17), (2.18) have the simple form

$$\frac{\partial \psi}{\partial x}\,|_{z=-H} = \frac{\partial \psi}{\partial x}\,|_{z=0} = 0. \tag{2.21}$$

As the stream function is defined with an accuracy to an arbitrary function of time, the boundary conditions (2.21) may be rewritten as

$$\psi\,|_{z=0} = 0, \quad \psi\,|_{z=-H} = C(t), \tag{2.22}$$

where $C(t)$ has a meaning of the total fluid flow through a 'cross-section of the ocean'.

$$\int_{-H}^{0} u\,dz = \int_{-H}^{0} \frac{\partial \psi}{\partial z}\,dz = \psi\,|_{z=0} - \psi\,|_{z=-H} = -C(t).$$

For wave propagation in undisturbed water, i.e., in the absence of the background flow caused by external sources, the total flow of incompressible fluid through the cross-section of the ocean is independent of t. Moreover, damping of the fluid's motion should be assumed at infinity, i.e., $\{u, w, \rho, p\} \to 0$ when $|x| \to \infty$. These requirements define the function $C(t)$ at once in the boundary conditions (2.22) to $C = 0$. Then conditions (2.22) become

$$\psi\,|_{z=-H} = \psi\,|_{z=0} = 0. \tag{2.23}$$

Eqs. (2.17), (2.18) are considerably simplified while considering stationary waves. Stationary waves (waves of stationary type) mean waves
propagating without variation of shape. The stream function ψ, the
density ρ, and the pressure p for such waves may be expressed as:

$$\psi = \psi(\vartheta, z); \quad \rho = \rho(\vartheta, z); \quad p = p(\vartheta, z), \qquad (2.24)$$

$\vartheta = x - c_f t$ and c_f is the phase velocity. Substituting (2.24) in the
equation of isopycnicity (2.18) we obtain $J(\rho, S) = 0$ with $S = \psi - c_f z$.
Then

$$\rho = \rho(S) = \rho(\psi - c_f z), \qquad (2.25)$$

where ρ is an arbitrary function of S. After substitution of (2.25) to
equation (2.17) and some transformations we obtain

$$J\left\{ \Delta S + \varphi'(S) \left[\frac{1}{2}\left(\frac{\partial S}{\partial \vartheta}\right)^2 + \frac{1}{2}\left(\frac{\partial S}{\partial z}\right)^2 + gz \right], S \right\} = 0, \qquad (2.26)$$

where

$$\Delta = \left\{ \frac{\partial^2}{\partial \vartheta^2} + \frac{\partial^2}{\partial z^2} \right\},$$

$$\varphi'(S) = \frac{d}{dS}\ln[\rho(S)] = \frac{d}{d\psi}\ln[\rho(\psi - c_f z)]. \qquad (2.27)$$

From (2.26) it follows that

$$\Delta\psi + \varphi'(\psi - c_f z)\left[\frac{1}{2}\left(\frac{\partial\psi}{\partial\vartheta}\right)^2 + \frac{1}{2}\left(\frac{\partial\psi}{\partial z} - c_f\right)^2 + gz \right] = \Phi'(\psi - c_f z).$$

$$(2.28)$$

$\Phi(\psi - c_f z) = \Phi(S)$ is an arbitrary function of S. In this case the pressure
is expressed by the stream function ψ and an arbitrary functions φ and
Φ by the formula (the Bernoulli integral)

$$p = \Phi(\psi - c_f z) - g\varphi(\psi - c_f z)z - \frac{1}{2}\rho(\psi - c_f z)\left[\left(\frac{\partial\psi}{\partial\vartheta}\right)^2 + \left(\frac{\partial\psi}{\partial z} - c_f\right)^2 \right].$$

$$(2.29)$$

Eq. (2.28) was derived for $c_f = 0$ by Dubreil-Jacotin [66] and then
independently by Long [215]. For $c_f \neq 0$ it was derived first, apparently,
by Magaard [226].

For equation (2.28) the boundary condition at the bottom, (2.20), will not change and the kinematic condition at the fluid's free surface is:

$$-\frac{\partial \psi}{\partial \vartheta} = \left(\frac{\partial \psi}{\partial z} - c_f\right) \frac{\partial \zeta}{\partial \vartheta} \quad \text{when} \quad z = \zeta(\vartheta). \tag{2.30}$$

The dynamic condition at the ocean's free surface is obtained by substituting (2.29) in (2.20b) for $z = \zeta$.

Eqs. (2.28) and (2.29) were studied in [20], [21], [58], [215], [226] [350], [351]. Some general results on stationary nonlinear internal waves behaviour were obtained in these works. However, these equations contain two arbitrary unknown functions, $\rho(S)$ and $\Phi(S)$, which makes a study quite difficult. Following Leonov and Miropol'sky [206], [207], we simplify these equations starting from some physical assumptions.

Let us consider the simplest case of solitary internal waves (or solitons) on quiet water for which all of the wave motions damp when $|\vartheta| \to \infty$:

$$\psi \big|_{|\vartheta| \to \infty} \to 0, \quad \zeta \big|_{|\vartheta| \to \infty} \to 0, \quad \rho \big|_{|\vartheta| \to \infty} \to \rho_0(z).$$

$\rho_0(z)$ is the distribution of the fluid density, undisturbed by a wave and assumed to be known. Then it follows ($s = z - \psi/c_f$) that

$$\rho(s) = \rho_0 \left(z - \frac{\psi}{c_f}\right),$$

i.e., the function ρ does not depend on any other arguments except for $z - \psi/c_f$; so it is a one-valued function. Assuming that when $|\vartheta| \to \infty$ the ψ derivatives vanish also, we derive the function Φ directly from (2.28)

$$\frac{\partial \Phi}{\partial z} = \frac{d\rho_0}{dz}\left(\frac{c_f^2}{2} + gz\right) = -\rho_0(z) N^2(z)\left(z + \frac{c_f^2}{2g}\right), \tag{2.31}$$

where

$$N^2(z) = -g \frac{d \ln \rho_0(z)}{dz}.$$

From (2.31) it follows that the functions Φ $\left(\text{and } \frac{d\Phi}{ds}\right)$ are one-valued functions:

$$\frac{d\Phi}{ds} = -\rho_0(s)\, N^2(s)\left(s + \frac{c_f^2}{2g}\right);$$

$$\Phi = -\int \rho_0(s)\, N^2(s)\left(s + \frac{c_f^2}{2g}\right) ds.$$

Substituting these expressions into Eqs. (2.28), (2.29) yields the equations describing solitary waves or wave trains damping at infinity [207]:

$$\Delta\psi + \left\{\frac{\psi}{c_f^2} + \frac{1}{2g\,c_f}\left[\left(\frac{\partial\psi}{\partial\vartheta}\right)^2 + \left(\frac{\partial\psi}{\partial z} - c_f\right)^2 - c_f^2\right]\right\}$$

$$\times N^2\left(z - \frac{\psi}{c_f}\right) = 0;$$

$$p - p_0 = -g\left\{\rho_0(s)\frac{\psi}{c_f} + \int_0^s \rho_0(\alpha)\,d\alpha\right\}$$

$$+\rho_0(s)\left\{c_f\frac{\partial\psi}{\partial z} - \frac{1}{2}\left[\left(\frac{\partial\psi}{\partial\vartheta}\right)^2 + \left(\frac{\partial\psi}{\partial z}\right)^2\right]\right\};$$

$$s = z - \frac{\psi}{c_f}.$$

$$(2.32)$$

The first equation in (2.32) is very simple under the Boussinesq approximation:

$$\Delta\psi + \frac{\psi}{c_f^2}N^2\left(z - \frac{\psi}{c_f}\right) = 0. \qquad (2.33)$$

Eqs. (2.32) or (2.33) are appropriate for working out a steady theory of nonlinear internal waves and were used in [206]–[208].

3.   EQUATIONS OF LINEAR INTERNAL WAVES THEORY

Basic combined equations for small amplitudes may be linearized relatively to some quasi-stationary state. Let us study two such cases.

3.1   LINEARIZATION WITH RESPECT TO STATE OF REST

Consider the ocean to be at rest without waves and the undisturbed density $\rho_0(z)$ and $c(z)$ may depend on the depth only. Then, linearizing equations (2.13), (2.14), (2.15a) yields:

$$\frac{\partial \mathbf{v}}{\partial t} + f\,\mathbf{k} \times \mathbf{v} = -\frac{\nabla p}{\rho_0} + g\,\frac{\rho}{\rho_0}, \tag{2.34}$$

$$\operatorname{div} \mathbf{v} = 0, \tag{2.35}$$

$$\frac{\partial p}{\partial t} + g\rho_0 w = c^2 \left(\frac{\partial \rho}{\partial t} + w\,\frac{d\rho_0}{dz} \right), \tag{2.36}$$

where p and ρ are used instead of $\hat{p}$ and $\hat{\rho}$ to simplify designations, see Eqs. (1.40). The linearized boundary conditions, taking into consideration the hydrostatic equation $g\rho_0 = \frac{\partial p}{\partial z}$, are worked out as having the following form

$$\frac{\partial p}{\partial t} + g\rho_0 w = \frac{\partial p_a}{\partial t} \quad \text{at} \quad z = 0, \tag{2.37}$$

$$w = \frac{\partial \zeta}{\partial t} \quad \text{at} \quad z = 0, \tag{2.38}$$

$$w = u\,\frac{\partial H}{\partial x} + v\,\frac{\partial H}{\partial y} \quad \text{at} \quad z = -H(x,y). \tag{2.39}$$

For free internal waves of small amplitude in an ocean of constant depth H, described by Eqs. (2.34)–(2.36) for $p_a = 0$ and $H = \mathrm{const}$, an energy integral exists [262]

$$E = \frac{1}{2}\rho_0 g\,\zeta^2 + \int_0^H \left[\frac{u^2 + v^2 + w^2}{2} + \frac{g^2}{2}\left(\frac{p - c^2\rho}{\rho_0 c^2 N} \right) \right] \rho_n\,dz, \tag{2.40}$$

where the first term in brackets is the kinetic energy and the second term is the potential energy of internal waves.

By means of equations (2.34)–(2.36) one can make sure that the horizontal average of E is time independent. Expression (2.40) is free of terms containing f because the Coriolis force does not contribute any work. In the simplified boundary condition $w = 0$, when $z = 0$ the first term in (2.40) vanishes. Using the isopycnicity approximation $(c \to \infty)$ and the Boussinesq approximation, expression (2.40) becomes

$$E = \frac{1}{2}\rho_0 g\,\zeta^2 + \rho_0 \int_0^H \left[\frac{u^2 + v^2 + w^2}{2} + \frac{g^2}{2}\frac{\rho^2}{\rho_0\,N^2} \right] dz, \qquad (2.41)$$

where $N^2 = (g/\rho_0)\,(d\rho_0/dz)$. The potential energy of the internal waves may be expressed as $(g^2/2)\,(\rho^2/\rho_0^2 N^2) = \frac{1}{2}N^2\xi^2$, where

$$\xi(\mathbf{x}, z, t) = \frac{\rho(\mathbf{x}, z, t)}{d\rho_0/dz}$$

is the displacement of the undisturbed level in internal waves[3].

For a flat bottom Eqs. (2.34)–(2.36) with the boundary conditions (2.37)–(2.39) yield by cross derivation a single equation for the vertical velocity w in an internal wave:

$$\frac{\partial^2}{\partial t^2}\left[\Delta w - \frac{N^2}{g}\frac{\partial w}{\partial z} \right] + f^2\left[\frac{\partial^2 w}{\partial z^2} - \frac{N^2}{g}\frac{\partial w}{\partial z} \right] + N^2(z)\Delta_h w = 0, \quad (2.42)$$

$$\left\{ \frac{\partial^2}{\partial t^2}\left[\frac{\partial w}{\partial z} \right] + f^2\frac{\partial w}{\partial z} - g\Delta_h w \right\}\bigg|_{z=0} = \frac{\partial}{\partial t}\Delta_h p_a \qquad (2.43)$$

$$w\,|_{z=-H} = 0, \qquad (2.44)$$

where Δ is the three-dimensional and Δ_h the plane Laplacian. Thus, the Brunt–Väisälä frequency appears to be the single characteristic of the ocean's stratification limiting the behaviour of internal waves. Under the Boussinesq approximation all terms proportional to N^2/g are omitted (these terms are really negligible relative to other terms in (2.42)).

All the remaining velocity components in a wave and p, ρ, and ξ may be obtained from a known w field by solving the differential equations:

$$\frac{\partial}{\partial t} \Delta_h u = -\left(\frac{\partial^2}{\partial t\, \partial x} + f \frac{\partial}{\partial y} \right) \frac{\partial w}{\partial z};$$

$$\frac{\partial}{\partial t} \Delta_h v = -\left(\frac{\partial^2}{\partial t\, \partial y} - f \frac{\partial}{\partial x} \right) \frac{\partial w}{\partial z};$$

$$\frac{\partial}{\partial t} \Delta_h p = \rho_0 \left(\frac{\partial^2}{\partial t^2} + f^2 \right) \frac{\partial w}{\partial z};$$

$$\frac{\partial}{\partial t} \Delta_h \rho = \frac{\rho_0}{c^2} \left(\frac{\partial^2}{\partial t^2} + f^2 \right) \frac{\partial w}{\partial z} - \frac{\rho_0 N^2}{g} \Delta_h w; \qquad \frac{\partial \xi}{\partial t} = w \qquad (2.45)$$

with the corresponding boundary conditions.

3.2 LINEARIZATION WITH RESPECT TO BACKGROUND FLOW

Consider internal waves propagating in the presence of a large scale background flow $\mathbf{U}(\mathbf{x}, z, t) = \{ U(\mathbf{x}, z, t), V(\mathbf{x}, z, t), W(\mathbf{x}, z, t) \}$. Then the total velocity field will be $\mathcal{U} = \mathbf{U} + \mathbf{v}$. Assuming $|\mathbf{v}| \ll |\mathbf{U}|$ and linearizing the basic equations in the isopycnicity and Boussinesq approximations (2.14), (2.14), (2.15a) relative to an intermediate state we derive:

$$\frac{Du}{Dt} + u \frac{\partial U}{\partial x} + v \frac{\partial U}{\partial y} + w \frac{\partial U}{\partial z} + f\,(V + v) = -\frac{1}{\rho_0} \frac{\partial p}{\partial x}; \qquad (2.46)$$

$$\frac{Dv}{Dt} + u \frac{\partial V}{\partial x} + v \frac{\partial V}{\partial y} + w \frac{\partial V}{\partial z} - f\,(U + u) = -\frac{1}{\rho_0} \frac{\partial p}{\partial y}; \qquad (2.47)$$

$$\frac{Dw}{Dt} + u \frac{\partial W}{\partial x} + v \frac{\partial W}{\partial y} + w \frac{\partial W}{\partial z} = -\frac{1}{\rho_0} \frac{\partial p}{\partial z} - g \frac{\rho}{\rho_0}; \qquad (2.48)$$

$$\frac{D\rho}{Dt} + w \frac{\partial \rho_0}{\partial z} = 0; \quad \operatorname{div} \mathbf{v} = 0, \qquad (2.49)$$

where

$$\frac{D}{Dt} \equiv \frac{\partial}{\partial t} + U \frac{\partial}{\partial x} + V \frac{\partial}{\partial y} + W \frac{\partial}{\partial z}.$$

The boundary conditions (1.29), (1.30), (1.27) become:

$$w = \frac{\partial \zeta}{\partial t} + U \frac{\partial \zeta}{\partial x} + V \frac{\partial \zeta}{\partial y} \quad \text{at} \quad z = \zeta, \tag{2.50}$$

$$\frac{Dp}{Dt} = \frac{Dp_a}{Dt} \quad \text{at} \quad z = \zeta, \tag{2.51}$$

$$w = u \frac{\partial H}{\partial x} + v \frac{\partial H}{\partial y} \quad \text{at} \quad z = -H(x,y). \tag{2.52}$$

The boundary conditions (2.50)–(2.52) assume that the large scale flow satisfies

$$W = 0 \quad \text{at} \quad z = \zeta,$$

$$W = U \frac{\partial H}{\partial x} + V \frac{\partial H}{\partial y} \quad \text{at} \quad z = -H(x,y).$$

In an important particular case when $\mathbf{U} = \{U(z), V(z), 0\}$ and $f \equiv 0$, Eqs. (2.46)–(2.49) have the form

$$\frac{Du}{Dt} + w \frac{dU}{dz} + \frac{1}{\rho_0} \frac{\partial p}{\partial x} = 0; \quad \frac{Dv}{Dt} + w \frac{dV}{dz} + \frac{1}{\rho_0} \frac{\partial p}{\partial y} = 0; \tag{2.53}$$

$$\frac{Dw}{Dt} + \frac{1}{\rho_0} \frac{\partial p}{\partial z} + g \frac{\rho}{\rho_0} = 0; \quad g \frac{D}{Dt} \left(\frac{\rho}{\rho_0} \right) - N^2(z) w = 0; \quad \text{div } \mathbf{v} = 0,$$

where

$$\frac{D}{Dt} \equiv \frac{\partial}{\partial t} + U \frac{\partial}{\partial x} + V \frac{\partial}{\partial y}.$$

The boundary conditions (2.50)–(2.52) will not change. The energy conservation equation may be worked out from (2.52) as usual

$$\frac{\partial E}{\partial t} + \nabla \mathbf{F} = -R, \tag{2.54}$$

where the total waves energy E, the wave energy flow $\mathbf{F}$ and the R value are given by

$$E = \tfrac{1}{2} \rho_0 [\mathbf{v}^2 + N^2 \xi^2]; \quad \mathbf{F} = \mathbf{U}E + \mathbf{v}p;$$

$$R = \rho_0 \left[uw \frac{dU}{dz} + vw \frac{dV}{dz} \right]. \tag{2.55}$$

Obviously from (2.55) R describes the interaction of the background flow components with the Reynolds stress. Integrating (2.54) over a closed volume G fixed in space and restricted by a surface Σ and using the Gauss theorem yields

$$\frac{\partial}{\partial t} \int_G E \, dG + \int_\Sigma \mathbf{F} \, d\Sigma = - \int_G R \, dG. \qquad (2.56)$$

From (2.56) it follows that even if the energy flow F_n through the surface Σ is absent, the integral wave energy is not conserved because of a term characterising waves' interaction with the background flow through the Reynolds stress. Internal waves in such a flow may become unsteady and grow in time.

Eqs. (2.52) yield by cross derivation a single equation for the vertical velocity component w

$$\frac{D^2}{Dt^2} \Delta w - \frac{D}{Dt} \left[\frac{d^2 U}{dz^2} \frac{\partial w}{\partial x} + \frac{d^2 V}{dz^2} \frac{\partial w}{\partial y} \right] + N^2(z) \Delta_h w = 0, \qquad (2.57)$$

which may be solved, e.g., with the simplified boundary conditions $w = 0$ at $z = 0$, $w = 0$ at $z = -H = \text{const}$. All the remaining velocity components in a wave and p, ρ, and ξ may be derived from the known w by solving the differential equations:

$$\left.\begin{aligned}
\frac{D}{Dt}\left(\frac{\partial}{\partial z}\frac{D}{Dt}u\right) &= -\left\{\frac{D}{Dt}\left(\frac{\partial}{\partial z}w\frac{dU}{dz}\right) - \frac{\partial}{\partial x}\left[\left(\frac{D^2}{Dt^2}+N^2\right)w\right]\right\}; \\[2mm]
\frac{D}{Dt}\left(\frac{\partial}{\partial z}\frac{D}{Dt}v\right) &= -\left\{\frac{D}{Dt}\left(\frac{\partial}{\partial z}w\frac{dV}{dz}\right) - \frac{\partial}{\partial y}\left[\left(\frac{D^2}{Dt^2}+N^2\right)w\right]\right\}; \\[2mm]
\frac{D}{Dt}\frac{\partial p}{\partial z} &= -\rho_0\left(\frac{D^2}{Dt^2}+N^2\right)w; \quad g\frac{D}{Dt}\left(\frac{\rho}{\rho_0}\right) = N^2 w; \quad \frac{D}{Dt}\xi = w
\end{aligned}\right\}$$
$$(2.58)$$

with the corresponding initial and boundary conditions.

Some peculiarities of mathematical problems associated with solutions of the system of equations describing oceanographic fields are discussed in [274].

Notes

1 For details see [292], where a weakly nonlinear theory of internal waves propagation in the atmosphere is considered.
2 In the literature this approximation is also called the Boussinesq approximation [217]. However, the term 'free convection approximation' is more convenient for us.
3 By definition $\zeta = \xi(\mathbf{x}, 0, t)$.

II

LINEAR THEORY OF INTERNAL WAVES

Chapter 3

LINEAR THEORY OF PROPAGATION OF INTERNAL WAVES IN THE UNDISTURBED HORIZONTALLY HOMOGENEOUS OCEAN

This chapter considers the simplest example of free harmonic internal waves of small amplitude propagating in an undisturbed horizontally homogeneous ocean. A detailed analysis of the basic boundary value problem for amplitude functions of these internal waves for a finite depth is given in Section 3.1. Examples of solutions of the basic boundary value problem for typical vertically undisturbed density distributions in the ocean are presented in Section 3.2. If the vertical scale of internal waves is small, the ocean may be considered as infinitely deep. In this case internal waves propagate not only horizontally, but also in the vertical direction. This case is examined in Section 3.3. A theory of the propagation of linear internal wave packets is worked out in Section 3.4 based on geometric optics' approximations. The influence of a fine structure of the ocean's density field on the propagation of free infinitesimal internal waves is studied in Section 3.5. The influence of the Earth's rotation (taking into account all components of the Coriolis force) on the propagation of free internal waves is considered in Section 3.6.

1.   BASIC BOUNDARY VALUE PROBLEM

To study the propagation of internal waves in the undisturbed horizontally homogeneous ocean, we will use Eqs. (2.34)–(2.36) and (2.37)–(2.39) with $H = $ const (see Section 2.3). The homogeneity of these equations and of boundary conditions with respect to variables $\mathbf{x}$, t al-

45

lows us to try elementary wave solutions in the form of plane waves

$$\{u, v, w, \rho, p, \zeta\}$$
$$= \{U(z),\ V(z),\ W(z),\ R(z),\ P(z),\ Z(z)\} \tag{3.1}$$
$$\times \exp\{i(\mathbf{kx} - \omega t)\}.$$

Here $\mathbf{k} = \{k_1,\ k_2\}$ is a wave vector, $\mathbf{x} = \{x,\ y\}$, and ω is the frequency of oscillations.

With the help of the elementary wave solutions (3.1) it is possible also to work out a solution of the boundary value problem with the initial data (2.34)–(2.39) by a Fourier transformation in terms of variables $\mathbf{x}$, t.

Using (3.1) and relations (2.45) we can derive expressions for the amplitude functions on the r.h.s. of (3.1) through $W(z)$

$$\left.\begin{aligned}
i\omega k^2 U(z) &= (k_1\omega - ik_2 f)\,W', \\
i\omega k^2 V(z) &= (k_2\omega + ik_1 f)\,W', \\
i\omega k^2 P(z) &= \rho_0(\omega^2 - f^2)\,W', \\
i\omega Z(z) &= W(0), \\
-i\omega R(z) &= \frac{\rho_0}{g} N^2(z)\,W + \frac{\rho_0(\omega^2 - f^2)}{c^2 k^2}\,W', \\
k^2 &= k_1^2 + k_2^2 = |\mathbf{k}|^2.
\end{aligned}\right\} \tag{3.2}$$

For $W(z)$, using equation (2.42), the boundary conditions (2.43) for $p_a = 0$ or $W|_{z=0} = 0$ (the 'rigid lid' approximation) and (2.44), we will obtain the following boundary value problem (a prime means derivation with respect to z):

$$W'' - \frac{N^2(z)}{g} W' + \frac{N^2(z) - \omega^2}{w^2 - f^2}\, k^2 W = 0, \tag{3.3}$$

$$W(-H) = 0, \quad W'(0) = \frac{gk^2}{w^2 - f^2}\, W(0), \tag{3.4}$$

$$W(-H) = 0, \quad W(0) = 0. \tag{3.4a}$$

Further, we will study a boundary value problem for equation (3.3) taking into account both a total boundary condition at the free surface (3.4), thus considering both surface and internal waves (the basic boundary value problem **A**), and filtering surface waves with the boundary condition (3.4a) (the basic boundary value problem **B**).

With

$$N^2(z) \equiv \mu(z); \quad \rho_*(z) = \rho_0(z) \exp\left\{ -g \int_z^0 \frac{dz}{c^2(z)} \right\}; \tag{3.5}$$

$$\lambda = \frac{k^2}{w^2 - f^2}; \quad \rho_*(0) = \rho_0(0) = \rho_a.$$

We obtain from (3.3), (3.4) (taking into consideration a definition for $\rho_*(z)$):

$$\mu(z) \equiv N^2(z) \equiv -\left(\frac{g}{\rho_0} \frac{d\rho_0}{dz} + \frac{g^2}{c^2(z)} \right) = -\frac{g}{\rho_*(z)} \frac{d\rho_*(z)}{dz}. \tag{3.6}$$

Using (3.5) and (3.6), we rewrite the problem (3.3), (3.4) and (3.4a) in a form more convenient for further analysis

$$(\rho_* W')' + [\lambda(\mu(z) - f^2) - k^2]\rho_*(z) W = 0, \tag{3.7}$$

$$W(-H) = 0, \quad W'(0) = \lambda g W(0), \tag{3.8}$$

$$W(-H) = 0, \quad W(0) = 0. \tag{3.8a}$$

Now $f^2 \sim 10^{-8}$ s^{-2} which is at least three orders less than $g^2/c^2 \approx 5 \times 10^{-5}$ s^{-2}. So we can omit f^2 in (3.7) in comparison with $\min \mu(z)$ and to consider in any case

$$\min \mu(z) > f^2. \tag{3.9}$$

The inequality (3.9) allows a complete analysis of the boundary value problem (3.7)–(3.8a).

Consider parameter k to be real ($k^2 > 0$). Then the basic boundary value problem (**A** and **B**) is reduced to the derivation of nontrivial solutions of (3.7)–(3.8a), i.e., of eigenfunctions $W(k, \lambda, z)$ and eigen-curves

$\lambda(k)$, which allows us to obtain a dispersion equation $\omega = \omega(k)$ from (3.5).

The spectral problem described corresponds to the derivation of eigenfunctions $W_n(k,z)$ and eigenvalues (frequency) ω_n for every fixed value of wave number k when $W_n(k,z) \equiv W(k,\omega_n(k),z)$. Another state of this spectral problem is possible when the eigenfunction $\widetilde{W}_n(\omega, z)$ and eigenvalues $k_n(\omega)$ for every fixed value of frequency ω are derived and $\widetilde{W}_n(\omega, z) \equiv \widetilde{W}(\omega, k_n(\omega), z)$.

Although dispersion curves $\omega_n(k)$ or $k_n(\omega)$ are defined with the same relations in both states, the eigenfunctions $W_n(k, z)$ and $\widetilde{W}_n(k, z)$ are different.

We shall study in detail only the first of the problems mentioned, i.e., we shall consider k to be given and $\omega_n(k)$ and $W_n(k,z)$ to be derived, as this case is more frequent for applied problems. The case when ω is given and $k_n(\omega)$ and $\widetilde{W}_n(\omega, z)$ are unknown will be briefly discussed at the end of the section.

We state some simple features of the problem (3.7)–(3.8a)[1].

1. The boundary value problem (3.7)–(3.8a) is self adjoint under condition (3.9).

To prove this, we will subtract Eq. (3.7) multiplied by W from (3.7) written for complex conjugate values $(\lambda^*, W^*))$ and multiplied by W^*. Integrating the obtained expression over z from $-H$ to 0 and using (3.8) yields the integral equation

$$(\lambda - \lambda^*) \left\{ \rho_a g \, |W(0)|^2 + \int_{-H}^{0} [\mu(z) - f^2] \rho_*(z) \, |W|^2 \, dz \right\} = 0, \quad (3.10)$$

from which it follows that $\lambda = \lambda^*$ under condition (3.9). Thus also the eigenfunctions W of the basic boundary value problem are real.

2. Eigen-curves of the boundary value problem (3.7)–(3.8a) $\lambda(k^2) > 0$ under condition (3.9).

A proof is based on the integral equation

$$
\lambda \left\{ \rho_a g W^2(0) + \int_{-H}^{0} (\mu - f^2) \rho_* W^2 dz \right\}
$$

$$
= k^2 \int_{-H}^{0} \rho_* W^2 dz + \int_{-H}^{0} \rho_* W'^2 dz,
$$

(3.11)

which may be derived multiplying (3.7) to W and integrating over z from $-H$ to 0, taking into account (3.8). As $W = \text{const}$ is not an eigenfunction of the problem (3.7)–(3.8a), from (3.11) it follows that for any k^2 $\lambda(k^2) \equiv k^2/(\omega^2 - f^2) > 0$, i.e., $\lambda \equiv 0$ is not an eigen-curve of the problem (3.7)–(3.8a).

Hence there follows the important inequality

$$
\omega^2 > f^2 > 0,
$$

(3.12)

showing that eigenfrequencies (or dispersion curves) $\omega(k)$ are real.

Generally speaking, a reality of $\omega(k)$ follows directly from the energy conservation law (2.40), [70]. But if inequality (3.9) is not satisfied, the boundary value problem (3.7)–(3.8a) may have both real and complex eigenfrequencies $\omega(k)$, as it can not be self-adjoint in this case.

3. If there are at least two eigen-curves $\lambda_m(k^2)$ or $\lambda_j(k^2)$, and besides $\lambda_m \neq \lambda_j$, eigenfunctions $W_m(z)$ and $W_j(z)$ corresponding to these eigen-curves satisfy the orthogonality condition

$$
\rho_a g W_m(0) W_j(0) + \int_{-H}^{0} \rho_* (\mu - f^2) W_m W_j \, dz = 0.
$$

(3.13)

To prove (3.13) let us carry out the following procedure. Write equation (3.7) for W_m, λ_m and multiply it to W_j. Then write equation (3.7) for W_j, λ_j and multiply it to W_m. And, finally, subtract the second combination from the first one. Integrating the equation obtained over z from $-H$ to 0, and taking into consideration (3.8), we shall obtain (3.13).

Condition (3.9) is not required to satisfy equation (3.13).

4. For the boundary value problem (3.7)–(3.8a), when satisfying (3.9), there is a denumerable set of eigen-curves $\{\lambda_m(k)\}$ and a corresponding denumerable set of eigenfunctions $\{W_m(z)\}$ $(m = 0, 1, 2, \ldots)$, complete in some Hilbert space.

Here $\lambda_0(k)$ and $W_0(z)$ appear in the boundary value problem **A** only, and for problems **A** and **B** $\{W_m(z)\}$ $(m = 0, 1, 2, \ldots)$ corresponds for every eigen-curve $\{\lambda_m(k)\}$.

To prove this important statement we examine a differential operator

$$\mathcal{L}(y) \equiv [\rho_*(z)\, y']' - k^2 \rho_*(z)\, y$$

for monotonic functions, defined in the range $-H \leq z \leq 0$, and obtain its Green's function [331]

$$G(z,\ \xi) = \begin{cases} y_1(z)\, y_2(\xi) & (z \leq \xi) \\ y_2(z)\, y_1(\xi) & (z \geq \xi) \end{cases} \quad (-H \leq z,\ \xi \leq 0).$$

Here $y_i(z)$ $(i = 1, 2)$ are two fundamental solutions of the equation $\mathcal{L}(y) = 0$ chosen so that $y_1(-H) = 0$, $y_2'(0) = 0$, $y_1' y_2 - y_2' y_1 = [\rho_*(z)]^{-1}$, which is always possible for the Wronskian y_1, y_2. As $\lambda = 0$ is not an eigen-curve of the boundary value problem (3.7)–(3.8a), $y_i(z)$ are not eigenfunctions of $\mathcal{L}(y)$. Thence

$$y_1'(0) \neq 0, \qquad y_2'(-H) \neq 0.$$

Using $G(z,\ \xi)$ we can reduce the problem (3.7)–(3.8a) to a corresponding integral equation with a positively defined symmetric kernel

$$\left.\begin{array}{c} \widetilde{W}(z) = \lambda \displaystyle\int_{-H}^{0} \widetilde{G}(z,\ \xi)\, \widetilde{W}(\xi)\, d\xi \\[2mm] + \lambda\, g\, W(0)\, y_1(z)\, \dfrac{\sqrt{\mu(z)-f^2}}{y_1'(0)}; \\[4mm] \widetilde{W}(z) = W(z)\, \sqrt{\mu(z) - f^2}; \\[2mm] \widetilde{G}(z,\xi) = \sqrt{(\mu - f^2)\, (\mu(\xi) - f^2)}\, G(z,\xi). \end{array}\right\} \qquad (3.14)$$

Eq. (3.14) belongs to the class of "oaded' integral equation studied in [169]. The Hilbert–Schmidt and Mercer theorems [331], proving the completeness of the corresponding denumerable $\{W_m(z)\}$ and $\{\lambda_m(k)\}$ systems, remain valid for such equations.

Eqs. (3.11), (3.13) under condition (3.9) allow us to introduce a Hilbert space G of real functions defined in the range $[-H,\,0]$ and summed with a square, for which a scalar product and a valuation are obtained as

$$
\left.
\begin{aligned}
(F,\varphi) &= \rho_a\, g\, F(0)\,\varphi(0) + \int_{-H}^{0} \rho_*(\mu - f^2)\, F\varphi\, dz; \\[2em]
\|F\|^2 &= \rho_a\, g\, F^2(0) + \int_{-H}^{0} \rho_*(\mu - f^2)\, F^2\, dz; \qquad F,\varphi \in G.
\end{aligned}
\right\}
\tag{3.15}
$$

The complete combined functions $\{W_m(z)\} \in G$ present an orthogonal basis in G, for which

$$
\left.
\begin{aligned}
(W_m, W_j) &\equiv \rho_a\, g\, W_m(0)\, W_j(0) \\
&\quad + \int_{-H}^{0} \rho_*(\mu - f^2) W_m W_j\, dz = 0 \quad (m \neq j), \\
(W_m, W_m) &\equiv \|W_m\|^2 = \rho_a\, g\, W_m^2(0) \\
&\quad + \int_{-H}^{0} \rho_*(\mu - f^2) W_m^2\, dz > 0.
\end{aligned}
\right\}
\tag{3.16}
$$

In particular, using the second equation of (3.16), it is possible to introduce an orthonormalized system $\{W_m^0(z)\}$, for which

$$
(W_m^0,\, W_j^0) = \delta_{mj}, \qquad W_m^0(z) = W_m(z)\,\|W_m\|^{-1}.
$$

Here δ_{mj} is the Kroneker symbol.

From the completeness of the system $\{W_m(z)\}$ it also follows that any function $\varphi(z)$ continuous in the range $[-H,\,0]$ and satisfying the condition $\varphi(-H) = 0$ may be developed as a Fourier series in terms of $\{W_m(z)\}$

$$
\varphi(z) = \sum_{m=0}^{\infty} \varphi_m W_m(z); \qquad \varphi_m = \|W_m\|^{-2}(\varphi,\, W_m),
\tag{3.17}
$$

which converges in $[-H, 0]$ to $\varphi(z)$ absolutely and uniformly. For every function summed with a square $\varphi \in G$, a series (3.17) converges to $\varphi(z)$ in a mean-square value. We obtain direct expressions for Fourier coefficients φ_m using definitions (3.15), (3.16)

$$\varphi = \frac{(\varphi,\ W_m)}{\|W_m\|^2} = \frac{\rho_a\, g\, \varphi(0)\, W_m(0) + \int\limits_{-H}^{0} \rho_*(\mu - f^{\,2})\, \varphi W_m\, dz}{\rho_a\, g\, W_m^2(0) + \int\limits_{-H}^{0} \rho_*(\mu - f^{\,2}) W_m^2\, dz}. \tag{3.18}$$

5. A solution of the inhomogeneous boundary value problem under condition (3.9) is:

$$(\rho_* F')' + (\lambda\tilde{\mu} - k^{\,2})\,\rho_* F = \rho_* \chi_1(z), \qquad (\tilde{\mu} \equiv \mu - f^{\,2}), \tag{3.19}$$

$$F(-H) = 0, \qquad F'(0) - \lambda g\, F(0) = \chi_0, \tag{3.20}$$

$$F(-H) = 0, \qquad F(0) = 0, \tag{3.20a}$$

where λ does not belong to any of the eigen-curves $\lambda_m(k)$ of the boundary value problem (3.7)–(3.8a) and looks like

$$F = \sum_{m=0}^{\infty} F_m W_m(z); \qquad F_m = \frac{-\rho_a\, \chi_0\, W_m(0) + \int\limits_{-H}^{0} \rho_* \chi_1 W_m\, dz}{(\lambda - \lambda_m)\|W_m\|^2}. \tag{3.21}$$

Here $W_m(z)$ is an eigenfunction of the problem (3.7)–(3.8a) corresponding to an eigen-curve $\lambda_m(k)$.

If $\lambda = \lambda_m(k)$, the problem (3.19), (3.20) may be solved when

$$-\rho_a\, \chi_0\, W_m(0) + \int\limits_{-H}^{0} \rho_* \chi_1 W_m\, dz = 0. \tag{3.22}$$

The solution (3.21) is obtained by a standard Fourier transformation with a $W_m(z)$ series expansion of F. Eqs. (3.21), (3.22) are written for a solution of the boundary value problem **A**. Corresponding formulae for the inhomogeneous problem **B** are obtained from (3.21), (3.22) when $W_m(0) = 0$, here $W_0(z) \equiv 0$.

6. Consider a variational problem [64] for the boundary value problem (3.7)–(3.8a).

$$(\rho_* \delta W')' + (\lambda \tilde{\mu} - k^2) \rho_* \delta W + (\tilde{\mu} \, d\lambda - dk^2) \rho_* W = 0, \qquad (3.23)$$

$$\delta W(-H) = 0, \qquad \delta W'(0) - \lambda g \, \delta W(0) = g \, W(0) d\lambda, \qquad (3.24)$$

$$\delta W(-H) = 0, \qquad \delta W(0) = 0. \qquad (3.24a)$$

Here a variation of an eigenfunction δW corresponds to a shift of a point λ, k^2 along an eigen-curve $\lambda(k^2)$ by values $d\lambda$, dk^2. I.e., for $W = W(\lambda(k^2), k^2, z)$ we have

$$\delta W = \left(\frac{\partial W}{\partial k^2} + \frac{\partial W}{\partial \lambda} \frac{d\lambda}{dk^2} \right) dk^2.$$

The inhomogeneous boundary value problem (3.23), (3.24) for δW corresponds to (3.19), (3.20) if we consider $\chi_1(z) = -(\tilde{\mu} d\lambda - dk^2) W$, $\chi_0 = g W(0) \, d\lambda$. Then, as $\lambda(k^2)$ is an eigen-curve of the boundary value problem (3.7)–(3.8a), to solve (3.23), (3.24) it is necessary and sufficient to meet condition (3.22), from which follows

$$\frac{d\lambda}{dk^2} = \frac{1}{\|W\|^2} \int_{-H}^{0} \rho_* W^2 \, dz > 0. \qquad (3.25)$$

Here $\|W\|^2$ is defined by the second formula in (3.16).

Eq. (3.25) indicates a monotonicity of eigenfunctions $\lambda(k^2)$.

7. When considering Eq. (3.11) to be a functional, which determines the dependence $\lambda = \lambda(W, k^2)$, varying (3.25), using an integral identity obtained by multiplication of (3.23) to δW, integrating the resulting equation between $[-H, 0]$, and taking into account (3.8) and (3.24) yields finally

$$\frac{d^2\lambda}{(dk^2)^2} = \|W\|^{-2} \left\| \frac{\partial W}{\partial k^2} \right\| \left\{ \lambda(W, k^2) - \lambda\left(\frac{\partial W}{\partial k^2}, k^2 \right) \right\},$$

$$\left(\frac{\partial W}{\partial k^2} = \frac{\partial W}{\partial k^2} + \frac{\partial W}{\partial \lambda}\frac{\partial \lambda}{\partial k^2}\right).$$

8. For evaluation of eigenfunctions $\lambda\,(k^2)$ and of correspondent dispersion curves $\omega\,(k)$ the following identity [154] will be useful

$$\left(g^2\lambda - \frac{k^2}{\lambda^2} - f^2\right)\int_{-H}^{0}\rho_*W^2\,dz = \lambda\int_{-H}^{0}\rho_*^{-1}\varphi^2\,dz;$$

$$\varphi = \rho_*(gW - \lambda^{-1}W'). \tag{3.26}$$

Identity (3.26) may be obtained when considering instead of (3.7)–(3.8a) a boundary value problem for the following combined equations

$$W' + \lambda\rho_*^{-1}\varphi - g\,\lambda\,W = 0;$$

$$\varphi' + g\lambda\varphi + (f^2 + k^2\lambda^{-1} - g^2\lambda)\,\rho_*W = 0;$$

$$W\,(-H) = 0, \quad \varphi\,(0) = 0\ (A) \quad W\,(0) = 0\ (B),$$

and then to multiply the first equation by φ, the second equation by W, to sum the results of multiplication and to integrate the latter between limits $[-H,\,0]$.

From (3.26) and (3.12), taking into account that $\lambda = k^2/(\omega^2 - f^2)$, we obtain an evaluation for dispersion curves

$$f^2 < \omega^2 < \frac{1}{2}f^2 + \sqrt{1/4\,f^2 + g^2k^2}. \tag{3.27}$$

9. Now we give without proof basic properties of the dispersion curves and the eigenfunctions of the basic boundary value problem (3.7)–(3.8a) **A** and **B**. A deduction of these results is found in [154] based on the Sturm theorem and the comparison theorem for (3.7)–(3.8a).

The eigenfunction $W_0(z)$, appearing in the problem (3.7), (3.8) only, decreases monotonically (and rapidly) with depth from $W_0(0) \neq 0$ to $W_0(-H) = 0$. A corresponding dispersion curve $\omega_0^2\,(k^2)$ increases monotonically from f^2 when k^2 grows from $k^2 = 0$, and $\omega^2 \to g\,|k|$ for $k^2 \to \infty$. So inequality (3.27) gives an exact asymptotic evaluation of ω^2 when $k^2 \to \infty$. An eigenfunction $W_0(z)$ and a dispersion curve $\omega_0^2\,(k^2)$ correspond to surface waves; these functions weakly depend on stratification parameters.

For $n \geq 1$, which corresponds to normal modes of internal waves, dispersion curves $\omega_{A,n}$ for problem (3.7), (3.8) (with the exact edge condition at the free surface) and $\omega_{B,n}$ for problem (3.7), (3.8a) (the 'rigid lid' approximation) satisfy inequalities

$$\omega_{A,n}^2 < \omega_{B,n}^2 < \omega_{A,\,n-1}^2, \qquad n = 1,\ 2,\ \dots. \tag{3.28}$$

The n-th eigenfunction $W_n(z)$ for problems **A** and **B** has $n - 1$ zeros inside the range $[-H,\ 0]$, and $W_{A,n}(0)$ decreases rapidly while n increases.

The condition of monotonicity (3.25) of eigenfunctions $\lambda_n\,(k^2)$ for boundary value problems (**A**) and (**B**) leads to a monotonic increase of dispersion curves $\omega_{A,n}^2$ and $\omega_{B,n}^2$ while k^2 increases, and:

$$\lim_{k^2 \to 0} \omega_{A,n}^2 = \lim_{k^2 \to 0} \omega_{B,n}^2 = f^2; \quad \lim_{k^2 \to \infty} \omega_{A,n}^2 = \lim_{k^2 \to \infty} \omega_{B,n}^2 = \mu_{\max}. \tag{3.29}$$

Bilateral evaluations of dispersion curves

$$\frac{k^2 H^2 \mu_{\min} + n^2 \pi^2 l f^2}{k^2 H^2 + n^2 \pi^2 l^2} < \omega_{B,n}^2 < \frac{l k^2 H^2 \mu_{\max} + n^2 \pi^2 f^2}{l k^2 H^2 + n^2 \pi^2} \tag{3.30}$$

$$\left(l = \frac{\rho_{*\ \max}}{\rho_{*\ \min}} \right)$$

yield that while n increases in a restricted variation range of k, dispersion curves $\omega_{A,n}\,(k^2)$ and $\omega_{B,n}\,(k^2)$ are 'flattened' against the k axis, which also follows from (3.28); so when k^2 is fixed the point f^2 is a limiting point for ω_n^2 (**A** and **B**).

In (3.29) and (3.30) the values $\rho_{*\ \max}$, $\rho_{*\ \min}$, $\mu_{\max}$, $\mu_{\min}$ correspond to maximum (minimum) values of functions $\rho_*\,(z)$ and $\mu\,(z)$ in the range $[-H,\ 0]$. Fig. 3.1 illustrates the behaviour of the dispersion curves. For $n > 0$, the dispersion curves for the problems **A** and **B** are close to each other.

10. Let us consider some special cases. All equations in this section become especially simple for the boundary value problem **B** (the 'rigid

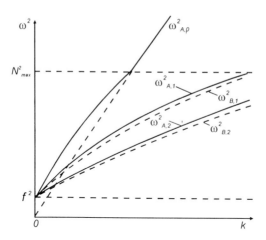

Figure 3.1.   Qualitative behaviour of dispersion curves for the basic boundary value problem.

lid' approximation) when surface waves are filtered. Then terms outside the integral in all of integral identities vanish.

If the internal waves are short enough, so that the Earth's rotation may be neglected, we may consider $f \equiv 0$ in all equations in this section. In such a case $\lambda = k^2/\omega^2 = [c_f(k)]^{-2}$, where c_f is the phase velocity of a wave, $\mu\rho_* = -g\rho'_* > 0$, and Eq. (3.11) gives a definition of the phase velocity of the m-th mode

$$c^2_{f,m} = g \, \frac{\rho_a W^2_m(0) + \int\limits_{-H}^{0} \rho'_* W^2_m \, dz}{k^2 \int\limits_{-H}^{0} \rho_* W^2_m \, dz + \int\limits_{-H}^{0} \rho_* W'^2_m \, dz}; \qquad m = 0, 1, 2 \dots . \quad (3.31)$$

For the group velocity of m-th mode $c_{g,m} = d\omega_m/dk$ from (3.25) and (3.31), taking into account that $\lambda = c_f^{-2}$, we have

$$c_{g,m} = c_{f,m} \, \frac{\int\limits_{-H}^{0} \rho_* W'^2_m \, dz}{\int\limits_{-H}^{0} \rho_* \left(k^2 W^2_m + W'^2_m\right) \, dz} < c_{f,m}. \qquad (3.32)$$

So the phase velocity is always greater than the group velocity[2].

For an isopycnic approximation $(c \to \infty) \sim (d\rho/dt = 0)$, it is necessary to replace ρ_* by $\rho_0(z)$ in all equations of this section and to omit the term f^2 in the expression $\mu(z) - f^2$, as $g^2/c^2 \gg f^2$. Thus, $\rho_* (\mu(z) - f^2) \approx -g\,\rho_0'$. Nevertheless, the Coriolis parameter should be retained in the expression for $\lambda = k^2 (\omega^2 - f^2)^{-1}$.

Now consider long waves, when $k \to 0$ (exactly $kH \ll 1$). As $\omega^2 = \omega^2(k^2)$, $\lambda = \lambda(k^2)$, then taking into consideration results of paragraph **9**, we can find the dependence $\omega^2(k^2)$ in terms of the expansion

$$\omega^2 = f^2 + \frac{1}{\lambda^0} k^2 + \alpha k^4 + \mathcal{O}(k^6). \tag{3.33}$$

Then with $\lambda = k^2 (\omega^2 - f^2)^{-1}$ we have for $\lambda(k^2)$ the expansion:

$$\lambda = \lambda^0 - (\lambda^0)^2 k^2 + \mathcal{O}(k^4). \tag{3.34}$$

The value of λ^0 is an eigenvalue of a self-adjoint boundary value problem obtained from (3.7)–(3.8a) under condition (3.9) when $k^2 \to 0$:

$$(\rho_* W^{0\prime})\prime + \lambda^0 (\mu - f^2) \rho_* W^0 = 0; \tag{3.35}$$

$$W^0(-H) = 0, \quad W^{0\prime}(0) = \lambda^0 g W^0(0); \tag{3.36}$$

$$W^0(-H) = 0, \quad W^0(0) = 0. \tag{3.36a}$$

From the results of paragraph **4** it follows that for the boundary value problem (3.35)–(3.36a) there is a denumerable system of positive eigenvalues $\{\lambda_n^0\}$ and a corresponding denumerable system of eigenfunctions $\{W_n^0(z)\}$ $(n = 0, 1, 2, \ldots)$. All of the formulae remain the same when proceeding to the limit $k^2 \to 0$.

In particular, from (3.11) yields

$$\lambda_n^0 = \frac{1}{\|W_n^0\|^2} \int_{-H}^{0} \rho_*(W_n^{0\prime})^2 \, dz;$$

$$\|W_n^0\| = \rho_a g \left(W_n^0(0)\right)^2 + \int_{-H}^{0} (\mu - f^2) \rho_*(W_n^0)^2 \, dz. \tag{3.37}$$

The parameter α in expansions (3.33), (3.34) is easily obtained from (3.25):

$$\alpha = \frac{1}{(\lambda^0)^2 \|W^0\|^2} \int_{-H}^{0} \rho_* (W^{0\prime})^2 \, dz < 0. \tag{3.38}$$

Thus for every dispersion branch ($n = 0$, 1, 2, ...) parameters α_n are defined with formula (3.38), where $W^0 = W_n^0$, and besides it is essential that all $\alpha_n < 0$.

For $f = 0$ Eqs. (3.33)–(3.38) are slightly simplified. In particular, in Eq. (3.34) the value $\lambda_0 = (c_f^0)^{-2}$, therefore

$$\left. \begin{aligned}
\lambda &= \frac{1}{(c_f^0)^2} - \frac{\alpha}{(c_f^0)^4} k^2 + \mathcal{O}(k^4); \\
\omega &= c_f^0 k + \frac{\alpha}{2 c_f^0} k^3 + \mathcal{O}(k^5); \\
c_f &= c_f^0 + \frac{\alpha}{2 c_f^0} k^2 + \mathcal{O}(k^4); \\
c_g &= c_f^0 + \frac{3}{2} \frac{\alpha}{c_f^0} k^2 + \mathcal{O}(k^4).
\end{aligned} \right\} \tag{3.39}$$

Here c_f^0 is the wave phase velocity when $k \to 0$. For every normal mode ($n = 0$, 1, 2, ...) the values of λ_n, ω_n, $c_{f,n}$, $c_{g,n}$ are defined from (3.39) for $\alpha \to \alpha_n$, $c_f^0 \to c_{f,n}^0$.

Considering in (3.39) the coefficient at k^3 in the expansion for $\omega(k)$ to be d, we have

$$d \equiv \frac{\alpha}{2 c_f^0} = -\frac{(2 c_f^0)^3}{2 \|W^0\|^2} \int_{-H}^{0} \rho_* (W^0)^2 \, dz < 0. \tag{3.40}$$

Here $\|W^0\|^2$ is obtained from (3.37) when $f^2 = 0$. At least, in the 'rigid lid' approximation, we have

$$d = -\frac{(2 c_f^0)^3}{2} \int_{-H}^{0} \rho_* (W^0)^2 \, dz \left[\int_{-H}^{0} \rho_* \mu (W^0)^2 \, dz \right]^{-1}. \tag{3.41}$$

In the isopycnic approximation in Eqs. (3.40), (3.41) it is necessary to put $\rho_*(z)$ instead of $\rho_0(z)$, and $\rho_0 \mu = -g \rho_0'(z) > 0$ in addition.

A negative value of the parameter d in the dispersion relation (3.39), i.e., $\omega = c_f^0 k - |d| k^3 + \ldots$ indicates that in the approximation of weak dispersion ($k \to 0$) internal waves have negative dispersion. The latter means that in coordinates moving at the 'velocity of sound' c_f^0 an internal wave propagates in direction opposite to c_f^0 at the velocity $\Delta c_f = -|d| k^2$ [see (3.1) and (3.39)].

Let us now discuss briefly properties of the spectral problem (3.7)–(3.8a) when ω is given and $\omega_n(k)$, $\widetilde{W}_n(\omega, z)$ are unknown. In this the case dispersion curves $\omega_n(k)$ and $k_n(\omega)$ remain invariant because their properties do not depend on the method of analysis of the boundary value problem (3.7)–(3.8a). And eigenfunctions depend essentially on the fact which of the parameters is fixed: k_0 (eigenfunctions $W_n(k_0, z) \equiv W(k_0, \omega(k_0), z)$) or ω_0 (eigenfunctions $\widetilde{W}_n(\omega_0, z) \equiv \widetilde{W}(\omega_0, k_n(\omega_0), z)$).

All of the properties found for the eigenfunctions $W_n(z)$ are valid for the eigenfunction $\widetilde{W}_n(z)$ too. But the orthogonality conditions (3.16) vary:

$$(\widetilde{W}_m, \widetilde{W}_j) \equiv \rho_a g \, \widetilde{W}_m(0) \, \widetilde{W}_j(0) + \int_{-H}^{0} \rho_*(\mu - \omega^2) W_m W_j \, dz = 0;$$

$$(\widetilde{W}_m, \widetilde{W}_m) \equiv \|\widetilde{W}_m\|^2 = \rho_a g \, \widetilde{W}_m^2(0) + \int_{-H}^{0} \rho_*(\mu - \omega^2) W_m^2 \, dz > 0.$$

$$(3.16a)$$

Eq. (3.16a) may be easily obtained from (3.7)–(3.8a) directly. Eqs. (3.17) remain the same also, when taking into consideration (3.16a). If ω does not belong to any of dispersion curves $k_n(\omega)$ (or $\omega_n(k)$), a solution of the inhomogeneous boundary value problem (3.19), (3.20) is expressed by Eqs. (3.21), where $\|\widetilde{W}_m\|^2$ is obtained from (3.16a). And if $\omega = \omega_n(k)$, then a condition of solvability of (3.22) remains valid.

2. SOLUTIONS OF BASIC BOUNDARY VALUE PROBLEM FOR PARTICULAR DISTRIBUTIONS OF $N(Z)$

This problem has been studied in numerous works (e.g., [15], [65], [70], [179], [186], [213], [266], [364]). Therefore we consider here only examples which illustrate the basic statements of the theory, discussed in Section 3.1, and which are of special physical interest. Consider at first the simplest exact solutions of the basic boundary value problem (3.3)–(3.4a).

2.1   CONSTANT BRUNT–VÄISÄLÄ FREQUENCY ($N^2 = $ CONST $\equiv N_0^2 = \mu_0$)

According to inequality (3.9) we shall consider the problem under the condition $\mu_0 > f^2$. This case is invalid for examination of internal waves in upper ocean layers, because of a seasonal thermocline presence. But it gives a satisfactory approximation for deep layers and allows us to study (rather roughly) internal waves in shallow water. This easy example will also illustrate differences in solutions of the boundary value problems **A** and **B** (Section 3.1).

Using the definition $N^2(z) \equiv \mu(z)$ from (3.6) we can obtain an expression for an undisturbed density distribution $\rho_0(z)$

$$\rho_0(z) = \rho_a \exp\left\{-\frac{\mu_0}{g}z + g\int_z^0 \frac{dz'}{c^2(z')}\right\}. \qquad (3.42)$$

The solution of Eq. (3.3) for $\mu = \mu_0 = $ const $= N_0^2$ is (multiplied by an arbitrary factor)

$$W = e^{\frac{\mu_0 z}{2g}}\frac{\sin\left[l\left(1 + \frac{z}{H}\right)\right]}{\sin l} \ ;$$

$$l^2 = \frac{\mu_0 - \omega^2}{\omega^2 - f^2}k_0^2 - \left(\frac{\mu_0 H}{2g}\right)^2 \ ; \qquad (k_0 = kH). \qquad (3.43)$$

Substituting (3.43) in the boundary conditions (3.4), (3.4a) we obtain

$$\operatorname{ctg} l = \alpha l + \beta/l \ ; \qquad \alpha = \frac{g}{H(\mu_0 - f^2)} \ ;$$

$$\beta = \alpha\left\{k_0^2 - \frac{\mu_0 H^2}{2g^2}\left(\frac{1}{2}\mu_0 - f^2\right)\right\} \ ; \qquad (3.44)$$

$$l = \pm n\pi \ , \qquad n = 1, \ 2, \ \dots \ . \qquad (3.45)$$

Let us examine the roots of Eq. (3.44). Besides a denumerable number of the roots l_n ($n = 1, 2, \ldots$), which are determined by the intercepts of cotangents with the r.h.s. of (3.44) and are very close to the values $n\pi$ for $\alpha \geq 1$, there is one more root $l_0(k)$ which will be studied below.

The values l_n ($n \geq 1$) are almost independent of the dimensionless wave number $k_0 = kH$, but l_0 depends considerably on k_0. The curves $l_n(n \geq 1)$ correspond to normal modes of internal waves. As mentioned before, it is possible to neglect the dependence l_n versus k_0 for correspondent dispersion curves $\omega_n(k_0)$. Then

$$\omega_n^2 = f_n^2 + \frac{(\mu_0 - f^2)\, k_0^2}{k_0^2 + l_n^2 + (\mu_0 H/2g)^2} \; ; \qquad l_n \approx n\pi. \tag{3.46}$$

Eq. (3.46) satisfies exactly the peculiarities of dispersion curves behaviour described in paragraph **9** of Section 3.1. The term $(\mu_0 H/2g)^2$ in the denominator goes to zero in the Boussinesq approximation.

The phase velocity of n-th mode $c_{f,n} = H\omega_n/k_0$ according to (3.46) is decreasing steadily from ∞ ($k_0 \to 0$) to 0 ($k_0 \to \infty$). For the group velocity modulus $c_{g,n}$ in virtue of (3.46) we have

$$c_{g,n} = \frac{H k_0\, (\mu_0 - f^2)\, [l_n^2 + (\mu_0 H/2g)^2]}{\omega_n\, [k_0^2 + l_n^2 + (\mu_0 H/2g)^2]}. \tag{3.47}$$

c_g has its maximum and $dc_{g,n}/dk > 0$ for $f \neq 0$ and when k is rather small. For $f = 0$ and $N = $ const the group velocity is steadily decreasing while k increases.

Let us now consider the dispersion curve $\omega_0(k)$. From Eqs. (3.44) it follows that the real root l_0 exists in the rather short wave range $\beta < 1$ only, i.e., when $k_0 > k_0^*$, where

$$(k_0^*)^2 = \frac{\mu_0 H}{2g^2} \left(\frac{1}{2} \mu_0 - f^2 \right) + \frac{H(\mu_0 - f^2)}{g}. \tag{3.48}$$

The dispersion curve is described by

$$\omega_0^2 = f^2 + \frac{(\mu_0 - f^2)\, k_0^2}{k_0^2 + l_0^2\, (k_0) + (\mu_0 H/2g)}. \tag{3.49}$$

While k_0 increases from zero to k_0^* the value of l_0 is monotonically going to zero (when $k_0 = k_0^*$).

A distribution of the vertical velocity component is determined with Eqs. (3.43), i.e., the maximum amplitude of such a wave occurs when $k_0 < k_0^*$ under the free surface, i.e., inside the fluid. Thus such waves may be defined as internal.

Owing to (3.43) the value of l_0 becomes imaginary when $k_0 > k_0^*$. Let us define $l_0 = i\,\hat{l}$. Then the distribution of $W(z)$ is

$$W = e^{\mu_o z/2g}\frac{\mathrm{sh}\left[\hat{l}\left(1+\frac{z}{H}\right)\right]}{\mathrm{sh}\,\hat{l}} \;;\; \hat{l}^2 = k_0^2 - \frac{\mu_0 - f^2}{\omega^2 - f^2}k_0^2 + \left(\frac{\mu_0 H}{2g}\right)^2 , \quad (3.50)$$

and the eigen-curve is described with the expression

$$\hat{l}\,\mathrm{cth}\,l = \frac{gk_0^2}{H(\omega^2 - f^2)} - \frac{\mu_0 H}{2g} . \quad (3.51)$$

There is always a single root for the eigen-curve $\hat{l}$ which defines an unknown dispersion relation. The $W(z)$ distribution shows that the studied wave has its maximum amplitude at the free surface and may be considered to be the surface wave.

If

$$k_0^2 \gg \frac{\mu_0 - f^2}{\omega^2 - f^2}k_0^2 - \left(\frac{\mu_0 H}{2g}\right)^2$$

then $\hat{l} \approx k_0$ and the dispersion relation for a surface wave will be $\omega^2 = gk\,\mathrm{th}\,kH + f^2$ which corresponds to free surface barotropic waves in rotating fluid. If additionally $f = 0$ then we have $\omega^2 = gk\,\mathrm{th}\,kH$, which is the well known relation for surface waves.

2.2   THREE-LAYER STRATIFICATION MODEL

Amongst piecewise constant distributions of Brunt–Väisälä frequency the most realistic application to the ocean is a three-layer model (Fig. 3.2). Here the density is constant in the upper layer of the thickness h modeling the quasi-homogeneous layer, it then varies in a layer of the thickness d modeling the upper thermocline, and then it roughly considered to be constant below the seasonal thermocline.

Let us study the boundary value problem (3.3)–(3.4a) in the isopycnic approximation neglecting the earth's rotation ($f = 0$) and considering

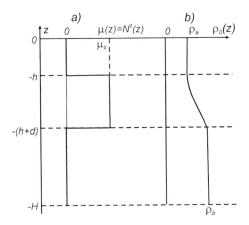

Figure 3.2. Distribution of the Brunt–Väisälä frequency (a) and of the density (b) for a three-layer stratification model.

the boundary condition (3.4a) at the free surface (the 'rigid lid' approximation).

$$W'' - \frac{\mu}{g} W' + \frac{\mu - \omega^2}{\omega^2} k^2 W = 0 \; ;$$

$$W(-H) = W(0) = 0 \; .$$

$$(3.52)$$

The distributions of $\mu(z) = N^2(z)$ and $\rho_0(z)$ are

$$\mu = \begin{cases} 0 & (0 \geq z > -H) \; , \\ \mu_0 & (-h \geq z \geq -(h+d)) \; , \\ 0 & (-(h+d) > z \geq -H) \; , \end{cases}$$

$$\rho_0(z) = \begin{cases} \rho_a & (0 \geq z \geq -H) \; , \\ \rho_a \, e^{\frac{-\mu_0}{g}(z+h)} & (-h \geq z \geq -h-d) \; , \\ \rho_b = \rho_a \, e^{\frac{-\mu_0 d}{g}} & (-h-d \geq z \geq -H) \; . \end{cases}$$

$$(3.53)$$

We can obtain the solution of the problem (3.52) from (3.53) taking into consideration the distribution of μ

$$W(z) = \begin{cases} A \operatorname{sh} kz & (0 \geq z \geq -H) , \\ e^{\frac{\mu}{2g} z}(C \sin lz + D \cos lz) & (-h \geq z \geq -h - d) , \\ B \operatorname{sh}[k(z + H)] & (-h - d \geq z \geq -H) . \end{cases} \quad (3.54)$$

An expression for l is defined below.

Conditions of continuity between functions (3.54) for $z = -h$ and $z = -(h + d)$ lead to the following dispersion equation

$$l \operatorname{ctg}(ld) = \frac{\mu_0 k}{2\omega^2} \left\{ \operatorname{cth}[k(H - d)] - \frac{\operatorname{ch}[k(H - d - 2h)]}{\operatorname{sh}[k(H - d)]} \right\}$$
$$- k \operatorname{cth}[k(H - d)] + \frac{\mu_0}{2g} \frac{\operatorname{sh}[k(H - d - 2h)]}{\operatorname{sh}[k(H - d)]} ; \quad (3.55)$$
$$l^2 = k^2 \frac{\mu_0 - \omega^2}{\omega^2} - \left(\frac{\mu_0}{2g}\right)^2 .$$

The Boussinesq approximation here means a neglect of the last terms in both Eqs. (3.55). The first of Eqs. (3.55) may be rewritten in the form

$$\operatorname{cth}(ld) = ld f_1(k) + \frac{f_2(k)}{ld} .$$

A study of this equation shows that there is a denumerable set of dispersion curves $\omega_n(k)$ whose qualitative behaviour is similar to that described in Section 3.1 (**2**) for $f = 0$. A denumerable set $W_n(z)$ of solutions of the boundary value problem (3.52) corresponds to $\omega_n(k)$. Frequencies $\omega_n(k)$ of internal waves vary from 0 to $\omega_{\max} = [\mu_0(1 + \mu_0^2/4g^2k^2)]^{1/2}$ (in the Boussinesq approximation $\omega_{\max} = \mu_0^{1/2}$). Additionally there are two internal waves propagating along the pycnocline borders $z = -h$ and $z = -(h + d)$ (are absent in the Boussinesq approximation).

Oceanic conditions are characterized by inequalities: when $h \sim d \gg H$ the most interesting are the waves for which $kH \gg 1$ and $kd \leq 1$ (waves in a narrow pycnocline). Eqs. (3.55) may be simplified slightly under these conditions and rewritten as follows (the stratification model in the infinitely deep ocean)

$$\operatorname{tg}(ld) = \frac{l/k\,(1 + \operatorname{cth} kh)}{(l/k)^2 - (1 + \mu_0/2gk)\,(\operatorname{cth} kh - \mu_0/2gk)} \; ;$$

$$\omega^2 = \mu_0 \left[1 + \frac{l^2}{k^2} + \left(\frac{\mu_0}{2gk}\right)^2 \right]^{-1} . \tag{3.56}$$

This dispersion equation is thoroughly studied in [262]. In particular, for waves long enough in comparison with the pycnocline's thickness (the lowest mode) when $kd \ll 1$ from (3.56) one can obtain ([262], Section 6.3)

$$\omega_0^2 = \frac{\Delta\rho}{\rho_a} \frac{gk}{1 + \operatorname{cth}(kh)} \; ; \quad (\Delta\rho = \rho_b - \rho_a) . \tag{3.57}$$

For $l \gg \mu_0/2g$ from the second (3.56) equation we have

$$\omega \approx \pm \frac{N_0 k}{\sqrt{k^2 + l^2}} = N_0 \cos\theta , \tag{3.58}$$

where θ is an angle included between the three-dimensional vector $\boldsymbol{æ} = \{k_1,\ k_2,\ l\}$ and a horizontal plane.

Owing to (3.58) high frequency internal waves ($k \gg l$) of the frequency close to N_0 propagate almost horizontally ($\theta \approx 0$), and low frequency waves propagate almost vertically.

Let us examine particular cases of (3.53)–(3.55) corresponding to more simple stratification models for the finite depth H.

1. A two-layer model with the 'lower' pycnocline may be formed from the studied three-layer model (see Fig. 3.2) when $d + h = H$. This case models the oceanic frequency distribution in the quasi-homogeneous layer and in the seasonal thermocline; the density variation in the mean thermocline is neglected. Distributions of μ, W are

$$0 \geq z > -h \quad \mu = 0; \qquad\qquad W = A\operatorname{sh} kz;$$

$$-h \geq z \geq -H \quad \mu = \mu_0 = \text{const}; \quad W = C\,e^{\frac{\mu z}{2g}} \sin\left[l\,(z + H)\right] . \tag{3.59}$$

The dispersion relation (3.55) turns into

$$l \operatorname{ctg}(ld) = -\left(\frac{\mu_0 H}{2g} + k \operatorname{cth} kh\right) ;$$

$$l^2 = k^2 \frac{\mu_0 - \omega^2}{\omega^2} - \left(\frac{\mu_0}{2g}\right)^2 . \qquad (3.60)$$

When $h \to 0$ we obtain from (3.60) the case studied above of the constant stratification $\mu_0 = \mathrm{const}$ $(0 \geq z \geq -H)$ in the isopycnic and the 'rigid lid' approximations, as well as for $f = 0$.

2. The two-layer stratification model of the 'upper' pycnocline is transformed from the studied three-layer model (see Fig. 3.2) when $h = 0$. This case models the natural density distribution in the mean thermocline layer and in the lower homogeneous ocean layer when neglecting the density variations in the upper seasonal thermocline[3].

Distributions of $\mu(z), W(z)$ in this case are

$$\left. \begin{array}{ll} 0 \geq z \geq -d & \mu = \mu_0; \quad C e^{\frac{\mu z}{2g}} \sin lz; \\ -d > z \geq -H & \mu = 0; \quad B \operatorname{sh}\left[k\left(z + H\right)\right] \end{array} \right\} \qquad (3.61)$$

The dispersion relation (3.55) is transformed into

$$\left. \begin{array}{l} l \operatorname{ctg} ld = -\left(1 - \frac{\mu_0}{2g\,k}\right) k \operatorname{cth}\left[k\left(H - d\right)\right] ; \\ l^2 = k^2 \frac{\mu_0 - \omega^2}{\omega^2} - \left(\frac{\mu_0}{2g}\right)^2 . \end{array} \right\} \qquad (3.62)$$

In the 'rigid lid' approximation, from (3.62) for $f = 0$ and $d = H$ we can obtain again a relation for the constant stratification.

3. The two-layer model of the stratification with the infinitely thin pycnocline may be obtained from the three-layer model (Fig. 3.2) when $d \to 0$ and corresponds to the simplest stratification model for compressible fluid with two homogeneous layers: $z > -h$ and $z < -h$.

Now we have the following distributions of parameters

$$\begin{array}{lll} 0 \geq z > -d & \rho = \rho_a; & W = A \operatorname{sh} kz; \\ -d > z \geq -H & \rho = \rho_b > \rho_a; & W = B \operatorname{sh}\left[k\left(z + H\right)\right] \end{array}$$

The dispersion relations corresponding to the latter case may be obtained from Eq. (3.55) by the limiting translation $d \to 0$, when

$$\mu_0 = \frac{g}{d} \ln \frac{\rho_b}{\rho_a} \; ; \quad l \approx id^{-1} \ln \frac{\rho_b}{\rho_a} \; ;$$

$$\omega^2 = kg \frac{\Delta\rho}{\rho_a} \left\{ \frac{\Delta\rho}{\rho_a} \operatorname{cth}\left[k\left(H - h\right)\right] + 1 + \operatorname{cth} kh \cdot \operatorname{th}\left[k\left(H - h\right)\right] \right\}^{-1},$$

$$(3.63)$$

where $\Delta\rho = \rho_b - \rho_a$. For $H \to \infty$ and $\Delta\rho/\rho_a \ll 1$ Eq. (3.63) goes to (3.57).

The dispersion relation (3.63) defines the wave propagating along the interface between two homogeneous fluids and models waves propagation in a very narrow oceanic pycnocline when the pycnocline layer oscillates as a whole.

In particular, from (3.63) for $h \ll H$, $\rho_a \to 0$ we have $\omega^2 \approx kg \operatorname{th} kH$, i.e., the dispersion relation for surface waves in a homogeneous layer of thickness H.

An investigation by analytical methods of natural stratification models more complicated than the studied above ones is rather difficult. So we have to use numerical methods for the solution of the basic boundary value problem [229], [103]. An essence of the simplest calculation method, suggested in [103], consists of a piecewise constant approximation of the Brunt–Väisälä frequency $N(z)$ by a finite number of layers p and in the use of an analytical solution of Eq. (3.3) for each of the layers with continuity at the joints. As a result of such an adjoin, at the $p + 1$ (the final) step corresponding to $z = -H$ we obtain the dispersion relation valid for the complete boundary condition (3.4). Corresponding eigenfunctions of the basic boundary value problem are found in this method using the normalization condition $\|W_n\|^2 = 1$ (see Eq. (3.16)). Increasing the number of layers p we can use this method for solution of the basic boundary value problem (3.3)–(3.4a), considering a fine structure of the Brunt–Väisälä frequency either.

When examining the basic boundary value problem the non-monotonic dependence of the group velocity c_g versus the wave number k is especially interesting, because in the vicinity of max c_g intensive wave trains (the Airy waves) may appear (see Section 5.3). For $N = \text{const}$, studied

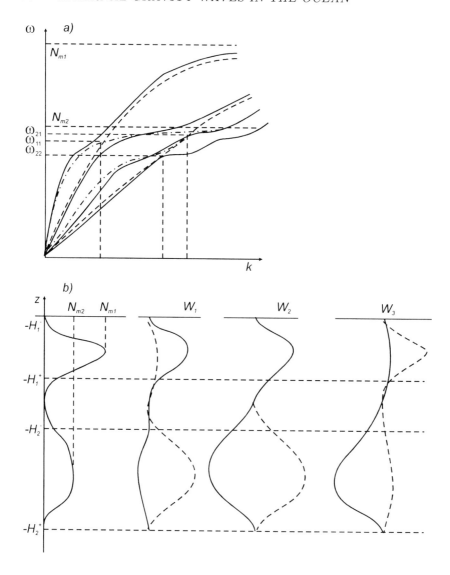

Figure 3.3. Dispersion curves for two independent waveguides (a) and their corresponding eigenfunctions (b) [103].

in the present paragraph, such a non-monotony occurs only when taking into consideration the Earth's rotation ($f \neq 0$). As taking into account the Coriolis parameter picks out for examination rather long waves ($\lambda \geq$ 30 km), the question arises whether local maxima of $c_g(k)$ dependence exist for short enough waves ($f = 0$) in stratification profiles $N(z)$ more complicated than those studied above.

The positive answer to this question, based on a simple example of almost two-layer stratification, was given in [103]. Results obtained by a numerical method suggested by the author are presented in Fig. 3.3 for two independent waveguides (pycnoclines) with the Brunt–Väisälä maxima N_{m1} and N_{m2}. Dispersion curves in Fig. 3.3 a correspond to the first modes. Dashed lines show dispersion curves for the first separate waveguide, dash-dotted lines for the second separate waveguide. The frequencies ω_{ij} in Fig. 3.3 a correspond to intersection points of the dispersion curves for separate waveguides (i number of a dispersion curve, j number of an intersection point). In the vicinity of the mentioned frequencies ω_{ij} the dispersion curves go smoothly from the curve for the first waveguide to that for the second. Inflection points of the dispersion curves $\omega_n(k)$ corresponding to local maxima of the dependences $c_{g,n}(k)$ are disposed, in fact, in the vicinity of the frequencies ω_{ij}.

The corresponding eigenfunctions $W_n(z)$ are shown in Fig. 3.3 b. When eigenfrequencies are far from the critical frequencies $\omega_{i,j}$ the eigenfunctions are localized in the respective waveguides (dashes). When $\omega \sim \omega_{n,j}$ (for example $\omega = \omega_{11}$) the first mode in the second (lower) waveguide is almost suppressed, the same occurs for the third mode in the upper waveguide.

These results can not only explain the appearance of local maximums of $c_{g,n}(k)$ dependences, but also allow us to suggest a simple physical mechanism for an energy redistribution of internal waves between mean and seasonal thermoclines [103] (which will be discussed in Section 5.2).

3. VERTICALLY PROPAGATING INTERNAL WAVES. BASIC BOUNDARY VALUE PROBLEM IN THE WKB APPROXIMATION

When the vertical scale of fluid motions is small compared to the thickness of the considered stratified ocean layer (this occurs for high number modes $n \gg 1$ or for the density stratification model (3.53) when $l > \mu_0/2g$), the parameter N in the basic equation (3.7) may be considered to be a slowly varying function of z. The bottom and surface reflection of internal waves may be neglected as a first approximation and the ocean may be considered to be infinitely deep.

First let us consider a case when $N(z)$ is constant in a local region, i.e., $N(z) = N_0 = \text{const}$ (see, e.g., [299] or [35]). Then Eq. (3.7) for $c = \infty$ and $f = 0$ in the Boussinesq approximation becomes

$$W'' + k^2 \left(\frac{N_0^2}{\omega^2} - 1 \right) W = 0 , \qquad (3.64)$$

where $k^2 = k_1^2 + k_2^2$, with its solution

$$W(z) = A \exp(i\,lz) + B \exp(-i\,lz). \qquad (3.65)$$

The vertical wave number component l

$$l = \frac{k}{\omega} (N_0^2 - \omega^2)^{1/2} . \qquad (3.66)$$

(3.66) yields a dispersion relation

$$\omega = \pm N_0 k (k^2 + l^2)^{-1/2} = \pm N_0 \mathbf{i} \cdot \nu = \pm N_0 \sin\theta , \qquad (3.67)$$

where $\mathbf{i}$ is a unit vector, directed along the line of the total wave vector $\mathbf{æ} = \{k_1, k_2, l\}$; ν a horizontal unit vector, θ an angle included between $\mathbf{æ}$ and the vertical z axis. With Eqs. (3.1) and (3.65) it is easy to obtain a general solution of the problem studied:

$$W \sim \exp\{i(\mathbf{kx} + lz - \omega t)\} = \exp\{i(\mathbf{æX} - \omega t)\} , \qquad (3.68)$$

where $\mathbf{X} = \{x, y, z\}$ is a three-component vector in a physical space. Naturally, the dispersion relation (3.67) may be obtained by a direct substitution of (3.68) into the common equation (2.42) for $f = 0$ and in the Boussinesq approximation. The solution (3.68) is a three-dimensional monochromatic internal wave running in all directions, including the vertical. The free waves studied in Sections 3.1, 3.2 were running in the horizontal direction only, and were standing in the vertical direction as a consequence of the boundary conditions at the free ocean surface.

The dispersion relation (3.67) suggests that internal waves similar to (3.68) are anisotropic, and that their frequency does not depend on the wave number $æ = |æ| = \sqrt{k^2 + l^2}$ and depends on the direction of the wave vector $æ$ only (on the angle θ). From (3.67) it is obvious that internal waves exist for $\omega < N_0$ only, for the given propagation angle θ the frequency ω is defined by Eq. (3.67). The wavelength and the phase velocity modulus

$$\mathbf{c}_f = \frac{\omega}{æ^2} \, æ = \left\{ \frac{\omega}{k_1}, \frac{\omega}{k_2}, \frac{\omega}{l} \right\} = |\mathbf{c}_f| \frac{æ}{æ}$$

$$|\mathbf{c}_f| = c_f = \frac{\omega}{æ} = \pm N_0 k \left(k^2 + l^2 \right)^{-1} = \pm N_0 \cos \theta \, , \qquad (3.69)$$

may be arbitrary. The group velocity of the internal waves is $\mathbf{c}_g = \nabla_{æ} \omega \, (æ)$ or

$$\begin{aligned}
\mathbf{c}_g &= \left\{ \frac{\partial \omega}{\partial k_1}, \frac{\partial \omega}{\partial k_2}, \frac{\partial \omega}{\partial l} \right\} = \pm N_0 \frac{\partial}{\partial æ_i} \left(\frac{\nu_j æ_j}{æ} \right) \\
&= \pm \{ N_0 k_1 l^2 q^{-1}, \; N_0 k_2 l^2 q^{-1}, \; N_0 k^2 l q^{-1} \} ; \\
q &= \left(k^2 + l^2 \right)^{3/2} k \, .
\end{aligned} \qquad (3.70)$$

$\mathbf{c}_g æ = 0$ and therefore $\mathbf{c}_f \mathbf{c}_g = 0$. Thus $\mathbf{c}_g$ is directed at right angles to $æ$ and to the phase velocity vector. The directions of $æ$, $\mathbf{c}_g$ and $\mathbf{c}_f$ for waves propagating in the positive z-direction [a plus sign in Eqs. (3.67), (3.69), (3.70)] and in the negative z-direction [a minus sign in Eqs. (3.67), (3.69), (3.70)] are shown in Fig. 3.4 a,b. From Eqs. (3.2), (3.67), (3.68) we have

$$u = -\frac{k_1 l}{k^2} \, w; \quad v = -\frac{k_2 l}{k^2} \, w; \quad p = -\rho_0 \frac{l}{k^2} \, w; \quad \rho = \mathbf{i} \frac{\rho_0 N_0^2}{g \omega} \, w \, , \qquad (3.71)$$

where w is defined by Eq. (3.68). From (3.71) we obtain $\nabla p = -i æ p$, i.e., the pressure gradient is directed along the vector $æ$. Eqs. (3.68) and (3.71) yield that $\mathbf{v} \times æ = 0$, i.e., fluid particles trajectories lie in the plane containing the vector $æ$ and the z axis, and streamlines are normal to $æ$ (a transverse wave). From the dispersion relation (3.67), the group velocity expression (3.70), and the equation $æ \times \mathbf{v} = 0$, it

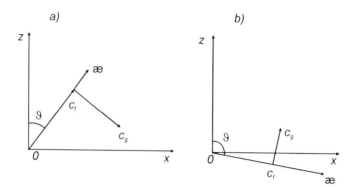

Figure 3.4. Directions of the wave vectors æ, the group $\mathbf{c}_g$ and the phase $\mathbf{c}_f$ velocities for three-dimensional internal waves: a — propagating upwards, b — propagating downwards.

follows that when $\omega \to N_0$ the wave vector æ is almost horizontal, the group velocity goes to zero $\mathbf{c}_g \to 0$, and the fluid particles oscillate in the vertical direction relative to their equilibrium levels. While ω decreases the inclination of the wave vector decreases too; for $\omega \to 0$ the wave vector becomes vertical, the group velocity goes to zero $(\mathbf{c}_g \to 0)$; and the wave motion degenerates to relative horizontal slip of layers. A time average energy flow $\mathbf{I}$ is defined by

$$\mathbf{I} = \overline{p\mathbf{v}} = \mathbf{c}_g \left[\frac{1}{2} \rho_0 \overline{\mathbf{v}^2} + \frac{1}{2} g^2 \frac{\overline{\rho^2}}{\rho_0^2 N_0^2} \right] . \tag{3.72}$$

From (3.72) it follows that the energy flow is directed along the line of the group velocity vector. When a wave propagates upwards the energy flow will go downwards, and *vice versa*, Fig. 3.4 a,b.

Let us define the conditions of reflection from a rigid obstacle (e.g., from an inclined bottom) for such three-dimensional waves. When reflecting from an obstacle the wave's frequency should remain constant, so from (3.67) it follows that the wave vector of a reflecting wave $æ_0$ makes with the vertical angle θ which is the same for a direct wave. If β is the bottom inclination (see Fig. 3.5) the angles of arrival and reflection of a wave with respect to the bottom surface are $\frac{\pi}{2} - (\theta - \beta)$ and $\frac{\pi}{2} - \theta + \beta$, respectively. As the normal velocity component at the bottom is equal to zero the normal components of both wave vectors should be equal. When a normal to the rigid surface and the wave vector of the

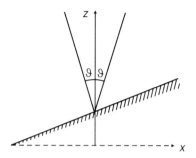

Figure 3.5. A scheme of reflection of a three-dimensional internal wave from an inclined bottom for $N = $ const.

forward wave æ_n lie in the picture's plane we have

$$\text{æ}_n \sin (\theta - \beta) = \text{æ}_0 \sin (\theta + \beta) . \qquad (3.73)$$

Thus when reflecting from the obstacle the internal waves' frequency remains constant and the wavelength (the wave number) varies.

Everything said above relates to the case when $N = $ const. If $N(z)$ is a slowly varying function of the depth, i.e., $N^{-1}(dN/dz) \approx \varepsilon \ll 1$, the approximate WKB method (the Wentzell–Kramers–Brillouin method) may be used for the analysis of Eq. (3.64)[4]. Let us consider a solution of Eq. (3.64) in the WKB approximation. When N is a slowly varying function of z we may write $N = N(\varepsilon z) = N(\eta)$, where $\eta = \varepsilon z$, and $\varepsilon \ll 1$ is a small parameter. Substituting η to (3.64) we obtain

$$W'' + \lambda^2 \Omega^2 (\eta) W = 0; \quad \lambda = \frac{k}{\varepsilon}; \quad \Omega^2 = \left(\frac{N^2}{\omega^2} - 1 \right) , \qquad (3.74)$$

where a prime means partial derivative with respect to η. We will consider $\lambda = k/\varepsilon \gg 1$ (owing to $\varepsilon \ll 1$ the condition $\lambda \gg 1$ will be satisfied for moderate k). For $N \approx $ const (or $\Omega \approx $ const) the solutions (3.74) will be $W = \exp \{\pm i\lambda\Omega\eta\}$. For $\lambda \gg 1$ the solution for $\Omega = \Omega(\eta)$ should be well described with the function $\exp \{\pm i\lambda \int_0^z \Omega(\eta) \, d\eta\}$. So we will find a solution of (3.74) in the form

$$W_{1,2} = A (\eta, \ \lambda) \exp \left\{ \pm i\lambda \int_0^z \Omega (\eta) \, d\eta \right\} . \qquad (3.75)$$

Substituting (3.75) to (3.74) we have

$$\frac{1}{\lambda} A'' \pm 2i\Omega A' \pm i\Omega' A = 0.$$

(3.76)

Neglecting the small value A''/λ in (3.76) we obtain $A(\eta) \sim [\Omega(\eta)]^{-1/2}$. Then solution (3.75) with respect to the previous variable looks like

$$W_{1,2} = \frac{C_{1,2}}{\left(\frac{N^2}{\omega^2} - 1\right)^{1/4}} \exp\left\{\pm ik \int_0^z \sqrt{\frac{N^2}{\omega^2} - 1} \, dz'\right\},$$

(3.77)

or in the trigonometric form

$$W = \left[\frac{N^2}{\omega^2} - 1\right]^{-1/4} \left\{ A\cos\left[k \int_0^z \sqrt{\frac{N^2}{\omega^2} - 1} \, dz'\right]\right.$$

$$\left. + B\sin\left[k \int_0^z \sqrt{\frac{N^2}{\omega^2} - 1} \, dz'\right]\right\}.$$

(3.78)

These solutions are valid with an accuracy to $O(1/\lambda)$, see e.g. [260]. Solutions (3.77) and (3.78), as well as (3.65), describe waves propagating in the positive and negative z-axis directions. Using the solution (3.78) we may satisfy the free surface and the bottom boundary conditions and to obtain the wave standing in z direction. Satisfying conditions $W = 0$ and $z = 0, -H$ from (3.78) we have $A = 0$ and

$$W = \left[\frac{N^2}{\omega^2} - 1\right]^{-1/4} B\sin\left[k \int_0^z \sqrt{\frac{N^2}{\omega^2} - 1} \, dz'\right]$$

(3.79)

when satisfying equations

$$k \int_0^{-H} \sqrt{\frac{N^2(z)}{\omega^2} - 1} \, dz' = n\pi, \quad (n = 1, 2, \ldots).$$

(3.80)

Eq. (3.80) is a dispersion relation of ω and k.

From expressions (3.78), (3.79) it is obvious that $|W| \to \infty$ when $N(z) \to \omega$. In the vicinity of the point $z = z_0$ where $N(z_0) = \omega$

the WKB approximation is invalid. The point $z = z_0$ is called the turning point because the solution (3.78) is oscillating for $N(z) > \omega$ and exponential when $N(z) < \omega$. In the vicinity of the turning point it is necessary to use methods more exact than WKB methods (see [270]). Usually in the vicinity of the turning point a solution is approximated by Airy functions. In their application to internal waves such solutions are worked out in [63], where the behaviour of the eigenfunctions of (3.64) is studied thoroughly in the vicinity of z_0 and $N \sim \omega$.

4. PROPAGATION OF WAVE PACKETS. RAY APPROXIMATION

Elementary wave solutions of basic equations of the dynamics of internal waves in the form $w \sim e^{i(\mathbf{kx} - \omega t)}$ were examined earlier. A general solution of any problem, e.g., the Cauchy problem, may be obtained by a superposition of elementary wave solutions

$$ w = \int\limits_{-\infty}^{\infty} A \, e^{i(\mathbf{kx} - \omega t)} \, d\mathbf{k} \ . $$

Such a solution will represent a wave packet, i.e., all wave characteristics (amplitude, frequency, wave number etc.) will vary in time and space. Examples of such an approach to the solution of the Cauchy problem for internal waves are given in Chapter 6. Now we will focus our attention on the 'ray method', a method of approximately describing wave packages' propagation. This method is valid for a study of both linear and nonlinear wave packets propagating both in homogeneous and inhomogeneous media when all the packet characteristics vary slowly over distances close to the wavelength and over times close to the wave period. All modulations caused by slow variations of medium parameters (e.g., density) lead to such conditions with characteristics slowly varying about the wavelength. When modulations occur as a result of wave generation by a modulated source (with frequency and amplitude variations) the ray method is valid if the modulations are relatively slow compared to the wave period[5].

Let us consider internal waves in the non-rotating ocean in the Boussinesq approximation. Waves may be described by a single equation for the vertical velocity component

$$\frac{\partial^2}{\partial t^2} \Delta w + N^2 (z) \Delta_h w = 0 \ , \tag{3.81}$$

where $N^2 (z)$ is a slowly varying function of z, i.e., $N^2 (z) = N^2 (\varepsilon z)$, $\varepsilon \ll 1$. We consider waves with a small vertical scale, neglecting the vertical boundary conditions. Propagation of the wave packet modulated slowly in time and space will be studied. In the absence of modulations and for $N^2 (z) = $ const, a solution of Eq. (3.81) is $w \sim A e^{i\theta}$, where $\theta = \mathbf{kx} + lz - \omega t$, and $\mathbf{x}$ and $\mathbf{k}$ are two-dimensional vectors. In the presence of slow modulations the amplitude and the phase of the wave will be slow functions of time and space. So the solution of (3.81) may be found as:

$$w (\mathbf{x}, z, t) = A (\xi, \eta, \tau) e^{i \varepsilon^{-1} \theta (\xi, \eta, \tau)} \ , \tag{3.82}$$

where $\xi = \varepsilon \mathbf{x}$, $\eta = \varepsilon z$, $\tau = \varepsilon t$, A is an amplitude, θ a slow phase, and $\hat{\theta} = \varepsilon^{-1} \theta$ a fast phase. The amplitude may be expanded as a power series of ε

$$A (\xi, \eta, \tau, \varepsilon) = \sum_{n=0}^{\infty} \varepsilon^n A_n (\xi, \eta, \tau) \ . \tag{3.83}$$

The frequency and the wave numbers may be defined, as well as in a homogeneous case, through the phase:

$$\omega = -\frac{\partial \theta}{\partial \tau}; \quad \text{æ} = \nabla \theta, \quad \mathbf{k} = \nabla_h \theta \ , \tag{3.84}$$

where

$$\nabla = \left\{ \mathbf{i} \frac{\partial}{\partial \xi_1} + \mathbf{j} \frac{\partial}{\partial \xi_2} + \mathbf{e} \frac{\partial}{\partial \eta} \right\} \ ;$$

$$\nabla_h = \left\{ \mathbf{i} \frac{\partial}{\partial \xi_1} + \mathbf{j} \frac{\partial}{\partial \xi_2} \right\} \ ;$$

$$\text{æ} = \{k_1, k_2, l\}; \quad \mathbf{k} = \{k_1, k_2\} \ .$$

From (3.84) it is easy to obtain the kinematic correlation between ω and æ, and the correlation between $æ_i$ $(i = 1, 2, 3)$

$$\frac{\partial \text{æ}}{\partial \tau} + \nabla \omega = 0; \quad \frac{\partial æ_i}{\partial x_j} - \frac{\partial æ_j}{\partial x_i} \quad x_i = \{\xi_1, \ \xi_2, \ \eta\} \ . \tag{3.85}$$

Substituting (3.82), (3.83) into (3.81) and equating to each other the terms of order $\mathcal{O}(1)$ and $\mathcal{O}(\varepsilon)$ we obtain

$$\omega^2 \ae^2 A_0 - N^2(\eta) k^2 A_0 = 0 , \tag{3.86}$$

$$2\omega \frac{\partial}{\partial \tau} (\ae^2 A_0) + \frac{\partial \omega}{\partial \tau} \ae^2 A_0 - \omega^2 2\ae \nabla A_0 - \omega^2 A_0 \nabla \ae$$
$$+ N^2 2\mathbf{k} \nabla_h A_0 + N^2 A_0 \nabla_h \mathbf{k} = 0 . \tag{3.87}$$

Here we limit ourselves to the derivation of A_0, the main term of the expansion (3.83). From (3.86) we have the dispersion relation

$$\omega = \Omega(\ae) = \frac{N^2(\eta) k^2}{\ae^2} . \tag{3.88}$$

This equation looks like the dispersion relation (3.67), but all of the parameters contained in this equation are functions of slow spatial coordinates and time. Multiplying (3.87) by A_0 and using (3.88) yields

$$\frac{\partial}{\partial \tau} [\omega \ae^4 A_0^2] = N^2 \{k^2 \nabla (A_0^2 \ae) - \ae^2 \nabla_h (A_0^2 \mathbf{k})\} . \tag{3.89}$$

Using the group velocity relation $\mathbf{c}_g = \partial \omega / \partial \ae$ (coinciding with (3.70)) and (3.88), we can reduce (3.89) to

$$\frac{\partial A_0^2}{\partial \tau} + \nabla (\mathbf{c}_g A_0^2) = 0 . \tag{3.90}$$

Writing the complex amplitude in the form $A_0 = a e^{i\beta}$, where a and β are real values, from (3.90) we have:

$$\frac{\partial a^2}{\partial \tau} + \nabla (\mathbf{c}_g a^2) = 0 ; \tag{3.91}$$

$$\frac{\partial \beta}{\partial \tau} + \mathbf{c}_g \nabla \beta = 0 . \tag{3.92}$$

As $E \sim a^2$ Eq. (3.91) represents an energy conservation law for a wave packet. Equation (3.92) describes a variation of the small addition β

to the total wave phase. It may be neglected in further considerations ($\beta \ll \varepsilon^{-1}\theta$).

Eqs. (3.85), (3.91), together with (3.88), form a closed system and define completely the behaviour of the wave packet (3.82). Similar correlations, differing just in the type of dispersion relation, describe internal waves of any nature. Let us analyze the closed system of equations following [391]. The first equation in (3.85) using the relation $\omega = \Omega(æ_i)$ may be rewritten as

$$\frac{\partial æ_i}{\partial \tau} + \frac{\partial \Omega}{\partial æ_j}\frac{\partial æ_j}{\partial x_i} = -\frac{\partial \Omega}{\partial x_i} ; \quad x_i = \{\xi_1, \xi_2, \eta\} . \tag{3.93}$$

As $\partial k_j / \partial x_i = \partial k_i / \partial x_j$, Eq. (3.93) yields:

$$\frac{\partial æ_i}{\partial \tau} + c_{gi}\frac{\partial æ_i}{\partial x_j} = -\frac{\partial \Omega}{\partial x_i} , \tag{3.94}$$

where $c_{gi} = \partial \Omega / \partial x_i$ is a component of the group velocity. Therefore the group velocity $\mathbf{c}_g(æ, \mathbf{x}, t)$ is the propagation velocity of perturbations of the wave vector $æ$. Eq. (3.94) may be rewritten in a characteristic form

$$\frac{dæ_i}{d\tau} = -\frac{\partial \Omega}{\partial x_i} \quad \text{when} \quad \frac{dx_i}{d\tau} = \frac{\partial \Omega}{\partial æ_i} = c_{gi}(æ) . \tag{3.95}$$

As $\Omega = \Omega(æ, \eta)$, values of the vector $æ$ vary when propagating in the line of characteristics. The characteristics itself represent curves in the space (ξ, η, τ). Using (3.85) we can show that

$$\frac{d\omega}{d\tau} = \frac{\partial \omega}{\partial d\tau} + c_{gi}\frac{\partial \omega}{\partial x_i} = \frac{\partial \Omega}{\partial \tau} , \tag{3.96}$$

and since Ω is independent of τ in our case, the frequency is constant for each of the characteristics.

The conservation equation for a square amplitude (3.91) shows that a total energy (proportional to a^2) in any volume, with all of the points moving at the group velocity, remains constant. Eq. (3.91) in the characteristic form looks like

$$\frac{da^2}{d\tau} = -\frac{\partial c_{gi}}{\partial x_i}a^2 \quad \text{when} \quad \frac{dx_i}{d\tau} = c_{gi}(æ) . \tag{3.97}$$

From Eqs. (3.97) and (3.95) it is obvious that the group velocity is here a double characteristic velocity, as far as both the wave number perturbations and the wave energy are transferred at this velocity. The Cauchy problem for a wave packet may be solved by means of the characteristics method using relations (3.95) and (3.97).

The main equations of the ray method were introduced above for the case in which wave characteristics varied slowly over space and time. But when the Brunt–Väisälä frequency varies considerably in the vertical direction (in the pycnocline) the approximation of geometric optics is valid under the condition of slow variation of parameters over only the horizontal coordinates and time.

Let us consider Eq. (3.81) when $N^2(z)$ is an arbitrary function of z and when taking into account the vertical boundary conditions $w\,|_{z=0} = 0$ (the 'rigid lid' approximation) and $w\,|_{z=-H} = 0$. Following the ray method, we will find a solution of (3.81) in the form

$$w(\mathbf{x}, z, t) = e^{i\varepsilon^{-1}\theta(\xi,\tau)} \sum_{n=0}^{\infty} \varepsilon^n \, \Phi_n(\xi, z, \tau) \, . \qquad (3.98)$$

According to stated earlier $\omega = -\partial\theta/\partial\tau$; $\mathbf{k} = \nabla_h\theta$ and (3.85), we have

$$\frac{\partial\mathbf{k}}{\partial\tau} + \nabla_h\omega = 0; \quad \frac{\partial k_i}{\partial\xi_j} - \frac{\partial k_j}{\partial\xi_i} = 0; \quad (i, j = 1,\, 2) \, . \qquad (3.99)$$

Substituting (3.98) into (3.81) and equating to each other the terms of order $\mathcal{O}(1)$ and $\mathcal{O}(\varepsilon)$ we obtain (a prime means derivation with respect to z):

$$\Phi_0'' + \frac{(N^2 - \omega^2)}{\omega^2} k^2 \Phi_0 = 0 \, , \quad \Phi_0|_{z=0,-H} = 0 \, ; \qquad (3.100)$$

$$\Phi_1'' + \frac{(N^2 - \omega^2)}{\omega^2} k^2 \Phi_1 = 0 \, , \quad \Phi_1|_{z=0,-H} = 0 \, , \qquad (3.101)$$

where

$$Q = -\frac{1}{\omega^2} \{ 2\omega \, (k^2 \Phi_0)_\tau + \omega_\tau k^2 \Phi_0 - 2\omega\mathbf{k}\,\nabla_h\Phi_0 - \omega\Phi_0\nabla_h\mathbf{k}$$

$$-\omega\Phi_{0\tau}'' - (\Phi_0''\omega)_\tau + N^2 2\mathbf{k}\nabla_h\Phi_0 + N^2\Phi_0\nabla_h\mathbf{k} \} \, . \qquad (3.102)$$

The homogeneous equation (3.101) admits solutions in the form

$$\Phi_0 (\xi, \tau, z) = A_0 (\xi, \tau) \varphi_0 (z),$$

where $\varphi_0 (z)$ is an eigenfunction of the boundary value problem

$$\varphi_0'' + \frac{(N^2 - \omega^2)}{\omega^2} k^2 \varphi_0 = 0, \quad \varphi_0|_{z=0,-H} = 0. \tag{3.103}$$

The boundary value problem (3.103) yields a series of eigenfunctions φ_0 and dispersion relations $\omega = \Omega_n (k)$ corresponding to normal modes. Thus the general solution is a superposition of all modes, i.e., $\Phi_0 = \sum_n A_{0n} \varphi_{0n}$. Let us define A_{0n} for each mode omitting the subscript n. To do this we examine equation (3.101) which solution may be presented in the form $\Phi_1 = A_0 (\xi, \tau) \varphi_1 (z)$, where $\varphi_1 (z)$ is the solution of the inhomogeneous boundary value problem

$$\varphi_1'' + \frac{(N^2 - \omega^2)}{\omega^2} k^2 \varphi_1 = Q (\varphi_0, A_0) A_0^{-1}; \quad \varphi_1|_{z=0,-H} = 0. \tag{3.104}$$

The inhomogeneous boundary value problem (3.104) has a solution if its r.h.s. is orthogonal to eigenfunctions of the corresponding homogeneous boundary value problem (3.103). The orthogonality condition is

$$\int_{-H}^{0} Q (\varphi_0, A_0) \varphi_0 (z) \, dz = 0. \tag{3.105}$$

Eq. (3.105) is an additional equation allowing to obtain $A_0 (\xi, \tau)$. Substituting (3.102) into (3.105), using (3.103) and the group velocity expressed through eigenfunctions $\varphi_0 (z)$ (3.32), yields the conservation law for the adiabatic invariant I (derivations see in [382])

$$\frac{\partial I}{\partial \tau} + \nabla_h \mathbf{c}_g I = 0, \tag{3.106}$$

where

$$I = A_0^2 \left[\int_{-H}^{0} \frac{N^2}{\omega^3} \varphi_0^2 \, dz \right];$$

$$\mathbf{c}_g I = A_0^2 \left[\int_{-H}^{0} \frac{\mathbf{k}}{k^2} \left(\frac{N^2}{\omega^2} - 1 \right) \varphi_0^2 \, dz \right], \qquad (3.107)$$

from the properties of the boundary value problem (3.103) (see Section 3.1) it may be easily shown that the quotient of Eqs. (3.107) is the group velocity $\mathbf{c}_g = \partial \Omega / \partial \mathbf{k}$. An expression for I may be written in the form $I = f(k) A_0^2$, where $f(k)$ is a bracketed functional from the first Eq. (3.107). So Eq. (3.106) may be written as

$$f(k) \left\{ \frac{\partial A_0^2}{\partial \tau} + \nabla_h \left(\mathbf{c}_g A_0^2 \right) \right\} + f'(k) A_0^2 \left\{ \frac{\partial \mathbf{k}}{\partial \tau} + \mathbf{c}_g \nabla_h \mathbf{k} \right\} = 0. \quad (3.108)$$

Since from (3.99) it follows that

$$\frac{\partial \mathbf{k}}{\partial \tau} + \mathbf{c}_g \nabla_h \mathbf{k} = 0, \qquad (3.109)$$

then Eq. (3.108), while separating the real part $|A_0| = a$ and using (3.109), may be rewritten as

$$\frac{\partial a^2}{\partial \tau} + \nabla_h \left(\mathbf{c}_g a^2 \right) = 0. \qquad (3.110)$$

In this case the problem was reduced to the basic equations of the ray approximation (3.99), (3.110). But unlike the previous case, when the dispersion relation had the evident form (3.88), here it is found as an eigen-curve of the boundary value problem (3.103) for a given distribution $N(z)$. All basic features of the ray method equations [see (3.93)–(3.97)] are valid for the system (3.99), (3.110). But the latter is simpler owing to its two-dimensionality and, moreover, its characteristics are straight lines in the $\{\xi, \tau\}$-space (the dispersion relation $\omega = \Omega(k)$ is independent of both ξ and τ).

5.   INTERNAL WAVES IN THE PRESENCE OF FINE VERTICAL STRUCTURE OF OCEANIC DENSITY FIELD

As a rule in the real ocean, variations with depth of temperature, salinity, density, and other characteristics are not smooth, but have a lot of inhomogeneities. Layers with high gradients of the parameters mentioned alternate with homogeneous layers, and the thickness of such 'steps' varies from centimeters to dozens of meters. Characteristics of such a 'fine structure' vary arbitrarily with the depth; that is why methods of the random functions theory are used for their study. Experimental data show that the fine structure of the density field retains for a long time (of about one day) and for a long distances (10–20 km). As the oceanic internal waves structure is defined by the density field disturbance, it is interesting to consider an influence of small vertical oscillations of the density field on internal waves propagation. This problem is being studied in the present section.

The ocean's fine structure can be induced by internal waves owed to nonlinear effects (such a theory is presented in Chapter 11). But the mutual influence of the waves and the fine structure may be neglected in a linear statement of the problem. The propagation of internal waves will be studied for a given vertical fine structure.

An undisturbed state of the density field in the ocean (in the absence of waves) in the fine structure presence may be described by

$$\rho_0\left(\mathbf{x}, z, t\right) = \langle \rho\left(z\right)\rangle + \rho_f\left(\mathbf{x}, z, t\right) , \qquad (3.111)$$

where $\langle \rho\left(z\right)\rangle$ is a mean density profile (angle brackets mean the statistical or space time average); ρ_f denotes density deflections from the mean value owed to the fine structure. Thus ρ_f is not the total density fluctuations spectrum, but its low frequency part corresponding to the fine structure. An expression for the Brunt–Väisälä frequency $N^2 = (g/\langle\rho\rangle)(\partial\rho_0/\partial z)$ while taking into account (3.111), and the property that in the real ocean $\rho_f/\langle\rho\rangle \ll 1$, we may write

$$N^2\left(\mathbf{x}, \ z, \ t\right) = \frac{g}{\langle\rho\rangle}\frac{\partial\langle\rho\rangle}{\partial z} + \frac{g}{\langle\rho\rangle}\frac{\partial\rho_f}{\partial z} = \langle N^2\left(z\right)\rangle + M^2\left(\mathbf{x}, \ z, \ t\right) . \quad (3.112)$$

Here $\langle N^2\rangle$ is the mean Brunt–Väisälä frequency, M^2 the deflection of the density field from $\langle N^2\rangle$ owed to the fine structure. We shall consider internal waves of small amplitude with periods τ and lengths λ satisfying conditions $\tau \ll T$, $\lambda \ll L$, where T is the characteristic time of the variation of the fine structure parameters, L the horizontal scale of the density field inhomogeneities. Under the above mentioned conditions (which are to some extent the fine structure definition) we may neglect variations of N^2 versus horizontal coordinates and time:

$$N^2 (\mathbf{x}, \ z, \ t) \approx N^2 (z) = \langle N^2 (z) \rangle + M^2 (z) .$$

In this case, for the description of internal waves $W = \varphi (z) \exp \{i (\mathbf{k}\mathbf{x} - \omega t)\}$ in the Boussinesq approximation for $f = 0$ in an ocean of finite depth we have the following equation [see (3.64)]

$$\varphi'' + \{\langle N^2 (\xi) \rangle [1 + \mu (\xi)] \omega^{-2} - 1\} H^2 k^2 \varphi = 0 \ ;$$

$$\varphi (0) = \varphi (-1) = 0 \ , \tag{3.113}$$

where a prime means derivation with respect to the dimensionless coordinate $\xi = z/H$ and the function $\mu (\xi) = M^2 (\xi)/\langle N^2 (\xi) \rangle$ defines the fine structure of the density field.

Let us assume that $\mu (\xi)$ is a homogeneous random function of ξ and $|\mu (\xi)| \ll 1$ (i.e., the Brunt–Väisälä frequency deflections owed to the fine structure of the density field are small in comparison with $\langle N^2 \rangle$). The latter is quite frequent for the ocean, at least for a weakly pronounced fine structure and a strong thermocline.

For $|\mu (\xi)| \ll 1$ we use the perturbation method to solve problem (3.113), [248], in the following form:

$$\varphi = \varphi_0 + \varphi_1 + \varphi_2 + \dots \ ; \quad k^2 = k_0^2 + k_1^2 + \dots \ . \tag{3.114}$$

Substituting (3.114) into (3.113) we obtain

$$\varphi_0'' + \left(\frac{\langle N^2 \rangle}{\omega^2} - 1 \right) H^2 k_0^2 \varphi_0 = 0 \ ; \quad \varphi_0 (0) = \varphi_0 (-1) = 0 \ ; \tag{3.115}$$

$$\varphi_1'' + \left(\frac{\langle N^2 \rangle}{\omega^2} - 1 \right) H^2 k_0^2 \varphi_1 = -H^2 \left[\left(\frac{\langle N^2 \rangle}{\omega^2} - 1 \right) \varphi_0 k_1^2 + \frac{\langle N^2 \rangle}{\omega^2} \mu k_0^2 \varphi_0 \right] \ ;$$

$$\varphi_1 (0) = \varphi_1 (-H) = 0 \tag{3.116}$$

. . . .

Giving the distribution $\langle N^2(\xi) \rangle$ we can derive undisturbed eigenfunctions and eigenvalues (or a dispersion relation) for (3.115) which describe free internal waves in the absence of a fine structure. Examples of such solutions are presented in Section 3.2. We solve the problem (3.116) for arbitrary $\langle N^2(\xi) \rangle$. Let us consider $\varphi_{0n}(\xi)$ $(n = 1, 2, ...)$ to be orthonormalized eigenfunctions of (3.115) and the n-th eigenvalue of (3.115) defines the dispersion relation $k_{0n}^2 = k_{0n}^2(\omega, H, \langle N^2 \rangle)$. Then for solvability of the inhomogeneous equation (3.116) its r.h.s. must be orthogonal to each of the eigenfunctions of the corresponding homogeneous equation:

$$\left(\frac{H^2}{\omega^2}\right) [k_{1n}^2 \mathcal{L}_{nm} + k_{0n}^2 J_{nm}] = 0 , \qquad (3.117)$$

where

$$\mathcal{L}_{nm} = \int_0^1 [\langle N^2 \rangle - \omega^2] \varphi_{0n} \varphi_{0m} d\xi ; \qquad J_{nm} = \int_0^1 \langle N^2 \rangle \mu(\xi) \varphi_{0n} \varphi_{0m} d\xi .$$

Further, from an orthogonality of functions φ_{0n} with the weight $H^2 \times \times (\langle N^2 \rangle / \omega^2 - 1)$ and under the appropriate choice of normalization, it follows that $\mathcal{L}_{nm} = \delta_{nm}$, where δ_{nm} is the Kronecker symbol. Then from (3.117) we derive

$$k_{1n}^2 = -k_{0n}^2 J_{nn} . \qquad (3.118)$$

Let us find a correction for a first approximation of eigenfunctions. We solve (3.116) as a series in terms of eigenfunctions for undisturbed Eq. (3.115), i.e.,

$$\varphi_{1n}(\xi) = \sum_m A_m^n \varphi_{0m}(\xi) . \qquad (3.119)$$

Substituting (3.119) into (3.116), multiplying (3.116) by $\varphi_{0n}(\xi)$ and integrating over ξ between the limits 0 and 1, we derive an expression for A_m^n. And using (3.112) we have finally

$$\varphi_{1n}(\xi) = \sum_{m \neq n} \frac{k_{0n}^2 J_{nm}}{(k_{0m}^2 - k_{0n}^2)} \varphi_{0m}(\xi) . \qquad (3.120)$$

A validity condition for the suggested perturbation theory method follows from (3.120): $|J_{mn}| \ll |k_{0m}^2 / k_{0n}^2 - 1|$.

Corrections for a second approximation of the dispersion relation and eigenfunctions may be calculated in a similar way. Omitting intermediate derivations, we write an expression for a mean square wave number $\langle k_n^2 \rangle$ and for an averaged eigenfunction $\langle \varphi(\xi) \rangle$ correct to second order

$$\langle k_n^2 \rangle = k_{0n}^2 \left[1 + \langle J_{nn}^2 \rangle + \sum_{m \neq n} \frac{k_{0n}^2 \langle J_{nm}^2 \rangle}{(k_{0n}^2 - k_{0m}^2)} \right] ; \tag{3.121}$$

$$\langle \varphi(\xi) \rangle = \sum_n [c_{0n} + \langle c_{2n} \rangle] \varphi_{0n}(\xi) , \tag{3.122}$$

where

$$\langle c_{2n} \rangle = \sum_{l \neq n} c_{0l} \frac{k_{0l}^2}{(k_{0l}^2 - k_{0n}^2)} \left[\sum_{j \neq l} \frac{k_{0l}^2 \langle J_{lj} J_{jn} \rangle}{k_{0l}^2 - k_{0j}^2} + \frac{k_{0n}^2 \langle J_{ll} J_{ln} \rangle}{k_{0n}^2 - k_{0l}^2} \right] . \tag{3.123}$$

While deriving (3.121), (3.122) we have taken into consideration that $\langle \mu(\xi) \rangle \equiv 0$. Here c_{0n} is an arbitrary constant.

We can make some quantitative conclusions from the relations obtained. E.g., Eq. (3.121) shows that while internal waves propagate in the ocean with random density field inhomogeneities, the dispersion relation (between k^2 and ω) and the wave phase also become random. From Eq. (3.123) it is obvious that the coefficient c_{2n} characterizes an undisturbed modes interaction caused by random fluctuations of the density field. It also follows from this expression that a coherence of a vertical velocity component decreases at different depths owing to influence of the fine structure. It occurs even for the lowest internal mode because additions owed to a fine structure includes a sum of all the rest higher modes.

The variables $\langle J_{nn}^2 \rangle, \langle J_{nm}^2 \rangle, \langle J_{lj} J_{jn} \rangle$, etc., in (3.121), (3.122) may be expressed through statistical characteristics of the Brunt–Väisälä frequency fluctuations, e.g.,

$$\langle J_{nm}^2 \rangle = \int_0^1 \int_0^1 [\varphi_{0n}(\xi) \varphi_{0m}(\xi) \varphi_{0n}(\xi') \varphi_{0m}(\xi') B(\xi, \xi')] \, d\xi \, d\xi' ,$$

where $B(\xi, \xi') = \langle \mu(\xi), \mu(\xi') \rangle$ is a normalized correlation function of fluctuations $M^2(\xi)$. Thus, having fixed profiles of $B(\xi, \xi')$ and $\langle N^2(\xi) \rangle$

and using (3.121)–(3.123), one can evaluate quantitatively an influence of the density field microstructure on the propagation of free internal waves. Such an influence in an infinitely deep ocean when $\langle N^2 \rangle = $ const (internal waves have an inclined wave vector) was studied in [234].

In the case studied above it was assumed that the Brunt–Väisälä frequency deflections owed to the fine structure are small in comparison with the mean frequency. Otherwise we have to make assumptions concerning the behaviour of the function $M^2(\xi)$. If $\mu(\xi)$ is a periodic function of the depth and $\langle N^2 \rangle = $ const, Eq. (3.113) may be analyzed quite thoroughly. The latter was carried out in the work of Ostrovskiy [283]. Let us introduce the following form for the Brunt–Väisälä frequency $N^2 = N_0^2 + Qf(z)$, where N_0^2 is the mean value of the Brunt–Väisälä frequency with respect to z (considered to be constant), Q is a constant ($|Q| < N_0^2$ means a stable stratification), $f(z)$ is a periodic function with a period l ($|f| \leq 1$, the mean f with respect to z over the range l is zero). Then in the Boussinesq approximation the basic boundary value problem for normal internal modes, instead of (3.113), is reduced to:

$$\varphi'' + [a + qf(z)]\varphi = 0 \; ; \quad \varphi(0) = \varphi\left(\pi \frac{H}{l}\right) = 0 \; , \qquad (3.124)$$

where a prime means derivation with respect to the dimensionless coordinate $\xi = \pi z/l$, and the parameters a and q are connected with the frequency and the wave number by:

$$\Omega = \left(1 - \frac{a\,s}{q}\right)^{1/2} \; ; \quad \text{æ} = \left(\frac{q}{s} - a\right)^{1/2} = \Omega\left(\frac{q}{s}\right)^{1/2} \; ;$$

$$\Omega = \frac{\omega}{N_0} \; ; \quad \text{æ} = \frac{kl}{\pi} \; ; \quad s = \frac{Q}{N_0^2} \; . \qquad (3.125)$$

Eq. (3.124) is the well known Gill equation, but the boundary conditions of the present problem are given in an interval different from the value of π or 2π, which makes our study more complicated. The behaviour of eigenfunctions and eigenvalues of problem (3.124) was analyzed completely by means of the Floquet theorem in [283]. For small s and small wave numbers (when $q > a$, i.e., $\Omega > (1-s)^{1/2}$, $\text{æ} \geq (s^{-1} - 1)^{1/2}$), and fixed Ω the values of æ are gathered in narrow 'clusters' each containing about H/l modes.

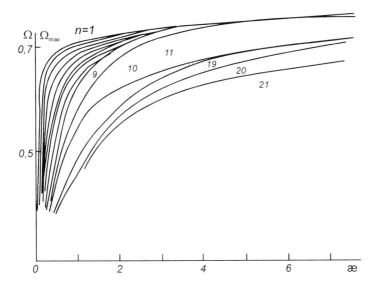

Figure 3.6. Dispersion curves for the first 21 modes of internal waves for $(N/N_{\mathrm{max}})^2 = 1 + 0.5 \sin (2\pi z/l)$, $H = 10\,l$ [283].

Fig. 3.6 (taken from [283]) shows the results of the numerical calculation of dispersion curves in the case $f = \sin 2pz$, $H = 10l$, $l = \pi/p$. The dispersion curves are grouped into clusters containing nine modes. Estimations showed that for natural oceanic conditions $s \approx 0.1$–0.3, and in this case a cluster character of dispersion (dispersion curves grouping) is considerable for frequencies $\omega \geq 0.9 N_{\mathrm{max}}$ and horizontal wave lengths $\lambda \leq l \approx 10$–$20$ cm. I.e., rather short and high frequency internal waves will be localized in a wave number – frequency space in groups.

6. INFLUENCE OF THE EARTH'S ROTATION ON FREE INTERNAL WAVES

In previous sections the influence of the Earth's rotation on oceanic internal waves was analyzed in the 'traditional' approximation neglecting a horizontal component of the earth rotation velocity. Here short enough internal waves are considered for which a 'plane' description is valid (see Chapters 1, 2). In this section the validity of this approximation is examined along the lines of the work by Kamenkovich and Kulakov [153].

As the basic simultaneous equations describing the propagation of infinitesimal internal waves we consider Eqs. (2.13), (2.14), (2.15a), linearized relative to a state of rest (see Section 2.3) with the standard state $\rho_0 = \rho_0(z)$ and $p_0 = p_0(z)$. Within the 'plane' description x is directed east, y north, and z upwards from an undisturbed free ocean surface. We consider free internal waves propagating in an ocean layer of constant depth H with a wavy free surface.

For elementary plane waves written in the form (3.1) all variables are expressed through an amplitude function of a vertical component of the velocity W [compare with (3.2)]

$$\left.\begin{aligned}
U,(z) &= \frac{1}{\omega^2 - f_z^2}\left[-i\omega f_y W + (\omega k_1 + i f_z k_2)\frac{P}{\rho_0}\right] ; \\
V(z) &= \frac{1}{\omega^2 - f_z^2}\left[f_y f_z W + (i f_z k_1 - \omega k_2)\frac{P}{\rho_0}\right] ; \\
\frac{P(z)}{\rho_0(z)} &= -\frac{i}{\omega k^2}\left[(\omega^2 - f_z^2)W' + f_y(\omega k_1 - i f_z k_2)W\right] ; \\
R(z) &= (i\rho_0 N^2/g\omega)W .
\end{aligned}\right\} \quad (3.126)$$

Here $f_y = 2\Omega\cos\varphi$, $f_z = 2\Omega\sin\varphi$ are components of the doubled earth rotational velocity vector Ω, φ is a latitude, N the Brunt–Väisälä frequency (in a further analysis considered to be constant), and $k^2 = k_1^2 + k_2^2$.

For W we have the following boundary value problem [compare with (3.3)–(3.4a)]:

$$(\omega^2 - f_z^2)W'' - \left[2ik_2 f_y f_z + (\omega^2 - f_z^2)\frac{N^2}{g}\right]W'$$

$$- \left[(\omega^2 - N^2)k^2 - k_2^2 f_y^2 + \frac{f_y N^2}{g}(\omega k_1 - i f_z k_2)\right]W = 0; \quad (3.127)$$

$$(\omega^2 - f_z^2)W'(0) = \left[gk^2 - f_y(\omega k_1 - i f_z k_2)\right]W(0); \quad (3.128)$$

$$W|_{z=-H} = 0 .$$

Eq. (3.127) has constant coefficients and admits the simple solution $W \sim \exp\{ilz\}$. This allows us to derive easily a dispersion relation in an evident form. But formulae both for a solution and a dispersion relation are rather cumbersome. Estimations given in [153] show that terms divergent from the Boussinesq approximation (proportional to N^2/g) may be neglected in Eqs. (3.127), (3.128). In such a case we have, to a satisfactory accuracy (excluding a narrow equatorial zone, where the solution is derived separately),

$$W = \exp\left\{\frac{ik_2 f_y f_z (z + H)}{\omega^2 - f_z^2}\right\}$$

$$\times \sin\left\{\frac{k[(\omega_1^2 - \omega^2)(\omega^2 - \omega_2^2)]^{1/2}(z + H)}{\omega^2 - f_z^2}\right\} ; \qquad (3.129)$$

$$l_{1,2} = \frac{k_2 f_y f_z \pm [k_m^2(\omega_1^2 - \omega^2)(\omega^2 - \omega_2^2)]^{1/2}}{\omega^2 - f_z^2} ; \qquad (3.130)$$

$$k_0^2 = (\omega^2 - f_z^2)/gH ;$$

$$k_m = \frac{m\pi}{H}(\omega^2 - f_z^2)[(\omega_1^2 - \omega^2)(\omega^2 - \omega_2^2)]^{-1/2}, \quad \omega^2 > f_z^2 ; \qquad (3.131)$$

$$k_m = \frac{m\pi}{H}(f_z^2 - \omega^2)[(\omega_1^2 - \omega^2)(\omega^2 - \omega_2^2)]^{-1/2},$$

$$\omega^2 < f_z^2, \quad m = 1, 2, \ldots \qquad (3.131a)$$

Here $\omega_{1,2}^2$ is defined by

$$2\omega_{1,2}^2 = N^2 + f_z^2 + f_y^2 \sin^2\alpha \pm$$

$$\pm[(N^2 - f_z^2)^2 + f_y^4 \sin^4\alpha + 2f_y^2(N^2 + f_z^2)\sin^2\alpha]^{1/2} ; \qquad (3.132)$$

$$(k_1 = k\cos\alpha ; \quad k_2 = k\sin\alpha) .$$

Formulae for $k_0^2 = k_{10}^2 + k_{20}^2$ describe a surface wave and k_m are internal waves. They are valid when

$$\omega_2^2 < \omega^2 < \omega_1^2 \qquad (3.133)$$

(the case $\omega^2 > \omega_1^2$ corresponds to the presence of only surface waves, for $\omega^2 < f_z^2$ a surface wave is absent).

According to Eqs. (3.129)–(3.131) the existence of internal waves is independent of the presence of a free surface, and the condition (3.129) for $z = 0$ may be replaced with an enough accuracy by the 'rigid lid' condition $W(0) = 0$.

Eq. (3.132) yields the following inequalities:

$$\max\left(N^2,\ f_z^2\right) < \omega_1^2 < N^2 + f_z^2 + f_y^2 = N^2 + 4\Omega^2\ ;$$

$$\omega_2^2 < \min\left(N^2,\ f_z^2\right). \tag{3.134}$$

For $N/2\Omega \gg 1$ for any α we have from (3.131)

$$\omega_1^2 = N^2\left[1 + \frac{f_y^2\sin^2\alpha}{N^2} + \mathcal{O}\left(\frac{\Omega^4}{N^4}\right)\right]\ ;$$

$$\omega_2^2 = f_z^2\left[1 - \frac{f_y^2\sin^2\alpha}{N^2} + \mathcal{O}\left(\frac{\Omega^4}{N^4}\right)\right]. \tag{3.135}$$

Owing to Eqs. (3.135) and the inequalities (3.133), for high frequency internal waves (3.131) the Earth's rotation is essential only at low horizontal wave numbers k (long internal waves). Their frequencies, to an accuracy of $\mathcal{O}\left(\Omega^2/N^2\right)$, are within the range (f_z, N). In contrast, for low frequency internal waves (3.131a) rotation is essential for every k.[6]

Let us now estimate for $f_y = 0$ the 'traditional' approximation studied above. This approximation, as follows from (3.132), is valid exactly for $\alpha = 0$, i.e., when internal waves propagate in the latitude direction only (east–west), and $\omega_2 \equiv 0$. In Fig. 3.7 a,b,c, [153], the dispersion curves $\omega_m(k)$ for the first and the third internal wave modes are presented for given angles α of a horizontal wave vector inclination with respect to the latitude direction (figures at curves), for different values of the Brunt–Väisälä frequency N and for different latitudes φ (dispersion curves are symmetric with respect to ω-axis and only their right branches are shown). The dispersion curves divided by the line $\omega = f_z$ are two-valued and ω-axis scales for $\omega > f_z$ and $\omega < f_z$ are different.

Dispersion curves for $\omega > f_z$ ($N \gg \Omega$) differ slightly from those counted for the non-rotational case ($\Omega = 0$) (see dashed curves in

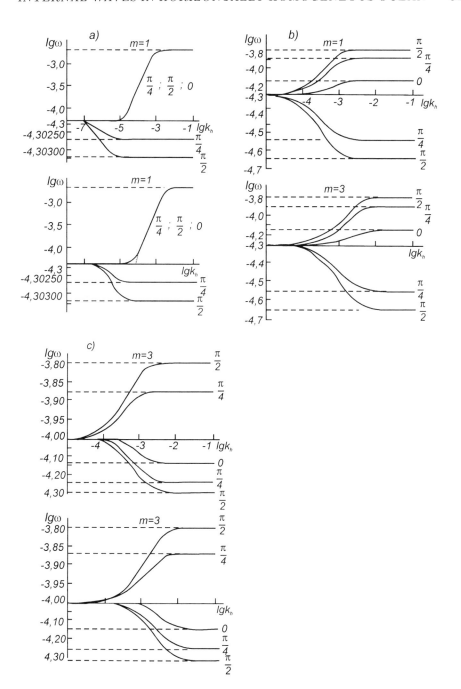

Figure 3.7. Dispersion curves $\omega(k)$ under given angles α for the 1st and the 3rd modes of internal waves at different N and φ ($\omega = 7.29 \times 10^{-5}$ 1/s): a) $N = 2 \times 10^{-3}$ 1/s, $\varphi = \pi/9$; b) $N = 7.29 \times 10^{-5}$ 1/s, $\varphi = \pi/9$ ($N > f_z$); c) $N = 7.29 \times 10^{-5}$ 1/s, $\varphi = \pi/4$ ($N < f_z$).

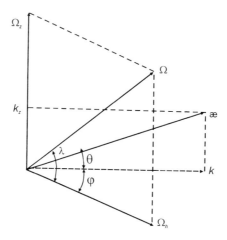

Figure 3.8. A diagram for the wave vector æ in the coordinates associated with the Earth.

Fig. 3.7 a). In this case the dispersion relation is isotropic (i.e., $\omega(k)$ is independent of the angle α). But for $\omega < f_z$ an anisotropy of the dispersion relation (weakly pronounced in Fig. 3.7 a) is observed, which is obvious from Fig. 3.7 b,c for $N \approx \Omega$.

From everything stated above it follows that the 'traditional' approximation for internal waves, when $f_y = 2\Omega \cos \varphi \approx 0$, describes correctly internal gravity waves (3.131) for $N \gg \Omega$. Internal waves (3.131a) vanish under this approximation (as illustrated by Fig. 3.7 a). For $N \leq \Omega$ the approximation $f_y = 0$ is invalid (see Fig. 3.7 b,c).

For the equatorial zone (where $\varphi = 0$), when assuming $f_z = 0$ and taking into account that from (3.132) $\omega_1^2 = N^2 + f_y^2 \sin^2 \alpha$ and $\omega_2^2 = 0$, (3.131) yields

$$k_0^2 = \frac{\omega^2}{gH}, \qquad k_m = \frac{m\pi}{H} \frac{|\omega|}{(N^2 + f_y^2 \sin^2\alpha - \omega^2)^{1/2}} . \qquad (3.136)$$

Unlike the dispersion curves (3.131) the curves (3.136) contain one branch only. Eq. (3.136) shows that the influence of the Earth's rotation on the propagation of internal waves may be neglected. The dispersion relation then becomes isotropic. Its anisotropy, following (3.136), becomes quite essential for $N \sim \Omega$.

Assuming $N \ll \Omega$ in all equations considered and letting $N \to 0$ yields the case of gyroscopic (or inertial) waves the appearance of which is independent of stratification, but is caused by the Earth's rotation only.

In a simplified state, when the vertical wave number satisfies the inequality $lH \gg 1$ [which is valid for the highest modes $m \gg 1$, see (3.130), (3.131)], the analysis simplifies considerably in the Boussinesq approximation. In this case, assuming the ocean to be infinitely deep and $W \sim \exp\{ilz\}$, from (3.127) for $N = $ const, neglecting the terms with N^2/g yields the simple expression

$$\omega^2 \ae^2 = (\mathbf{f}\ae)^2 + N^2 k^2 \,,$$

$$\ae^2 = k_1^2 + k_2^2 + l^2 \,, \quad k^2 = k_1^2 + k_2^2 \,,$$

$$\mathbf{f} = \{0, 2\Omega\cos\varphi, 2\Omega\sin\varphi\} \,. \tag{3.137}$$

Here $\ae$ is the wave vector modulus and k the horizontal wave vector modulus.

Now introducing the angles φ, θ, α (Fig. 3.8), where φ defines the latitude, θ the inclination, and α the azimuth for low frequency internal waves considered, then taking into consideration Fig. 3.8 we may rewrite (3.137) in the following form ([299], Section 5.7):

$$\omega^2 = 4\Omega^2 (\sin\varphi\sin\theta + \cos\varphi\cos\theta\sin\alpha)^2 + N^2\cos^2\theta \,. \tag{3.138}$$

Eq. (3.138) demonstrates an anisotropy of the dispersion relation for low frequency internal waves when the Earth's rotation is essential. In particular, if internal waves propagate along a line of latitude ($\alpha = 0$, $\alpha = \pi$) then (3.138) yields

$$\omega^2 = 4\Omega^2 \sin^2\varphi\sin^2\theta + N^2\cos^2\theta.$$

So, the most interesting case, when the latitude φ is fixed, reveals that for arbitrary variations of the wave inclination θ

$$4\Omega^2 \sin^2\varphi \leq \omega^2 \leq N^2 \,. \tag{3.139}$$

When internal waves propagate in a meridional direction (e.g., north, i.e., $\alpha = \pi/2$), Eq. (3.138) yields

$$\omega^2 = 4\Omega^2 \cos^2(\varphi - \theta) + N^2 \cos^2\theta.$$

Thus we may also conclude that in a thermocline zone where $N \gg 2\Omega$, we have

$$4\Omega^2 \sin^2\varphi \leq \omega^2 \leq N^2 + 4\Omega^2 \cos^2\varphi. \tag{3.140}$$

For the lowest frequencies we have again $\omega \geq f_z$, which has already been shown above with a detailed analysis (see Fig. 3.7 a).

Low frequency internal waves with the anisotropic dispersion relations mentioned are quite frequent in the ocean, although they are rarer in the open ocean than high frequency internal waves (which cover most of this book). Low frequency internal waves are basically generated while ocean tides propagate towards the continental shelf [309], [416], and therefore have pronounced half day and day periods. The dynamic role of these waves is still vague at present, although they contribute considerably to the total energy balance of internal waves in the ocean.

Notes

1 The problem (3.7)–(3.8a) has been solved by many authors (see, e.g., [402] for $f \equiv 0$, the review [403] and its bibliography).

2 Eqs. (3.31) and (3.32) define a modulus of the phase $|c_f|$ and of the group $|c_g|$ velocity vectors. Vectors for plane waves (3.1) are, accordingly, $c_f = (\omega/k^2)\,\mathbf{k}$, $c_g = (d\omega/dk)\,\mathbf{k}/k$.

3 The condition $f = 0$ selects for further consideration internal waves of moderate length.

4 This method is described in detail, e.g., in [270].

5 The propagation of the internal waves packet in the long wave (internal or Rossby type) field is considered for real oceanic conditions by the WKB approximation both numerically and theoretically. It was shown that the amplitude and the length of short waves may change several times in magnitude during propagation. The effect of large scale motions on the dispersion curves of short internal waves is also discussed.

6 Remember that $\Omega \approx 7.3 \times 10^{-5}$ 1/s. And in the approximation of N = const the mean Brunt–Väisälä frequency in the ocean $N \approx 10^{-3}$ 1/s which is a order lower than the value N in a seasonal thermocline.

Chapter 4

SHEAR FLOW INFLUENCE ON INTERNAL WAVES PROPAGATION

Shear flows slightly varying in time and horizontal coordinates are almost always found in the ocean[1]. Such shear flows are especially pronounced in the seasonal thermocline zone and may strongly influence the internal waves' dynamics. If a characteristic period of a shear flow variation is considerably greater than the internal waves' period and the horizontal variations are small, then the flow may be considered to be stationary and horizontally homogeneous.

1. BASIC BOUNDARY VALUE PROBLEM. STABILITY NUMBER

Consider the influence of a plane stationary shear flow on small amplitude internal waves. Assume that the mean velocity of the flow $\mathbf{U} = \{U(z); 0; 0\}$ is horizontal and depends on the depth only. We shall consider internal waves of relatively high frequencies ($\omega \gg f$) and shall not take into account the Earth's rotation using the free convection approximation. The basic equation system describing internal waves in this case will be Eqs. (2.52) for $V(z) \equiv 0$. In Section 2.3 it was shown that the integral wave energy of internal waves propagating on the background shear flow is not conserved owing to wave–flow interaction through the Reynolds stress. If the Reynolds stress increases the wave energy, waves become unsteady and will grow in time. We shall start by studying the stability of internal waves on a mean shear flow.

This problem described by Eq. (2.52) is quite complicated. It is the problem of the hydrodynamic steadiness of a flow with vertical gradients of velocity and density in the gravity field. According to Squire's theorem [192] two-dimensional disturbances tend more to instabilities than three-dimensional ones. So for a conservative estimate of stability, we may limit ourselves to plane internal waves $v \equiv 0$, $\partial/\partial y \equiv 0$. Eqs. (2.52) have elementary wave solutions in the form of plane waves. When assuming

$$\xi\,(x, z, t) = \varphi\,(z)\exp\left\{ik\,(x - c_f t)\right\},\qquad(4.1)$$

where $\xi = \rho\,/\,(\partial\rho_0/\partial z)$ is the particle displacement in a wave, k the real wave number, and c_f the (complex) phase velocity, Eqs. (2.52) for $V = v \equiv 0$ and $\partial/\partial y \equiv 0$ are reduced to the single equation for $\varphi\,(z)$:

$$\frac{d}{dz}\left[\rho_0\,(U - c_f)^2\,\frac{d\varphi}{dz}\right] + \rho_0\left[N^2\,(z) - k^2\,(U - c_f)^2\right]\varphi = 0\,.\qquad(4.2)$$

The remaining wave characteristics are expressed through $\varphi\,(z)$ by

$$u = -\frac{d}{dz}[(U - c_f)\,\varphi]\,;\quad w = ik\,(U - c_f)\,\varphi\,;\quad p = \rho_0\,(U - c_f)^2\,\frac{d\varphi}{dz}\,.\quad(4.3)$$

The boundary conditions for (4.2) are as a rule conditions in the 'rigid lid' approximation $\varphi\,(0) = \varphi\,(-H) = 0$. Eq. (4.2) has singularities for $U\,(z) = c_f$. Obviously a flow produces unstable (growing) waves when the eigenvalue problem (4.2) has nontrivial solutions with $\mathrm{Im}(c_f) > 0$. This problem was thoroughly studied by Miles [242] and Howard [130] (see also [243]). To define the instability conditions we shall assume (as in [130]) that $\varphi = F^{-n}Q$, where $F = U - c_f$ and φ corresponds to an unstable mode, so that $c_f = c_r + ic_i$ ($c_i > 0$). Multiplying (4.2) by Q^*, using the expression $\varphi = F^{-n}Q$, and integrating the resulting equation from 0 to $-H$, yields

$$\int\limits_{-H}^{0} \rho_0 F^{2\,(1-n)}\left[\left|\frac{dQ}{dz}\right|^2 + k^2|\,Q|^2\right]dz + n\int\limits_{-H}^{0} F^{1-2n}\,\frac{d}{dz}\left(\rho_0\frac{dU}{dz}\right)$$

$$\times|\,Q|^2\,dz + \int\limits_{-H}^{0} \rho_0 F^{-2n}\left[n\,(1 - n)\left(\frac{dU}{dz}\right)^2 - N^2\,(z)\right]|\,Q|^2\,dz = 0\,.\quad(4.4)$$

The integral relation (4.4) allows us to formulate some important theorems.

Consider, for example, $n = \frac{1}{2}$. Then the imaginary part in Eq. (4.4) must be zero for $c_i > 0$:

$$\int\limits_{-H}^{0} \rho_0 \left[\left| \frac{dQ}{dz} \right|^2 + k^2 |Q|^2 \right] dz + \int\limits_{-H}^{0} \rho_0 \left[N^2 - \frac{1}{4} \left(\frac{dU}{dz} \right)^2 \right] \left| \frac{Q}{F} \right|^2 dz = 0 .$$

$$(4.5)$$

This is impossible for $N^2 - \frac{1}{4} \left(\frac{dU}{dz} \right)^2 \geq 0$. Thus the necessary instability condition is $N^2 - \frac{1}{4} \left(\frac{dU}{dz} \right)^2 < 0$. When defining the local Richardson number as $Ri = N^2 / \left(\frac{dU}{dz} \right)^2$ we have the instability number $Ri < \frac{1}{4}$. This characteristic number was established by Miles [242].

Taking into account the obvious inequality $|F|^2 \leq c_i^{-2}$, we obtain from (4.5) an estimate for a rate of unstable waves growth:

$$k^2 c_i^2 \leq \max \left[\frac{1}{4} \left(\frac{dU}{dz} \right)^2 - N^2 \right] .$$

$$(4.6)$$

The important theorem on the complex phase velocity value for any unstable wave (initially proved by Howard [130]) may be obtained from (4.5) for $n = 0$. Assuming $n = 0$ in (4.4) and separating real and imaginary parts, yields for $c_i > 0$

$$\int\limits_{-H}^{0} U \tilde{Q} \, dz = c_r \int\limits_{-H}^{0} \tilde{Q} \, dz ,$$

$$\int\limits_{-H}^{0} U^2 \tilde{Q} \, dz = (c_r^2 + c_i^2) \int\limits_{-H}^{0} \tilde{Q} \, dz + \int\limits_{-H}^{0} N^2 \rho_0 |\varphi|^2 \, dz , \qquad (4.7)$$

where $\tilde{Q} = \left[\left| \frac{d\varphi}{dz} \right|^2 + k^2 |\varphi|^2 \right]$. Consider $a \leq U(z) \leq b$; then (4.7) and $N^2 \geq 0$, $\tilde{Q} > 0$ yield

$$\left[c_r - \frac{1}{2}(a+b) \right]^2 + c_i^2 \leq \left[\frac{1}{2}(a-b) \right]^2 .$$

$$(4.8)$$

Thus the complex phase velocity c_f for any unstable mode must be disposed in the upper half-plane inside a hemisphere of diameter equal to the variation range U.

For $n = 1$ from (3.4) we can derive a theorem disseminating the Rayleigh theorem on a stratified flow (see [243]). In line with this theorem an instability occurs when the value of

$$\frac{d}{dz}\left(\rho_0 \frac{dU}{dz}\right) - 2N^2\rho_0|F|^{-2}(U - c_r)$$

changes its sign within the range $[0, -H]$.

The conclusions of the linear stability theory, in particular the critical value of the stability parameter $(Ri = \frac{1}{4})$, have been validated in laboratory experiments [321], [354], [355], [358], [360]. Examples of solution of the boundary-value problem (4.2) for the certain $N^2(z)$ and $U(z)$ distributions are given in [243], [179], [359].

2.   INTERNAL WAVES' BEHAVIOUR IN THE VICINITY OF CRITICAL LAYER

As mentioned above, in the vicinity of $z = z_c$, where $U(z) = c_f$, the equations of internal waves' propagation in a shear flow have some peculiarities. Consider the behaviour of internal waves in the vicinity of the critical layer $z = z_c$ along the lines of Booker and Bretherton [28].

Examine Eq. (2.57) for an internal wave vertical velocity component when $V(z) \equiv 0$. We find its solution in the form

$$w = \mathrm{Re}\left[\widehat{W}(k, z, c_f)\exp\{ik(x - c_f t)\}\right],$$

This yields for $\widehat{W}$

$$\frac{d^2\widehat{W}}{dz^2} + \left\{\frac{N^2}{(U - c_f)^2} - \frac{1}{(U - c_f)}\frac{d^2U}{dz^2} - k^2\right\}\widehat{W} = 0. \qquad (4.9)$$

We define a critical level as:

$$U(z_c) - c_r = 0, \qquad (4.10)$$

where $c_f = c_r + ic_i$. If the wave phase velocity c_f is real $(c_i = 0)$ Eq. (4.9) has a singularity when satisfying equality (4.10). The condition $c_i = 0$

is met for stable waves, i.e., $Ri = N^2 / \left(\frac{dU}{dz}\right)^2 > 1/4$. A solution of
Eq. (4.9) on both sides of the singular point (for $z < z_c$ and $z > z_c$) was
derived by Miles [242]. To obtain a solution in the vicinity of $z = z_c$ we
consider c_i to be small

$$\left|\frac{d^2U}{dz^2}\right|\left[\frac{dU}{dz}\right]^{-2} c_i \ll 1 \qquad (4.11)$$

and $Ri > \frac{1}{4}$. Then to solve (4.9) the Frobenius method[2] may be used,
i.e., two linearly independent solutions may be expressed as a power
series of $z - z_c$. In the vicinity of the critical level this method gives the
following complete solution of Eq. (4.9)

$$\widehat{W} = A\left[(z - z_c) - \frac{ic_i}{dU/dz}\right]^{1/2+i\mu} + B\left[(z - z_c) - \frac{ic_i}{dU/dz}\right]^{1/2-i\mu},$$
$$(4.12)$$

where

$$\mu = (Ri - 1/4)^{1/2} > 0, \qquad (4.13)$$

A and B are arbitrary constants. For $c_i = 0$ each of the particular
solutions (4.12) have a branch point.

If $c_i > 0$, for $z - z_c$ decreasing from a positive value (greater than
$c_i / \frac{dU}{dz}$) to a negative one, the argument $(z - z_c) - c_i / \frac{dU}{dz}$ continuously
varies from 0 to $-\pi$. When fixing an exponent in the complex expression
(4.12) we have

$$(z - z_c)^{1/2+i\mu} = |z - z_c|^{1/2}e^{i\mu\,|z-z_c|} \quad \text{when } z > z_c, \qquad (4.14)$$

and further

$$\left.\begin{array}{l} (z - z_c)^{1/2+i\mu} = -ie^{\mu\pi}|z - z_c|^{1/2}e^{-i\mu\,|z-z_c|} \\[2mm] (z - z_c)^{1/2-i\mu} = -ie^{-\mu\pi}|z - z_c|^{1/2}e^{i\mu\,|z-z_c|} \end{array}\right\} \quad \text{when } z < z_c. \quad (4.15)$$

Eqs. (4.12), (4.15) show that $\widehat{W}$ rapidly oscillates for $c_i = 0$, $Ri > 1/4$
and $z \approx z_c$.

The expressions obtained are valid for arbitrary depth variations of $N^2(z)$ and $U(z)$ (N^2 and U are considered to be continuous and twice-differentiable functions). At first we consider the simplest case when $N^2(z)$ and $U(z)$ are piecewise constant functions (the case of slowly varying functions $N^2(z)$ and $U(z)$ is studied later).

When $N^2 \approx$ const and $U \approx$ const the solution of Eq. (4.9) is

$$\widehat{W} = Ae^{imz} + Be^{-imz} , \tag{4.16}$$

where the vertical wave number m is

$$m = \{N^2/(U - c_f)^2 - k^2\}^{1/2} . \tag{4.17}$$

We assume further $m_i > 0$ for $c_i > 0$. Then

$$\left.\begin{array}{ll} m \sim \{N/(U - c_f), & \text{when } k^2 \ll N^2/(U - c_f)^2 \\ m \sim ik, & \text{when } k^2 \gg N^2/(U - c_f)^2 \end{array}\right\} . \tag{4.18}$$

The first solution of Eq. (4.16) gives a complete vertical velocity distribution in the form

$$w = \mathrm{Re}\,\{A \exp [i\,(kx + mz - kc_f t)]\}$$

and represents a plane wave with the stationary phase front $kx + mz - kc_f t =$ const when kc_f, m, and c_f are real in a given frequency range. If $U - c_f$ is negative, i.e., a wave propagates in positive x direction, m is also negative, and stationary phase lines movie in the negative z direction, i.e., down. If $U - c_f$ is positive m is also positive, and stationary phase lines move upwards. Thus the first solution of Eq. (4.16) describes waves with a phase velocity component directed downwards and the second solution $\widehat{W} = Be^{-imz}$ describes waves having an upward phase velocity component. From Eq. (4.17) follows that $\omega = kc_f$ is expressed by

$$\omega = kU \pm \frac{kN}{(k^2 + m^2)^{1/2}} , \tag{4.19}$$

where, following (4.18), a plus sign is valid when $U - c_f$ and m are negative and a minus sign when $U - c_f$ and m are positive. From (4.19) for the vertical component of the phase velocity $w_g = \partial\omega/\partial m$ in the first term in (4.16) and the first inequality in (4.18) we have

$$w_g = \frac{\partial\omega}{\partial m} = \mp\frac{kNm}{(k^2+m^2)^{3/2}} = \frac{km\,(U-c_f)^3}{N^2}. \qquad (4.20)$$

This value is always positive. Thus the energy propagates in the positive z direction, i.e., upwards. For the second term in (4.16) $w_g < 0$, and energy is transferred vertically down. From the equation for the horizontal velocity component

$$(U-c_f)\,\frac{\partial u}{\partial x} + \frac{1}{\rho_0}\frac{\partial p}{\partial x} = 0$$

we can obtain the equation for the vertical energy flow

$$\overline{pw} = -\rho_0\,(U-c_f)\,\overline{uw}, \qquad (4.20a)$$

where overline means an averaging with respect to time and horizontal wave number. For the solution of (4.9) in the form $\exp\{imz\}$ the value $\overline{pw}$ is always positive and for $\exp\{-imz\}$ always negative, i.e., an energy flow is directed upwards and downwards, respectively. So for $N^2 =$ const and $U =$ const internal waves generated in a shear flow may be subdivided into waves carrying energy upwards (the first term in (4.16)) and down (the second term in (4.16)). Analogous subdivision may be performed for arbitrary depth variations of N^2 and U in the vicinity of the critical level $z \approx z_c$. Consider again the solution of (4.9) derived from (4.12). If $c_i = 0$ and $z - z_c$ is real and positive, then

$$(z-z_c)^{1/2+i\mu} = |z-z_c|^{1/2}[\cos\mu\,\log|z-z_c| + i\sin\mu\,\log|z-z_c|].$$

So as $z-z_c$ decreases, oscillations of $(z-z_c)^{1/2+i\mu}$ grow and its amplitude and wave number vary over the depth z. We define the local vertical wave number as

$$m = \frac{\mu}{z - z_c}. \qquad (4.21)$$

For large μ the vertical wave amplitude variations are small at distances of about m^{-1} and waves are locally sinusoidal. If $dU/dz > 0$ from (4.21) yields

$$m = \frac{\mu}{z - z_c} = \frac{N \left(1 - \frac{1}{4} Ri^{-1}\right)^{1/2}}{z - z_c} \sim \frac{N}{U(z) - c_f} \left(1 - \frac{1}{4} Ri^{-1}\right)^{1/2}.$$

$$(4.22)$$

When Ri is large enough Eq. (4.22) is similar to Eq. (4.18) for the wave number of waves moving upwards in a medium with $N^2 = $ const and $U = $ const. As Eq. (4.22) is valid on both sides of the critical level, the wave number will change its sign simultaneously with $U(z) - c_f$.

Consider further the Reynolds stresses and the vertical energy flow for both solutions of (4.16). Using (4.14) and (4.15), after simple calculations we derive:

$$\overline{uw} = -\frac{1}{4ik} \left(\widehat{W}^* \frac{d\widehat{W}}{dz} - \widehat{W} \frac{d\widehat{W}^*}{dz} \right)$$

$$= \begin{cases} -\dfrac{\mu}{2k} \{|A|^2 - |B|^2\} & \text{for } z > z_c; \\[2mm] \dfrac{\mu}{2k} \{|A|^2 e^{2\mu\pi} + |B|^2 e^{-2\mu\pi}\} & \text{for } z < z_c. \end{cases}$$

$$(4.23)$$

From Eq. (4.23) the Reynolds stresses have a discontinuity crossing the critical level $z = z_c$ with simultaneous change of both the sign and an absolute value. But the energy flow $-\rho_0 (U - c_f) \overline{uw}$ for the first solution (in the form $A(z - z_c)^{1/2 + i\mu}$) is positive both above and below the critical level, although it is also discontinuous at $z = z_c$. For the second solution (in the form $B(z - z_c)^{1/2 - i\mu}$) the energy flow is always negative. Thus waves in the first form transfer energy upwards and in the second form downwards. As the energy flow is discontinuous at $z = z_c$ (with the jump $e^{\pm\pi\mu}$) the wave energy absorption occurs in the critical level. This process in the absence of dissipation and nonlinearity leads to growth of energy of mean motion[3].

Furthermore, for $c_i = 0$ the vertical velocity w becomes rather small in the vicinity of $z \approx z_c$. But the horizontal velocity

$$u = \text{Re} \left\{ \frac{1}{ik} \frac{d\widehat{W}}{dz} e^{ik(x - c_f t)} \right\}$$

Table 4.1. Joint properties of two solutions of Eq. (4.9) in the vicinity of the critical level

$\hat{W}(z)$		Location	$\bar{c}_f$	$\dfrac{|\hat{W}|}{|z-z_c|^{\frac{1}{2}}}$	$\overline{uw}$	Sign of $\overline{pw}$	Direction
$\dfrac{dU}{dz} > 0$	$A\left(z - \dfrac{c_i}{U'}\right)^{\frac{1}{2}+i\mu}$	$z > z_c$	↙	$|A|$	$-\dfrac{\mu}{2k}|A|^2$	$+$	Up
		$z < z_c$	↗	$|A|e^{\mu\pi}$	$\dfrac{\mu}{2k}|A|^2 e^{2\mu\pi}$	$+$	Up
	$B\left(z - \dfrac{c_i}{U'}\right)^{\frac{1}{2}-i\mu}$	$z > z_c$	↖	$|B|$	$\dfrac{\mu}{2k}|B|^2$	$-$	Down
		$z < z_c$	↗	$|B|e^{-\mu\pi}$	$-\dfrac{\mu}{2k}|B|^2 e^{-2\mu\pi}$	$-$	Down
$\dfrac{dU}{dz} < 0$	$A\left(z - \dfrac{c_i}{U'}\right)^{\frac{1}{2}+i\mu}$	$z > z_c$	↗	$|A|$	$-\dfrac{\mu}{2k}|A|^2$	$-$	Down
		$z < z_c$	↖	$|A|e^{-\mu\pi}$	$\dfrac{\mu}{2k}|A|^2 e^{-2\mu\pi}$	$-$	Down
	$B\left(z - \dfrac{c_i}{U'}\right)^{\frac{1}{2}-i\mu}$	$z > z_c$	↘	$|B|$	$\dfrac{\mu}{2k}|B|^2$	$+$	Up
		$z < z_c$	↙	$|B|e^{\mu\pi}$	$-\dfrac{\mu}{2k}|B|^2 e^{2\mu\pi}$	$+$	Up

Note. The variable $\mu = (Ri - 1/4)^{1/2}$ is considered to be real and positive. Arrows indicate a quadrant of the (x,y)-plane for the phase velocity relatively to moving fluid at the level z. The wave front is always normal to the phase velocity.

may be quite large because $u \sim (z - z_c)^{-1/2}$. So the wave's particle motion becomes almost horizontal near $z = z_c$ and the wave frequency tends to zero accordingly.

In conclusion we present Table 4.1 [28] describing internal waves' behaviour for $c_i \neq 0$ in the vicinity of $z = z_c$.

Internal waves in a shear flow taking into account the Coriolis force were studied by Jones [145]. Fluid rotation causes the conservation with depth of the vertical flow of the angular momentum (contrary to the constancy of the horizontal flow of the angular momentum in a non-rotating fluid) out of the critical level which is defined by the equality $\omega - kU = \pm f$ or $U - c_f = \pm f/k$. The qualitative behaviour of internal waves in the vicinity of the critical level remains similar to that of a non-rotating fluid.

3.   EXAMPLE OF EXACT SOLUTION DESCRIBING INTERNAL WAVES IN MEAN FLOW

A clear particular solution of Eq. (2.57) for $Ri > \frac{1}{4}$, $\mathbf{U} = \{U, 0, 0\}$, $N^2 = \text{const}$, $\Gamma = dU/dz = \text{const}$ was given by Phillips [299]. His results with respect to internal waves in rotating fluid were generalized by Frankignoul [81]. Consider internal waves propagating in rotating infinitely deep fluid with a shear flow of a small vertical scale (large vertical wave numbers) and a plane mean flow $U(z)$. A local description (N and Γ are considered to be quasi-constant) is enough for internal waves with large vertical wave numbers. Then Eqs. (2.46)–(2.49) may be reduced to a single equation for w

$$\frac{d^3}{dt^3}(\Delta w) + \frac{d}{dt}\left(f^2 \frac{\partial^2}{\partial z^2} + N^2 \Delta_h\right) w - \Gamma f^2 \frac{\partial^2 w}{\partial x \, \partial z} = 0, \qquad (4.24)$$

where

$$\frac{d}{dt} = \frac{\partial}{\partial t} + U \frac{\partial}{\partial x}.$$

Assume that internal waves are stable and $Ri > \frac{1}{4}$. A common solution (4.24) may be found in the form

$$w = W \, \exp\{i\,(k_1 x + k_2 y - k_1 \Gamma z t)\}, \qquad (4.25)$$

where W is independent of z. Eq. (4.25) describes wave motion where the wave vector $\mathbf{k} = \{k_1, k_2, 0\}$ is horizontal at an initial time $t = 0$. The vertical wave vector component $m = -k_1\Gamma t$ appears with time and an internal wave has a three-dimensional wave vector $\ae = \{k_1, k_2, -k_1\Gamma t\}$, i.e., turns gradually from horizontal to vertical direction. Substituting (4.25) to (4.24) yields:

$$\frac{\partial^3}{\partial t^3}\left[(k_1^2 + k_2^2 + k_1^2\Gamma^2 t^2)\,W\right] + \frac{\partial}{\partial t}\left[\{(k_1^2 + k_2^2)\,N^2 + f^2 k_1^2\Gamma^2 t^2\}\,W\right]$$

$$+ f^2 k_1^2\Gamma^2 t W = 0 \qquad (4.26)$$

The following dimensionless variables will be convenient:

$$T = \Gamma t\,\cos\Phi; \quad F = f/N; \quad K^2 = N^2/\Gamma^2\,\cos^2\Phi = \mathrm{Ri}/\cos^2\Phi, \quad (4.27)$$

where $\cos\Phi = k_1^2/(k_1^2 + k_2^2)^{1/2}$ defines an initial horizontal orientation of the wave vector. Then Eq.(4.26) becomes

$$K^{-2}\frac{\partial^3}{\partial T^3}[(1 + T^2)\,W] + \frac{\partial}{\partial T}[(1 + F^2 T^2)\,W] + F^2 T W = 0. \qquad (4.28)$$

Solutions of this equation may be expressed through hypergeometric functions. But for $K^2 \gg 1$ an asymptotic solution of Eq. (4.28) may be derived also. A case with $K^2 \gg 1$ corresponds to either $Ri \gg 1$ (a large dynamic stability) or $\cos\Phi \ll 1$ (the wave vector is almost normal to a mean flow direction). For $K^2 \gg 1$ an asymptotic solution of (4.28) has the form

$$W \sim G\,(1 + T^2)^{-3/4}\exp\{\pm i\psi\,(T)\}; \qquad (4.29)$$

$$\psi\,(T) = K\int^T\left(\frac{1 + F^2 T^2}{1 + T^2}\right)^{1/2}dT, \qquad (4.30)$$

where G is an arbitrary constant. Using Eqs. (4.26)–(4.29) and the solution (4.29), (4.30) we obtain expressions for all of the velocity components and an elevation of a level in an internal wave, which are valid with an accuracy $\mathrm{O}\,(K^{-1})$:

$$w \sim G\,(1 + T^2)^{-3/4}\exp\{i\theta\}\exp\{-i\psi\,(T)\}, \qquad (4.31)$$

$$u \sim G \cos \Phi \left(1 + T^2\right)^{-3/4} T \exp\{i\theta\} \exp\{-i\psi\left(T\right)\}, \qquad (4.32)$$

$$v \sim G \frac{k_2}{k_1} \cos \Phi \left(1 + T^2\right)^{-3/4} T \exp\{i\theta\} \exp\{-i\psi\left(T\right)\}, \qquad (4.33)$$

$$\xi \sim -\frac{iG}{N} \left(1 + T^2\right)^{-1/4} \left(1 + F^2 T^2\right)^{-1/2} \exp\{i\theta\} \exp\{-i\psi\left(T\right)\}, \quad (4.34)$$

where

$$\theta = k_1 x + k_2 y - \frac{k_1 z t}{\cos \Phi}. \qquad (4.35)$$

These solutions (coinciding with solutions of Phillips [299] for the particular case $f \equiv F \equiv 0$) describe internal waves where amplitude, frequency, and propagation direction vary over time owing to a mean velocity gradient action. The wave number module $|\text{æ}| = (k_1^2 + k_2^2 + k_1^2 \Gamma^2 t^2)^{1/2}$

grows in time (the wave length decreases) and the angle $\tilde{\Phi}$, included between the total wave vector and a horizontal, is defined by

$$\cos \tilde{\Phi} = (k_1^2 + k_2^2)^{1/2}(k_1^2 + k_2^2 + k_1^2 \Gamma^2 t^2)^{-1/2} = (1 + \Gamma^2 t^2 \cos^2 \Phi)^{-1/2}$$

and grows with t because $\tilde{\Phi} \to \pi/2$ when $\Gamma t \to \infty$. Thus the mean velocity gradient stretches internal waves' stationary phase lines, making them more horizontal and reducing their interval.

A wave motion frequency ω, which may be defined as

$$\omega = \frac{d\psi}{dt} = N \left(1 + F^2 T^2\right)^{1/2}(1 + T^2)^{-1/2}, \qquad (4.36)$$

also varies persistently over time, and furthermore

$$\frac{\partial \omega}{\partial t} = -k_1 \Gamma c_g, \qquad (4.37)$$

where

$$c_g = \frac{\partial \omega}{\partial m} = -\frac{N}{k_1} \cos \Phi \left(F^2 - 1\right) T \left(1 + T^2\right)^{3/2}(1 + F^2 T^2)^{-1/2}$$

is the vertical component of the group velocity.

In the absence of velocity shear the dispersion relation has the form

$$(N^2 - \omega^2)(\omega^2 - f^2)^{-1} = m^2 (k_1^2 + k_2^2)^{-1}.$$

And $\textbf{æ} = \textbf{k}$, $\omega = N$, $\partial\omega/\partial t = 0$ when the vector $\textbf{æ}$ is horizontal. When $\textbf{æ}$ is vertical, $\partial\omega/\partial t = 0$ and $\omega \to f$. Since the wave vector turns to the vertical ($\tilde{\Phi} \to \pi/2$ for $\Gamma t \to \infty$), velocity shear redistributes energy from higher frequencies to lower frequencies. This follows directly from Eq. (4.36) for $\Gamma t \to \infty$. The mean kinetic energy density of wave motion

$$E = \frac{1}{2}\rho_0 (\bar{u}^2 + \bar{v}^2 + \bar{w}^2) = \frac{1}{4}\rho_0 G^2 (1 + T^2)^{-1/2}$$

decreases and the motion dampens when $\Gamma t \cos \Phi \to \infty$. Such energy loss in the absence of dissipation and nonlinearity must be restored by an increase in the mean motion energy. But this increase is not taken into consideration, because we assume $\Gamma = \text{const}$.

In the previous section internal waves of constant frequency ω propagating upwards and downwards were considered. The wave energy absorption in the critical level was similar to the wave energy transfer to the mean flow. In this case the critical level was defined in the non-rotating fluid as the level at which the wave phase velocity is equal to the flow velocity, and in the presence of rotation as the level at which the Doppler velocity is equal to the Coriolis parameter f. The present model also considers the wave–flow energy transfer, but the nature of the transfer is different. So far as the frequency and the phase velocity are now functions of time, a continuous 'spectrum' of critical levels appears (contrary to a single critical level for $\omega = \text{const}$). Thus the wave–flow energy transfer occurs totally inside the fluid.

If there is a stationary source of internal waves in a flow, then the evolution of waves' frequencies $\omega(t)$ and square amplitudes $|\xi(t)|^2$ leads to a formation of the stationary spectrum of vertical displacements of fluid particles $S_\xi(\omega)$. In fact, in the interval $[\omega, \omega + d\omega]$ $\overline{d\xi^2} = S_\xi(z, \omega)\,d\omega$, where $S_\xi(z, \omega)$ is the frequency spectrum of ξ at a fixed level z. Furthermore, we have

$$d\omega = \frac{\partial\omega}{\partial t}\,dt\,, \quad \overline{d\xi^2} = \overline{\dot{\xi}^2}\,dt\,,$$

$$\overline{\xi^2} \sim \frac{G^2}{2N^2}(1+T^2)^{-1/2}(1+F^2T^2)^{-1}.$$

Combining these relations and using (4.37) we derive

$$S_\xi(\omega) \sim (N^2-\omega^2)^{-1/2}\,\omega^{-1}. \qquad (4.38)$$

The frequency dependence $S_\xi(\omega)$ in the form (4.38) is observed in the ocean for relatively weak velocity shears [81].

4.   PROPAGATION OF INTERNAL WAVE PACKETS IN SHEAR FLOW

In this section we consider—following Bretherton [37]— the propagation of internal wave packets in a fluid with a slowly varying $N(z)$ and mean velocity components $\mathbf{U} = \{U(z), V(z), 0\}$ so that the Richardson number

$$Ri = \frac{N^2}{\left[\left(\frac{dU}{dz}\right)^2 + \left(\frac{dV}{dz}\right)^2\right]} \gg 1$$

is large because the vertical derivatives of U and V are small. In this sense the theory presented below differs from the theory of Section 4.2 where $Ri > \frac{1}{4}$, but it was not required that $Ri \gg 1$. Consider Eq. (2.57) to be the basic equation. Introduce a small parameter $\varepsilon \ll 1$ characterizing the slow variation of $N(z)$, $U(z)$ and $V(z)$. Then we may use the WKB approximation to solve Eq. (2.57). Following [37] we introduce the slow time

$$\tau = \varepsilon t \qquad (4.39)$$

and the solution of (2.57) is then found in the form

$$w = \mathrm{Re}\left\{(W + \varepsilon W_1 + \ldots)\, e^{i\varepsilon^{-1}\theta}\right\}, \qquad (4.40)$$

where $W(x, y, z, \tau)$ is the amplitude function of a sinusoidal wave packet with the local phase $\varepsilon^{-1}\theta(x, y, z, \tau)$. The terms $\varepsilon W_1 + \ldots$ are corrections to W for small, but finite ε. W is independent of ε, but depends on a time scale of the order ε^{-1}. Define in a usual way (see Section 3.4) the local frequency and the wave vector $\mathbf{\ae} = \{k, l, m\}$:

$$\omega = -\varepsilon^{-1}\frac{\partial\theta}{\partial t} = -\frac{\partial\theta}{\partial\tau} \qquad (4.41)$$

$$k = \varepsilon^{-1} \frac{\partial \theta}{\partial x} \; ; \quad l = \varepsilon^{-1} \frac{\partial \theta}{\partial y} \; ; \quad m = \varepsilon^{-1} \frac{\partial \theta}{\partial z} \; . \qquad (4.42)$$

Thus for small ε the wave numbers are large and the Bretherton theory in the form (4.39)–(4.42) describes a short wave asymptotics of Eq. (2.57). Before substituting (4.40) to (2.57) we define a velocity scale as

$$U = \varepsilon U^* , \quad V = \varepsilon V^* . \qquad (4.43)$$

So formally

$$Ri = N^2 \left[\left(\frac{dU}{dz} \right)^2 + \left(\frac{dV}{dz} \right)^2 \right]^{-1} = \frac{N^2}{\varepsilon^2} \left[\left(\frac{dU^*}{dz} \right)^2 + \left(\frac{dV^*}{dz} \right)^2 \right]^{-1} \to \infty$$

$$\text{for} \quad \varepsilon \to 0 .$$

Naturally, the theory is valid for large but finite values of $Ri \gg 1$. Substituting (4.39), (4.40), (4.43) in Eq. (2.57) and equating coefficients of the same powers of ε for the terms in the order of ε^{-2} yields

$$QW e^{i\varepsilon^{-1}\theta} = 0 , \qquad (4.44)$$

where

$$Q = [(\theta_\tau + U^*\theta_x + V^*\theta_y)^2 (\theta_x^2 + \theta_y^2 + \theta_z^2) - N^2(\theta_x^2 + \theta_y^2)] . \qquad (4.45)$$

Here a subscript means derivation with respect to an according variable. From Eq. (4.44) it follows that $Q = 0$, whence using (4.41) and (4.42) yields

$$\omega = kU + lV \pm N \left\{ \frac{K^2 + l^2}{k^2 + l^2 + m^2} \right\}^{1/2} . \qquad (4.46)$$

Eq. (4.46) looks like a dispersion relation for internal waves in a medium with $N = \text{const}$ moving at constant velocity $\mathbf{U} = \{U, V, 0\}$. But in the present case k, l, m, ω are functions of time and coordinates. We may rewrite Eq. (4.45) in the symbolic form

$$\omega = \Omega(k, l, m, z) \qquad (4.47)$$

(because U, V, N depend on z), and then define the group velocity $\mathbf{c_g}$ with the relation

$$\mathbf{c_g} = \{u_g, v_g, w_g\} = \left[\frac{\partial\Omega}{\partial k}, \frac{\partial\Omega}{\partial l}, \frac{\partial\Omega}{\partial m}\right]. \qquad (4.48)$$

Introduce a convective derivative moving at the local group velocity

$$\frac{D_g}{Dt} = \frac{\partial}{\partial t} + \mathbf{c_g}\nabla.$$

Then (4.41), (4.42), (4.47), (4.48) yield

$$k_t = \varepsilon^{-1}\theta_{xt} = -\omega_x = -\frac{\partial\Omega}{\partial k}k_x - \frac{\partial\Omega}{\partial l}l_x - \frac{\partial\Omega}{\partial m}m_x$$

$$= -u_g k_x - v_g l_x - w_g m_x = -u_g \theta_{xx} - v_g \theta_{xy} - w_g \theta_{xz}$$

$$= -u_g k_x - v_g k_y - w_g k_z. \qquad (4.49)$$

Then it follows that

$$\frac{D_g k}{Dt} = 0. \qquad (4.50)$$

By analogy with the previous derivation we can prove that

$$\frac{D_g \omega}{Dt} = 0; \quad \frac{D_g l}{Dt} = 0; \quad \frac{D_g m}{Dt} = -\frac{\partial\Omega}{\partial z}. \qquad (4.51)$$

Eqs. (4.50), (4.51) show that although the frequency and wave numbers are functions of time at a fixed point, they are constant in the coordinates moving at the group velocity. This is a common and purely kinematic result which is independent of any specific dispersion relation and is defined only by relations like (4.41), (4.42). The dispersion relation (4.46) yields

$$u_g = U \pm Nkm^2 q^{-1}, \quad v_g = V \pm Nlm^2 q^{-1},$$

$$w_g = \pm N\left(k^2 + l^2\right)mq^{-1}, \qquad (4.52)$$

where $q = (k^2 + l^2)^{1/2}(k^2 + l^2 + m^2)^{3/2}$. From (4.51) it follows that the total group velocity is the sum of the local flow velocity and the group velocity of a wave packet relative to a medium at rest.

To estimate variations of the wave packet amplitude we equate the terms of order ε^{-1} in Eq. (2.57) when substituting there Eqs. (4.39), (4.40), (4.43). The result is a complicated relation which is reduced, using Eqs. (4.44), (4.45), (4.41), (4.42), to

$$\frac{\partial Q}{\partial(-m)} W_t + \frac{\partial Q}{\partial k} W_x + \frac{\partial \Omega}{\partial z} W_y + \frac{\partial \Omega}{\partial z} W_z + PW = 0, \qquad (4.53)$$

where Q is defined with Eq. (4.44) and P has the form

$$P = -\left\{ \frac{d}{dt}\left(\omega' æ^2\right) + \omega' \frac{d}{dt} æ^2 + \omega' \nabla æ + N^2 \nabla_h æ \right\}, \qquad (4.54)$$

where

$$æ = |æ|, \qquad \frac{d}{dt} = \frac{\partial}{\partial t} + U \frac{\partial}{\partial x} + V \frac{\partial}{\partial y}$$

and $\omega' = \omega - kU - lV$ is the Doppler frequency.

Taking into account that $\Omega\left(\omega, k, l, m\right) = 0$ and

$$u_g = \frac{\partial \Omega}{\partial k} = -\frac{\partial \Omega}{\partial k} \Big/ \frac{\partial \Omega}{\partial \omega},$$

$$v_g = \frac{\partial \Omega}{\partial l} = -\frac{\partial \Omega}{\partial l} \Big/ \frac{\partial \Omega}{\partial \omega},$$

$$w_g = \frac{\partial \Omega}{\partial m} = -\frac{\partial \Omega}{\partial m} \Big/ \frac{\partial \Omega}{\partial \omega},$$

(4.53) yields

$$\frac{1}{W} \frac{D_g W}{Dt} = P \left(\frac{\partial \Omega}{\partial \omega}\right)^{-1}. \qquad (4.55)$$

Expressing the complex function W through the amplitude and the phase, i.e., $W = |W| e^{i\alpha}$, from (4.55) we have

$$\frac{D_g \alpha}{Dt} = 0; \qquad \frac{D_g \mathcal{F}}{Dt} + \left(u_{gx} + v_{gy} + w_{gz}\right) \mathcal{F} = 0, \qquad (4.56)$$

where

$$\mathcal{F} = |W|^2 \frac{\left(k^2 + l^2 + m^2\right)}{\omega' \left(k^2 + l^2\right)}. \qquad (4.57)$$

From the first Eq. (4.56) it follows that the phase $\varepsilon^{-1}\theta$ does not change while the wave packet propagates along the beam. Using Eq. (4.40) for the other wave components u, v, p, ρ, ξ we obtain as a first approximation a relation between W and amplitude functions of mentioned components $W_u, W_v, W_p, W_\rho, W_\xi$:

$$W_u = -\frac{km}{k^2 + l^2} W ; \quad W_v = -\frac{lm}{k^2 + l^2} W ; \quad W_p = -\frac{\rho_0 m^2 \omega'}{k^2 + l^2} W ;$$

$$W_s = i \frac{\rho_0 N^2}{g\omega'} W ; \quad W_\xi = i \frac{W}{\omega'} . \tag{4.58}$$

Then the expression for the period and wavelength average of the total internal waves' energy is

$$\overline{E} = \overline{\left(\frac{1}{2} \mathbf{u}^2 + N^2 \xi^2\right)} = |W|^2 \frac{(k^2 + l^2 + m^2)}{(k^2 + l^2)} .$$

Thence $\mathcal{F} = \overline{E}/\omega'$ and the second Eq. (4.56) may be rewritten in the divergence form:

$$\frac{\partial \overline{E}/\omega'}{\partial t} + \nabla \left(\mathbf{c}_g \frac{\overline{E}}{\omega'}\right) = 0 . \tag{4.59}$$

Eq. (4.59) states the fundamental fact that the shear flow does not retain the wave energy, but its relative value $\mathcal{F} = \overline{E}/\omega'$, frequently called the wave action. From Eq. (4.59) follows that the value of $\int_G \overline{E}/\omega' dG$ is constant in any containment volume moving at the local group velocity. Thus for small amplitude internal waves propagating in a moving medium, the wave action, i.e., the ratio of the wave energy to the Doppler frequency, will be an adiabatic invariant. Such a form of the conservation law is common for waves in a moving medium; it is defined for waves of different nature (e.g., surface, sound, etc., [38]).

Eq. (4.59) may be presented in another form

$$\frac{\partial \overline{E}}{\partial t} + \nabla (\mathbf{c}_g \overline{E}) = \frac{\overline{E}}{\omega'} \frac{D_g \omega'}{Dt} . \tag{4.60}$$

The l.h.s. of this equation relates to the wave energy conservation law in the absence of mean flow. So the r.h.s. defines the work of the

Reynolds stresses (or the radiation stresses as it is called in the wave approximation). Using (4.46), (4.52), (4.58), (4.59) yields

$$\rho_0 \left\{ \frac{dU}{dz}\, \overline{uw} + \frac{dV}{dz}\, \overline{vw} \right\} = \left[k\, \frac{dU}{dz} + l\, \frac{dV}{dz} \right] w_g\, \frac{\overline{E}}{\omega'} = -\frac{\overline{E}}{\omega'}\, \frac{D_g \omega'}{Dt}, \quad (4.61)$$

and the radiation stresses are equal to $\{\rho_0 \overline{uw}, \rho_0 \overline{vw}, 0\} = \{k\mathcal{F}w_g, l\mathcal{F}w_g, 0\}$. Eq. (4.60) follows from the conservation equation (2.54) for the wave energy in the mean flow in the WKB approximation. Thus Eq. (2.54) under the condition of slow variation of the medium characteristics (the WKB approximation) is reduced to the divergent form (4.59).

Eq. (4.59) has a singularity in the vicinity of z_c where $\omega' = \omega - kU - lV = 0$. Expression (4.52) for the vertical group velocity component w_g may be written in the form

$$w_g = \pm \left[1 - N^2 (\omega')^2 \right]^{1/2} N^{-1} (\omega')^2 (k^2 + l^2)^{1/2}.$$

In the vicinity of $z \approx z_c$ where $\omega' \approx 0$

$$w_g \sim \pm \left[k\, \frac{dU}{dz} + l\, \frac{dV}{dz} \right] N^{-1}\, \frac{(z - z_c)^2}{(k^2 + l^2)^{1/2}}. \quad (4.62)$$

From Eq. (4.62) follows an estimation of the time interval required for a wave group passage from the level z_1 to the level z_2 (where $z_1 > z_c > z_2$),

$$t_2 - t_1 \sim \frac{(k^2 + l^2)^{1/2}\, N}{\left(k\, \frac{dU}{dz} + l\, \frac{dV}{dz} \right)^2} \left[\frac{1}{(z_2 - z_c)} - \frac{1}{(z_1 - z_c)} \right]. \quad (4.63)$$

Thus for $z_2 \to z_c$ the passage time is growing infinitely. So the vertically moving wave group will never reach the critical level z_c. While drawing near z_c, the vertical wave number tends to zero [see (4.62)] and the horizontal component of the group velocity, normal to the wave front, looks like

$$ku_g + lv_g \sim kU(z_c) + lV(z_c) + \mathcal{O}\left[(z - z_c)^2 \right]$$

Both become independent of z for $z \sim z_c$. Such a situation corresponds to waves becoming 'trapped' in the vicinity of the critical level[4].

For $\omega' = \pm N$, w_g also goes to zero. In the vicinity of $z = z_m$, where $\omega' = \pm N$, we have the following estimate for the vertical wave number

$$m^2 \sim \pm (k^2 + l^2) \left[k \frac{dU}{dz} + l \frac{dV}{dz} \right] N^{-1} (z - z_m) \qquad (4.64)$$

For $z \sim z_m$ we have $m \sim 0$, i.e., the reflection of waves from the z_c level.

Bretherton [37] shows that the wave energy density grows while approaching z_c and the linear theory becomes invalid.

Notes

1 Comprehensive reviews of gravity waves and current interaction are given by Peregrine [295] and Jonsson [147].

2 According to the Frobenius method a solution of Eq. (4.9) is found in the form

$$\widehat{W} = A\tilde{z}^a \left[1 + a_1\tilde{z}^{-1} + a_2\tilde{z}^{-2} + \ldots\right] + B\tilde{z}^a \left[1 + b_1\tilde{z}^{-1} + b_2\tilde{z}^{-2} + \ldots\right],$$

where

$$\tilde{z} = (z - z_c) - ic_i \, / \, \frac{dU}{dz}$$

and constants a, b, a_1, $\ldots$, a_n; b_1, $\ldots$, b_n are obtained when equating coefficients of the same powers of $\tilde{z}^{-n}$.

3 Sutherland studied the influence of nonlinear effects on this problem [343]. The main result is that, at large amplitude, the wave packet may transmit energy across the level z_c through a nonlinear mechanism involving the generation of a mean flow interacting with the wave. The energy transmission is manifested as the generation of an internal wave packet of lower frequency. This means that, for instance, internal waves generated near the surface of equatorial oceans may propagate to a greater depth and constitute a significant momentum source for the deep equatorial countercurrent.

4 The trapping and reflection of gravity waves in the Gulf Stream and the Circumpolar Current were analysed theoretically by Kenyon [161] using the geometrical optics approximation. Kunze [188] used a ray tracing approach to show the behaviour of near inertial waves in a geostrophic jet. Trapped internal gravity waves in a geostrophic boundary current were studied by Ma [221].

Chapter 5

PROPAGATION OF INTERNAL WAVES IN HORIZONTALLY INHOMOGENEOUS OCEAN

The real ocean has, as a rule, different horizontal inhomogeneities which sometimes considerably affect internal waves' propagation. Among the most characteristic horizontal inhomogeneities are: bottom unevenness; horizontal inhomogeneities of the density field; and horizontal variations of the mean flows. The influence of these phenomena on internal waves' propagation will be studied in this chapter.

1. INTERNAL WAVES OVER AN UNEVEN BOTTOM

Consider propagation of small amplitude internal waves in the undisturbed ocean with a given bottom relief where the undisturbed density depends on the vertical coordinate z only. In this case we may use Eq. (2.42) for the vertical velocity of fluid particles in a wave with boundary conditions (2.39) and (2.43) for $p_a \equiv 0$. Consider harmonic waves

$$\{w, u, v\} = \{W, U, V\} \exp\{-i\omega t\}. \tag{5.1}$$

Substitution of (5.1) in (2.42), (2.43) and (2.39) yields

$$W'' - \frac{1}{\alpha^2}\Delta_h W = 0, \quad \alpha^2 = \frac{\omega^2 - f^2}{N^2 - \omega^2}, \tag{5.2}$$

$$(f^2 - \omega^2) W' + g\Delta_h W = 0 \quad \text{at} \quad z = 0, \tag{5.3}$$

119

$$W = U \frac{\partial H}{\partial x} + V \frac{\partial H}{\partial y} \quad \text{at} \quad z = -H(x, y). \tag{5.4}$$

While deriving Eq. (5.2) we use the Boussinesq approximation, i.e., small terms proportional to N^2/g are omitted (a prime means derivation with respect to z). In the general case the separation of variables in Eq. (5.2) is impossible under condition (5.4). The problem (5.2), (5.4) with the surface boundary conditions (5.3) (or with the simplified condition $W = 0$ at $z = 0$) can be solved only approximately. But there are particular cases in which an exact analytical solution can be obtained.

Following Wunsch [395] we consider plane internal waves ($U = \partial/\partial x \equiv \equiv 0$) under a plane slope $H(y) = \gamma y$ (where γ is an inclination) for constant Brunt–Väisälä frequency $N = $ const and under the 'rigid lid' condition at the free surface. Introduce the stream function ψ ($V = -\partial\psi/\partial z$, $W = \partial\psi/\partial y$). Then Eqs. (5.2)–(5.4) yield

$$\psi'' = \frac{1}{\alpha^2} \frac{\partial^2\psi}{\partial y^2}; \quad \left.\frac{\partial\psi}{\partial y}\right|_{z=0} = 0; \quad \left.\frac{\partial\psi}{\partial y}\right|_{z=\gamma y} = \gamma\psi', \tag{5.5}$$

where $\alpha^2 = $ const. The solution of problem (5.5) for $\gamma < \alpha$ looks like (for details see [395])

$$\psi = A\left\{\exp\left[iq\ln(\alpha y - z)\right] - \exp\left[iq\ln(\alpha y + z)\right]\right\}, \tag{5.6}$$

where

$$q = 2\pi n \left[\ln\left(\frac{\alpha - \gamma}{\alpha + \gamma}\right)\right]^{-1},$$

A is a constant, and n is the mode number. The velocity components in terms of derivatives of function (5.6) looks like

$$\left.\begin{aligned} V &= Aiq \left\{\frac{\exp\left[iq\ln(\alpha\gamma - z)\right]}{\alpha y - z} + \frac{\exp\left[iq\ln(\alpha\gamma + z)\right]}{\alpha y + z}\right\}; \\ W &= Aiq\,\alpha \left\{\frac{\exp\left[iq\ln(\alpha\gamma - z)\right]}{\alpha y - z} - \frac{\exp\left[iq\ln(\alpha\gamma + z)\right]}{\alpha y + z}\right\}. \end{aligned}\right\} \tag{5.7}$$

From Eqs. (5.6), (5.7) it is obvious that these solutions describe standing waves appearing while the incident waves reflect from the shore. Both solutions (5.6) and (5.7) have singularities at $\alpha = \gamma$. This equality determines the critical values of the internal waves propagation angle and the waves frequency $\alpha^2 = (\omega^2 - f^2)(N^2 - f^2)^{-1}$), and corresponds to an inclination equality of the bottom and a characteristics of Eq. (5.5). When $\alpha > 0$ Eq. (5.5) is hyperbolic. From solutions (5.6) and (5.7) it follows that decreasing the depth also decreases the internal wave length λ as $\lambda \sim \gamma |y|$. For $\gamma < \alpha$ the wave energy is transferred to the singular point (the energy density increases as $y \to -0$). We can show that for $\alpha > \gamma$ the energy reflects from the coastal slope (the energy density decreases towards the coast). For $\alpha = \gamma$ the stated linear theory yields an infinite velocity at the bottom. As mentioned in [395], to avoid a singularity one should take into account the bottom boundary layer ([396], [399]).

For internal waves study of more complicated functions $H(x, y)$ the method of characteristics may be used. For $\alpha > 0$ Eq. (5.2) is hyperbolic and therefore has real characteristics. E.g., for $H = H(y)$ and $N^2 = $ const ($\alpha = $ const) the common solution of Eq. (5.2) is

$$\mathbf{W} = \varphi(\xi) + g(\eta), \qquad (5.8)$$

where φ and g are arbitrary functions of the characteristic variables

$$\xi = z + \alpha y, \quad \eta = z - \alpha y. \qquad (5.9)$$

Functions φ and g can be obtained from the boundary conditions. Such a theory for different $H(y)$ with a thorough analysis of diffracting and reflected wave field was developed by Baines [12] and Sandström [319]. For $N^2 = N^2(z)$ the characteristics are presented along the lines

$$\xi = \int \sqrt{N^2(z) - \omega^2} \, dz + \sqrt{\omega^2 - f^2} \, y,$$

$$\eta = \int \sqrt{N^2(z) - \omega^2} \, dz - \sqrt{\omega^2 - f^2} \, y. \qquad (5.10)$$

Internal waves for some particular distributions $N^2(z)$ and $H(y)$ were calculated by the method of characteristics by Magaard [225] (and are presented by Krauss [179]).

Approximate methods should be used to calculate internal waves for more complicated functions $H(x, y)$. For small unevenness of the bottom, i.e., $H(x, y) = H_0 + \varepsilon H_1(x, y)$, where $\varepsilon \ll 1$, the perturbation method may be used while expanding an unknown function in the series $W = \sum_{n=0}^{\infty} \varepsilon^n W_n$. Similar calculations are presented in [179] where an effect of small corrections owed to the small bottom unevenness on a basic solution for $H = \text{const}$ is estimated.

If the ocean depth varies slowly compared to the characteristic internal waves' length (which is satisfied in the real ocean), then the ray approximation may be used for the problem's solution. Such a theory was worked out by Keller and Van Mow [160]. As the modification of the ray method for multi-mode systems, suggested in [160], will be used later on, we will discuss some details of this work.

Consider harmonic waves

$$\{u, v, w, p, \rho, \xi\} = \{U, V, W, P, \hat{\rho}, \hat{\xi}\} \exp\{-i\omega t\}, \qquad (5.11)$$

where ξ means a displacement of the fluid level in an internal wave. We examine internal waves propagation under an uneven bottom, omitting the Coriolis force, under the condition of isopycnic motion, for $\rho_0 = \rho_0(z)$ and without the Boussinesq approximation. Simultaneous equations (2.34)–(2.39), taking into account abovementioned assumptions, may be reduced to the single pressure equation

$$\omega^2 \rho_0 \left[\rho_0^{-1} P' \left(\omega^2 - N^2 \right)^{-1} \right]' + \Delta P = 0 ; \qquad (5.12)$$

$$P' = \frac{1}{g} \left(\omega^2 - N^2 \right) P \quad \text{at} \quad z = 0 ; \qquad (5.13)$$

$$\rho_0 P' + \rho_0 \left[1 - \frac{N^2}{\omega^2} \right] \nabla H \nabla P = 0 \quad \text{at} \quad z = -H(x, y), \qquad (5.14)$$

where

$$\nabla = \left\{ \frac{\partial}{\partial x}, \frac{\partial}{\partial y} \right\}, \quad \Delta = \nabla \nabla,$$

a prime means the derivative with respect to z. All the remaining wave parameters in terms of P yield:

$$U = -\frac{i}{\omega \rho_0} \frac{\partial P}{\partial x} ; \quad V = -\frac{i\omega}{\rho_0 (\omega^2 - N^2)} \frac{\partial P}{\partial y} ; \quad W = -\frac{i}{\rho_0 \omega} P' ;$$

$$\hat{\rho} = -\frac{N^2}{g\,(\omega^2 - N^2)}\,P'\,; \quad \hat{\xi} = \frac{P}{g\rho_0}\,. \tag{5.15}$$

Introduce the following dimensionless variables

$$\tilde{x} = xL^{-1}, \quad \tilde{y} = yL^{-1}, \quad \tilde{z} = \omega^2 z g^{-1}, \quad h = \omega^2 H g^{-1},$$

$$\alpha = 1 - \frac{N^2}{\omega^2}, \quad \beta = \omega^2 L g^{-1}, \tag{5.16}$$

where L is a characteristic horizontal scale of $H(x, y)$ variations. Eqs. (5.12)–(5.14), with (5.16), take on the form

$$\beta^2\left[P'' - \frac{(\alpha\rho_0)'}{\alpha\rho_0}\,P'\right] + \alpha\Delta P = 0\,; \tag{5.17}$$

$$P' - \alpha P = 0 \quad \text{at } z = 0\,; \tag{5.18}$$

$$\beta^2 P' + \alpha \nabla h \nabla P = 0 \quad \text{at } z = -h\,(\tilde{x},\,\tilde{y}) \tag{5.19}$$

Consider a case of large values $\beta \gg 1$ (for the typical frequency values of short period internal waves $\omega \sim 5 \times 10^{-3}$ s^{-1} and $g = 10^3$ cm $\times$ s^{-2} it corresponds to $L \gg 4 \times 10^7 = 400$ km). A solution of Eqs. (5.17)–(5.19) for greater β may be tried in the form typical for the ray method:

$$P\,(\mathbf{x},\,\beta) = \left[A_0\,(\mathbf{x}) + \sum_{m=1}^{\infty} (i\beta)^{-m}\,A_m\,(\mathbf{x},\,z)\right]\Phi\,(\mathbf{x},\,z)\,e^{i\beta S\,(\mathbf{x})}, \tag{5.20}$$

here $\mathbf{x} = \{x,\,y\}$ and the symbol $\sim$ will be omitted while considering all variables to be dimensionless. The amplitudes A_0, A_m and the phase $S\,(\mathbf{x})$ are obtained from Eqs. (5.17)–(5.19) and $\Phi\,(x, z)$ from the solution of the boundary value problem

$$\Phi'' - \frac{(\alpha\rho_0)'}{\alpha\rho_0}\,\Phi' - k^2\,(\mathbf{x})\,\alpha\,\Phi = 0\,; \quad \{\Phi' - \alpha\,\Phi\}|_{z=0} = 0\,; \quad \Phi'|_{z=-h\,(\mathbf{x})} = 0\,. \tag{5.21}$$

The boundary value problem (5.21) corresponds to internal waves propagating in the ocean with $h = $ const. For the given distribution $N^2\,(z)$ the solutions of (5.21) yield the eigenvalues $k^n\,(\mathbf{x})$ (k^n is the dimensionless wave number) and the eigenfunctions $\Phi^n(z, \mathbf{x})$ corresponding to normal vertical modes, and besides k^n and Φ^n depend on $\mathbf{x}$ parametrically. Thus the general solution may be written in the form $P = \sum_n \sum_m A_m^n \Phi^n e^{i\beta S^n}$. Assuming that $k^n(\mathbf{x})$ and $\Phi^n(z, \mathbf{x})$ are known for each of the modes we proceed to derive A_n and S^n for each of the modes (omitting the superscript and relating further equations to a certain mode).

Substituting Eq. (5.20) in Eqs. (5.17)–(5.19), using the boundary value problem (5.21), and equating the terms of the same powers of β, we have

$$[\nabla S]^2 = k^2(\mathbf{x}) \, ; \qquad (5.22)$$

$$\Phi A_m'' - \left[\Phi \frac{(\alpha \rho_0)'}{\alpha \rho_0} - 2\Phi'\right] A_m'$$

$$= \alpha \left[2 \nabla S \nabla (\Phi A_{m-1}) + \Phi A_{m-1} \Delta S\right] + \alpha \Delta (\Phi A_{m-2}) \, ; \quad (5.23)$$

$$A_m' = 0 \quad \text{at} \quad z = 0 \, ; \qquad (5.24)$$

$$\Phi A_m' = \alpha \Phi A_{m-1} \nabla S \nabla h + \alpha \nabla h \nabla (\Phi A_{m-2}) \quad \text{at} \quad z = -h(\mathbf{x}) \, . \quad (5.25)$$

Eq. (5.22) is well known as the geometric optics eikonal, where the modulus of the wave vector $k(\mathbf{x})$ derived from the boundary value problem (5.21) serves as a 'refraction coefficient'. Eqs. (5.23) allow us to obtain recursively the amplitude functions A_m starting with $m = 1$. For $m = 0$, Eq. (5.23) is satisfied trivially because $A_0' = A_0'' = 0$ (A_0 is independent of z) and $A_{-1} \equiv A_{-2} \equiv 0$. To obtain $A_0(\mathbf{x})$, which is the basic term of expansion (5.20), we multiply Eq. (5.23) for $m = 1$ by Φ and integrate over z from $-h$ to 0. Further, using Eqs. (5.21), (5.24), and (5.25) yields

$$2\psi \nabla S \nabla A_0 + A_0 \nabla S \nabla \psi + \psi A_0 \Delta S = 0 \, , \qquad (5.26)$$

where

$$\psi(\mathbf{x}) = \int_{-h}^{0} \frac{1}{\rho_0} \Phi^2(z, \mathbf{x}) \, dz \, . \qquad (5.27)$$

By analogy with the method mentioned, equations for A_m are derived from (5.23)–(5.25) (for details see [160]). Eqs. (5.22), (5.26) together with (5.21), (5.27) form a closed system for Φ, A_0 and S.

The eikonal equation (5.22) may be solved by ray theory methods, which present the characteristics of the partial first-order equation (5.22). Ray equations have the form (see, e.g., [31])

$$\frac{d(k\mathbf{J})}{d\sigma} = \nabla k ; \quad \frac{d\mathbf{x}}{d\sigma} = \mathbf{J} , \qquad (5.28)$$

where $\mathbf{x}(\sigma)$ is a ray, i.e., a vector normal to the wave front or to the stationary phase surfaces $\beta S = \text{const}$, $\mathbf{J}$ is a unit vector tangential to the curve $\mathbf{x}(\sigma)$. σ is a parameter determining the length of the curve $\mathbf{x}(\sigma)$, i.e., $d\sigma^2 = dx^2 + dy^2$. Eqs. (5.27) together with the initial conditions determine the curve $\mathbf{x} = \mathbf{x}(\sigma)$. After that, S may be tried in the form of the curvilinear integral

$$S = \int k\,[\,\mathbf{x}(\sigma')\,]\,d\sigma' = S\,[\,\mathbf{x}(\sigma_0)\,] + \int_{\sigma_0}^{\sigma} k\,[\,\mathbf{x}(\sigma')\,]\,d\sigma' , \qquad (5.29)$$

where σ_0 is the value of σ corresponding to the initial condition $S = S_0$ for $\mathbf{x} = \mathbf{x}_0$.

Eq. (5.26) may be written as a trivial differential equation along the ray because the derivative in the line of $\nabla S \nabla$ equals to $k\,(d/d\sigma)$, since the value ∇S is k, and the ray is directed towards a normal to the curve $S = \text{const}$ and is parallel to ∇S. Taking into consideration all the above, Eq. (5.26) yields

$$2k\,\frac{d}{d\sigma}\,[\,\psi^{1/2}A_0\,] + [\,\psi^{1/2}A_0\,]\,\Delta S = 0 . \qquad (5.30)$$

Solution (5.30) has the form

$$A_0(\sigma) = A_0(\sigma_0)\frac{\psi^{1/2}(\sigma_0)}{\psi^{1/2}(\sigma)}\,\exp\left\{ -\frac{1}{2}\int_{\sigma_0}^{\sigma} \frac{\Delta S}{k}\,d\sigma' \right\} . \qquad (5.31)$$

The exponent in (5.31) is written in the form (see [333])

$$\exp\left\{ -\frac{1}{2}\int_{\sigma_0}^{\sigma} \frac{\Delta S}{k}\,d\sigma' \right\} = \left[\frac{k(\sigma_0)\,da(\sigma_0)}{k(\sigma)\,da(\sigma)} \right]^{1/2} \qquad (5.32)$$

where $da(\sigma)$ means the breadth of the ray tube in the point $\mathbf{x}(\sigma)$ of the ray, and $da(\sigma_0)$ the breadth of the ray tube of the same ray for $\mathbf{x}(\sigma_0)$. The quotient $da(\sigma)/da(\sigma_0)$ is also the Jacobian of the wave front transformation along the ray between the points $\mathbf{x}(\sigma_0)$ and $\mathbf{x}(\sigma)$. Using (5.32) Eq. (5.31) may be rewritten in the form

$$k(\sigma)\, A_0^2(\sigma)\, \psi(\sigma)\, da(\sigma) = k(\sigma_0)\, A_0^2(\sigma_0)\, \psi(\sigma_0)\, da(\sigma_0). \qquad (5.33)$$

Eq. (5.33) is the conservation law for the wave energy flow proportional to $A^2 \psi k\, da$. The energy flow is constant along the ray tube.

Eqs. (5.20), (5.29), (5.33) for P yield the following expression

$$P\left[\mathbf{x}(\sigma),\, z,\, \beta\right] \sim A_0\left[\mathbf{x}(\sigma_0)\right] \left[\frac{\psi(\sigma_0)\, k(\sigma_0)\, da(\sigma_0)}{\psi(\sigma)\, k(\sigma)\, da(\sigma)}\right]^{1/2} \Phi(\mathbf{x}, z)$$

$$\times \exp\left[i\beta \left\{ S\left[\mathbf{x}(\sigma_0)\right] + \int_{\sigma_0}^{\sigma} k\left[\mathbf{x}(\sigma')\right] d\sigma' \right\}\right]. \qquad (5.34)$$

From Eqs. (5.15) and (5.34) we obtain

$$\{U, W\} = \frac{1}{i\omega\rho_0}\left\{\frac{\partial P}{\partial x}, \frac{\partial P}{\partial y}\right\} \sim \frac{\beta}{\omega\rho_0 L}\, P \nabla S; \quad V = -\frac{i\beta\Phi'}{\omega\rho_0 \alpha L\Phi}\, P;$$

$$\hat{\rho} \sim -\frac{\beta N^2 \Phi'}{g\alpha L\Phi}\, P; \quad \hat{\xi} \sim \frac{1}{g\rho_0}\, P. \qquad (5.35)$$

To illustrate the solution obtained, we consider the simplest example N^2 = const, i.e., $\alpha = 1 - N^2/\omega^2$ = const, following [160]. The solution of the boundary value problem (5.21) for N^2 = const under the 'rigid lid' approximation $\Phi' = 0$ for $z = 0$ (used here for simplification) yields

$$\Phi(z) = \frac{1}{\lambda_1}\exp\left[\lambda_1(z+h)\right] - \frac{1}{\lambda_2}\exp\left[\lambda_2(z+h)\right], \qquad (5.36)$$

where

$$\lambda_{1,2} = -\frac{N^2}{2\omega^2} \pm \frac{in\pi}{h}.$$

The dispersion relation has the form

$$k = \left(\frac{N^2}{\omega^2} - 1\right)^{1/2}\left[\left(\frac{n\pi}{h}\right)^2 + \frac{N^4}{4\omega^4}\right]^{1/2}. \qquad (5.37)$$

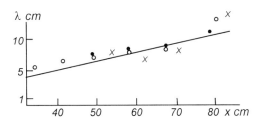

Figure 5.1.   Internal wave length λ vs. the distance x from the shore (a straight line — the theoretical result, experimental points — from [332]).

From Eqs. (5.16), (5.37) for the dimensional wavelength yield

$$\lambda = \frac{2H}{n}\left(\frac{N^2}{\omega^2} - 1\right)^{-1/2}\left[1 + \left(\frac{N^2 H}{2\pi n g}\right)^2\right]^{-1/2}. \qquad (5.38)$$

Keller and Van Mow [160] compared Eq. (5.38) with the experimental evidence by Wunsch [395] who measured internal waves on the plane slope for $N^2 = $ const. In his experiments $H = 0.112\,x$, where x is a distance from the shore, $\left(N^2/\omega^2 - 1\right)^{1/2} = 1.68$. Then for $n = 1$ (the first mode), (5.38) yields

$$\lambda = 0.133x\,[1 + 10^{-8}x^2]^{-1/2} \approx 0.133x \quad (x \ll 10^4). \qquad (5.39)$$

Eq. (5.39) is in line with the experimental data (Fig. 5.1).

Eqs. (5.27), (5.36) yield

$$\psi = \int_{-h}^{0} \frac{\Phi^2}{\rho_0}\,dz = \frac{-2h^3}{(n\pi)^2\rho_0}\exp\left(-\frac{N^2 h}{\omega^2}\right)\left[1 + \left(\frac{N^2 h}{2n\pi\omega^2}\right)^2\right]^{-1}. \qquad (5.40)$$

Then Eq. (5.34) will be transformed to the form

$$P \sim \frac{2\,(-1)^n}{in\pi}\,A_0 h_0\left(\frac{da_0}{da}\right)^{1/2}\cos\left(\frac{n\pi z}{h}\right)$$

$$\times \exp\left\{i\beta\left[S_0 + n\pi\left(\frac{N^2}{\omega^2} - 1\right)^{-1/2}\int_{\sigma_0}^{\sigma}\frac{d\sigma}{h}\right]\right\}. \qquad (5.41)$$

To simplify Eq. (5.41) we assume that $N^2 h/\omega^2 \ll 1$. In the two-dimensional case (i.e., when all of the functions depend only on x and z) $da_0/da = 1$ and for the plane slope we may assume $\sigma = x$ and $h = \gamma x$. With

$$P_0 = \exp\{i\beta S_0\} \, 2(-1)^n \frac{A_0 h_0}{n\pi i},$$

Eq. (5.41) yields

$$P \sim P_0 \cos\left(\frac{n\pi z}{\gamma x}\right) \left(\frac{x}{x_0}\right)^{i\beta \frac{n\pi}{\gamma} \left(\frac{N^2}{\omega^2}-1\right)^{-1/2}}. \tag{5.42}$$

Eqs. (5.41), (5.42) describe the wave moving in x-direction. The complex conjugate function P^* represents the wave moving in the opposite direction. A superposition of these waves may form a standing wave.

Some difficulties in studying internal waves on an uneven bottom appear in the presence of sharp relief deformations because the ray approximation is invalid in this case and singularities appear in the angular points while using the method of characteristics. In the simplest case of a vertical and infinitely thin bottom barrier this problem was analyzed by Larsen [195], who evaluated diffracting and reflected internal wave fields.

2.   INTERNAL WAVES IN THE PRESENCE OF HORIZONTAL INHOMOGENEITIES OF DENSITY FIELD

The cases in which ρ_0 and N depended on z and did not depend on horizontal coordinates, were considered above. This was because the vertical variations of N are more considerable than its horizontal variations. So the weak dependence of N versus **x** was neglected (see Fig. 5.2).

But because of weak damping of ocean waves and their subsequent long distance propagation the dependences of ρ_0 and N on **x** can considerably affect the wave's behaviour. Thus it seems to be interesting to consider the problem of the propagation of internal waves in the ocean with a mean density ρ_0 and N varying over all of the coordinates. It is difficult to solve such a problem in the linear approximation because the

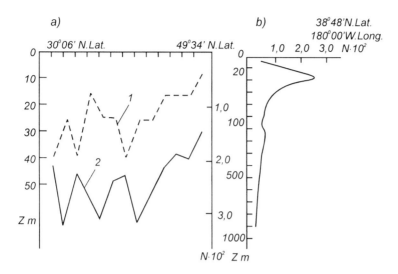

Figure 5.2.   An example of $N(z,x)$ distribution in the ocean: a — variations of h_{max} (1) and N_{max} (2) for a meridional cross-section; b — the characteristic distribution of N for this cross-section.

partial differential equations describing internal waves for $N = N(z, \mathbf{x})$ do not permit separation of variables. But the ray method, discussed in the previous section, may be used for the solution of the problem because the horizontal scale Λ of variations of ρ_0 and N is quite large compared to the characteristic length λ of high frequency internal waves. In the ocean Λ is some hundreds of kilometers and λ is less than ten kilometers (detailed evidence on the horizontal variations of N and its characteristic scales are presented in [259]). The effect of horizontal inhomogeneities of the density field on internal waves propagation in the ray approximation was studied in [250][1].

Consider relatively high frequency internal waves with $\omega \gg f$ in a non-rotating ocean of constant depth with $\rho_0 = \rho_0(z, \mathbf{x})$. Generally speaking, from the averaged hydrodynamics equations it follows that when the undisturbed density is a function of horizontal coordinates the stationary flows $\mathbf{U}(z, \mathbf{x})$ correspond to the stationary density distribution $\rho_0 = \rho_0(z, \mathbf{x})$. But these flows are quite slow and may be neglected in a first approximation, which was used in this section. In general, such a neglect of the stationary flow $\mathbf{U}(z, \mathbf{x})$ causes an absence of corresponding equations of motion for the density field $\rho_0 = \rho_0(z, \mathbf{x})$. Thus it is necessary to consider the oceanic density field $\rho_0 = \rho_0(z, \mathbf{x})$ to be the given background field (known from the evidence) formed by some mass forces and non-adiabatic sources. Such an *a priori* knowledge is rather frequent in oceanographic calculations.

Assume $\rho_0 = \rho_0(z, \mathbf{x})$ in the linearized equations of motion (2.35)–(2.37). We use further the Boussinesq and the isopycnic approximations. Consider harmonic waves

$$\{\mathbf{v}, \xi, \rho, P\} \sim \exp\{-i\omega t\}\{\mathbf{V}, \hat{\xi}, \hat{\rho}, P\}. \tag{5.43}$$

Then Eqs. (2.35)–(2.37) under the mentioned approximations may be reduced to a single equation for the function $\Pi = P/\rho_0$:

$$[\alpha^{-1}\Pi']' - \Delta\Pi = -[\alpha^{-1}\nabla\Pi\nabla\gamma]', \tag{5.44}$$

where Δ is the horizontal Laplacian, a prime means derivation with respect to z (the z-axis is directed downwards from the free surface), α and γ are defined by

$$\alpha(z, \mathbf{x}) = \frac{N^2(z, \mathbf{x})}{\omega^2} - 1; \quad \gamma(z, \mathbf{x}) = \frac{g}{\omega^2}\ln[\rho_0(z, \mathbf{x})]. \tag{5.45}$$

All of the unknown internal waves characteristics may be rewritten in the terms of Π:

$$\left.\begin{array}{l} \mathbf{U} = \{U, V\} = -\dfrac{i}{\omega}\nabla\Pi; \quad W = \dfrac{i}{\omega\alpha}[\Pi' + \nabla\Pi\nabla\gamma]; \\[2ex] \hat{\xi} = -\dfrac{1}{\omega^2\alpha}[\Pi' + \nabla\Pi\nabla\gamma]; \quad \hat{\rho}/\rho_0 = \dfrac{1}{g\alpha}\left[\dfrac{N^2}{\omega^2}\Pi' + \nabla\Pi\nabla\gamma\right]. \end{array}\right\} \tag{5.46}$$

Assume the vertical velocity to be zero at the free surface and the ocean bottom as the boundary conditions for Eq. (5.44). Taking (5.46) into account these conditions look like

$$\Pi' + \nabla\Pi\nabla\gamma = 0 \quad \text{at} \quad z = 0, H, \tag{5.47}$$

where H is the ocean's depth, which may be infinite. Introduce the dimensionless variables

$$\tilde{\mathbf{x}} = \mathbf{x}\Lambda^{-1}, \quad \tilde{z} = zl^{-1}, \tag{5.48}$$

where Λ is the characteristic scale of ρ_0 (or N) horizontal variations ($\Lambda \sim 10^5$ m), l is the characteristic vertical scale of N variations (it may be the pycnocline thickness or the ocean depth, so $l \leq 10^3$ m). Eqs. (5.44) and (5.47) in dimensionless variables[2] are

$$\beta^2 [\alpha^{-1} \Pi']' - \Delta\Pi = -[\alpha^{-1} \nabla\Pi\nabla\ae]' \qquad (5.49)$$

$$\beta^2 \Pi' + \nabla\Pi\nabla\ae = 0 \quad \text{at} \quad z = 0, H , \qquad (5.50)$$

where $\beta = \Lambda l^{-1}$ is the dimensionless parameter, $\beta \gg 1$, $\ae = \gamma l^{-1} = (g/\omega^2 l) \ln \rho_0$. For greater $\beta \gg 1$ the modified ray method, described in the previous section, may be used. We try the asymptotic solution of the problem (5.49), (5.50) in the form

$$\Pi(z, \mathbf{x}, \beta) = \left\{ A_0(\mathbf{x}) + \sum_{m=1}^{\infty} (i\beta)^{-m} A_m(z, \mathbf{x}) \right\} \Phi(z, \mathbf{x}) \exp\{i\beta S(\mathbf{x})\} , \qquad (5.51)$$

where $\Phi(z, \mathbf{x})$ is obtained from the boundary-value problem

$$[\alpha^{-1} \Phi']' + k^2 \Phi = 0, \quad \Phi' = 0 \quad \text{at} \quad z = 0, H . \qquad (5.52)$$

Substituting (5.51) and taking into account (5.52) from (5.49), (5.50), while equating terms at the same β powers, yields

$$[\nabla S]^2 = k^2(\mathbf{x}), \qquad (5.53)$$

$$\begin{aligned}
\Phi[\alpha^{-1} A_m']' &+ 2\alpha^{-1} A_m' \Phi \\
&= [\alpha^{-1} A_{m-1} \Phi \nabla \ae \nabla S]' \\
&\quad -[2\nabla(A_{m-1}\Phi)\nabla S + A_{m-1}\Phi\nabla S] \\
&\quad + [\alpha^{-1} \nabla \ae \nabla(A_{m-2}\Phi)]' - \Delta(A_{m-2}\Phi) ;
\end{aligned} \qquad (5.54)$$

$$[A_m\Phi]' = A_{m-1}\Phi\nabla\ae\nabla S + \nabla\ae\nabla(A_{m-2}\Phi) \quad \text{at} \quad z = 0, H . \quad (5.55)$$

Eqs. (5.54) and (5.55) are valid for $m \geq 1$, and $A_m \equiv 0$ when $m \leq 0$. To obtain $A_0(\mathbf{x})$, which is the main term of the expansion (5.51), we multiply (5.54) for $m = 1$ by Φ and integrate with respect to z between the limits 0 and H. Then we use the identity $\Phi\{\Phi\,[\,\alpha^{-1}A'_m]' + 2\alpha^{-1}A'_1\Phi' = [\alpha^{-1}A'_m\Phi^2]'$ and Eqs. (5.52), (5.55). It yields

$$2L\,\nabla\,S\,\nabla\,A_0 + LA_0\Delta S + A_0\,\nabla S\,[\nabla\,L + \mathbf{I}] = 0\,, \tag{5.56}$$

where

$$L = \int_0^H \Phi^2\,(z,\mathbf{x})\,dz\,; \quad \mathbf{I} = \frac{1}{2}\int_0^H \frac{\nabla\,\text{æ}}{\alpha}\,(\Phi^2)'\,dz\,. \tag{5.57}$$

If Φ, S and A_0 are obtained, then Π is defined by Eq. (5.51) and all the rest values are defined from (5.46):

$$\mathbf{U} = [\,\omega l\,]^{-1}\,[A_0\Phi\,\nabla\,S]\,e^{i\beta S}\,; \quad W = i\,[\,\omega\alpha l\,]^{-1}\,[A_0\Phi']\,e^{i\beta S}\,;$$

$$\hat{\xi} = -[\,\omega^2\alpha l\,]^{-1}\,[A_0\Phi']\,e^{i\beta S}\,; \quad \frac{\hat{\rho}}{\rho_0} = \frac{N^2}{\omega^2}\,[\,gl\alpha\,]^{-1}\,[A_0\Phi']\,e^{i\beta S}\,. \tag{5.58}$$

By analogy with Section 5.1, Eq. (5.56) may be written in the form of a differential equation along the ray $\mathbf{x}\,(\sigma)$, the solution of which yields

$$A_0^2\,(\sigma)\,L\,(\sigma)\,k\,(\sigma)\,da\,(\sigma) = A_0^2\,(\sigma_0)\,L\,(\sigma_0)\,k\,(\sigma_0)\,da\,(\sigma_0)$$

$$\times \exp\left\{-\int_{\sigma_0}^{\sigma} \mathbf{I}\,\nabla\,S\,(Lk)^{-1}\,d\sigma'\right\}. \tag{5.59}$$

From Eq. (5.59) it follows that, unlike in Section 5.1, the wave energy flow is not constant along the wave tube. This is caused by the medium 'instability': the undisturbed density field $\rho_0\,(z,\mathbf{x})$, in general, does not satisfy the hydrodynamics equations in the adiabatic approximation, but is given *a priori*, e.g., by an evidence.

Eqs. (5.52), (5.53),(5.56) are closed combined equations for the derivation of Φ, S and A_0 and may be solved along the rays by analogy with Section 5.1. But a particular case of the direct solution of these equations is more convenient for consideration.

Consider the case where $N^2 = N^2\,(z,x)$. Such a case is interesting for a study of long distance internal waves in the real ocean because longitudinal variations of ρ_0 are greater than latitudinal. If $N^2 = N^2\,(z,x)$ then

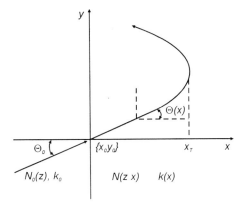

Figure 5.3. Inclined incidence of an internal wave on a layer. The incident wave has the form $\Pi = a_0 \Phi_0 \exp \{i\delta k_0 \left(x \cos \theta_0 + y \sin \theta_0\right)\}$ for $x \leq x_0$, $k = k_0$, $N_z = N_0 \left(z\right)$.

$k = k\left(x\right)$ and the rays' equations (5.28) (rays are the characteristics of the eikonal equation (5.53)) have the form

$$\frac{d}{d\sigma}\left[k J_x\right] = \frac{\partial k}{\partial x}\,; \quad \frac{dx}{d\sigma} = J_x = \cos\theta\left[x\left(\sigma\right)\right], \qquad (5.60)$$

$$\frac{d}{d\sigma}\left[k J_y\right] = 0\,; \quad \frac{dy}{d\sigma} = J_y = \sin\theta\left[x\left(\sigma\right)\right], \qquad (5.61)$$

where J_x and J_y are x and y components of the unit vector $\mathbf{J}$ tangential to the ray $\mathbf{x}\left(\sigma\right)$; θ is an angle included between the x-axis and the wave front's normal. We assume further that the ray leaves the medium with $N^2 = N_0^2\left(z\right)$ at the point $\{x_0; y_0\}$ at an angle θ_0 and has the given amplitude a_0 and wave number k_0. The ray geometry is presented in Fig. 5.3. Integrating Eq. (5.60) along a ray we obtain

$$k\left[x\left(\sigma\right)\right]\sin\theta\left[x\left(\sigma\right)\right] = k_0 \sin\theta_0 = \text{const}\,. \qquad (5.62)$$

Eq. (5.62) represents Snell's law (see, e.g., [31], [33]). Integration of Eq. (5.60) yields the consequence of Snell's law

$$\cos^2\theta = 1 - \left(k_0^2/k^2\right)\sin^2\theta_0\,.$$

Using (5.60)–(5.62) we may rewrite the explicit expression (i.e., free from σ) for the ray incident on an inhomogeneous medium in the point

$\{x_0; y_0\}$ at the angle θ_0:

$$y = y_0 + \int_{x_0}^{x} \text{tg}\, \theta(x)\, dx = y_0 + \int_{x_0}^{x} [E(x)]^{-1/2} k_0 \sin \theta_0\, dx, \qquad (5.63)$$

where

$$E(x) = k^2(x) - k_0^2 \sin^2 \theta_0; \quad \tan \theta(x) = [E(x)]^{-1/2} k_0 \sin \theta_0. \quad (5.64)$$

To derive S we use Eq. (5.53). As the r.h.s. of Eq. (5.53) is independent of y its solution may be tried in the form $S(x, y) = S_1(x) + S_2(y)$ by separation of variables. Omitting intermediate derivations we obtain the expression for S when $y \geq 0$ and the wave $\Pi = a_0 \Phi_0 \exp\{i\delta k_0 (x \cos \theta_0 + y \sin \theta_0)\}$ is given for $x = x_0$:

$$S(\mathbf{x}) = \pm \int_{x_0}^{x} [E(x)]^{1/2}\, dx + k_0 \sin \theta_0 y + X_0, \qquad (5.65)$$

where X_0 is an arbitrary constant. When $S(\mathbf{x})$ is known, $A_0(x)$ is derived from Eq. (5.56) for $N^2 = N^2(z, x)$. Omitting intermediate derivations (see [250]) we write this solution as

$$A_0(x) = \frac{a_0\, [L(x_0)]^{1/2}\, [k_0 \cos \theta_0]^{1/2}}{[L(x)]^{1/2}\, [E(x)]^{1/4}} \exp\{-\Omega\}, \qquad (5.66)$$

where

$$\Omega = \frac{1}{2} \int_{x_0}^{x} [L^{-1} I]\, dx. \qquad (5.67)$$

Using Eqs. (5.58), (5.65), and (5.66) it is possible to obtain expressions for all of the internal waves characteristics valid with an accuracy $\mathcal{O}[(i\beta)^{-1}]$. E.g., for the level elevation $\hat{\xi}$ in an internal wave we have

$$\hat{\xi} = -\frac{a_0 \Phi' L_0^{1/2}\, [k_0 \cos \theta_0]^{1/2}}{\omega^2 l \alpha L^{1/2}\, E^{1/4} \exp\{\Omega\}}$$

$$\times \exp\left\{ i\beta \left[\pm \int_{x_0}^{x} E^{1/2}\, dx + k_0 \sin \theta_0 y + X_0 \right] \right\}. \qquad (5.68)$$

Let us analyze this equation assuming for simplification that $\alpha\,(z, \mathbf{x}) \neq$ 0, i.e., $\omega^2 \neq N^2$. Eq. (5.68) has the solution in the form of two waves propagating in y-direction and in opposite x-directions (both positive and negative). According to (5.63) the rays have straight line trajectories when $\theta_0 = 0$ and the dependence on y vanishes. When $\theta_0 \neq 0$ the rays have curved trajectories, and in addition the rays turn towards decreasing $k\,(x)$. The solution of (5.68) always oscillates along the y-axis. The phase space is divided into two parts along the line of the x-axis: $E\,(x) > 0$ and $E\,(x) < 0$. When $E\,(x) > 0$ the wave phase is real and the solution of (5.68) oscillates along the x-axis. When $E\,(x) < 0$ the solution is growing or decreasing exponentially versus x. Thus the point $x = x_T$, where $E\,(x_T) = 0$ and $\theta\,(x_T) = \pi/2$, is according to (5.68) the turning point. The value of x_T is defined by the expression

$$E\,(x_T) = k^2\,(x_T) - k_0^2 \, \sin^2 \theta_0 = 0 \,, \tag{5.69}$$

where $k^2\,(x)$ and $k_0^2 = k^2\,(x_0)$ are defined through $N^2\,(z, x)$ and $N_0^2\,(z) = N^2\,(z, x_0)$ using (5.52). The solution obtained by the ray method is invalid in the turning point (see, e.g., [102]). The approximate criterion of the ray method validity [102] is

$$[k \, \cos\theta]'_x \, \{|\, k \, \cos\theta|\}^{-1} \ll \beta \,.$$

This condition is infringed for $\cos\theta \to 0$, i.e., in the vicinity of the turning point.

When the turning point is absent while the wave (5.68) propagates in x-direction, the wave will be running along the x-axis with the phase $\beta \int_{x_0}^{x} E^{1/2}\,dx$ and $X_0 = 0$. If $E\,(x)$ decreases as x increases and reaches the point $x = x_T$, where $E = 0$, then $X_0 = \pi/4$ and there are two waves — the incident and the reflecting — with corresponding phases $\pm\beta\,[\int_{x_0}^{x} E^{1/2}\,dx + \pi/4]$. The superposition of these waves may form the standing wave

$$\hat{\xi} \sim e^{i\beta k_0 \, \sin\theta_0 y} \, \cos\left[\beta\left(\int_{x_0}^{x} E^{1/2}\,dx + \pi/4\right)\right] \,.$$

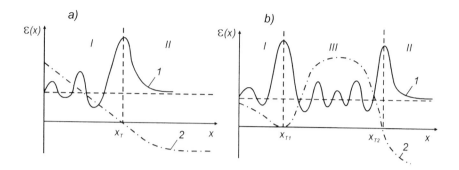

Figure 5.4. A scheme of internal waves' behaviour in the presence of one (a) and two (b) turning points: 1 — the wave amplitude; 2 — the dependence of E on x; I — the region of standing oscillating waves; II — the region of the exponential decay; III — the region of the standing waves with a discontinuous spectrum.

Thus in the presence of the turning point standing internal waves may be induced (and have been observed experimentally in the open ocean).

When $E(x)$ grows vs. x, the incident wave going from $x = +\infty$, reflects at the point $x = x_T$ towards increasing x values. In this case $X_0 = -\pi/4$.

If there are two turning points x_{T_1} and x_{T_2}, where $x_{T_2} > x_{T_1}$, then it yields $E > 0$ for $x_{T_1} < x < x_{T_2}$. The wave is oscillating, and is a standing wave along the x-axis, and decays exponentially out of this region. By analogy with the method of phase integrals [102] such a standing wave has a discontinuous spectrum of wave numbers k_s defined by the condition

$$\beta \int_{x_{T_1}}^{x_{T_2}} [E(x)]^{1/2} \, dx = \left(s + \frac{1}{2} \right) \pi . \qquad (5.70)$$

Eq. (5.70) together with the dispersion relation $\omega = \omega^n(k)$, obtained from the solution (5.52), defines the possible spectrum of frequencies ω_s and wave numbers k_s. A scheme of waves behaviour in the presence of one and two turning points is presented in Fig. 5.4.

Consider as an example a case $N^2 = N^2(x)$, i.e.,

$$\rho_0(z, x) = \rho_c(x) \exp \left\{ \frac{N^2(x)}{g} z \right\} .$$

Then from Eq. (5.52) this yields

$$\Phi^n = a_0^n \cos n\pi z, \quad n\pi = \alpha^{1/2} k^n (x).$$ (5.71)

Eq. (5.71) shows that the wavelength increases along x as N increases. Eqs. (5.57), (5.67) and (5.71) yield

$$L = a_0^2/2; \quad I = \frac{a_0^2}{4} [\ln \alpha]'_x; \quad \Omega = \frac{1}{4} \ln [\alpha/\alpha_0].$$

Then for $\hat{\xi}/\hat{\xi}_0$, where $\hat{\xi}_0 = \hat{\xi}(x \le x_0)$ is the incident wave and $\alpha_0 = \alpha(x \le x_0)$, we have

$$\frac{\hat{\xi}}{\hat{\xi}_0} = \frac{\alpha_0^{5/4} [\cos \theta_0]^{1/2}}{\alpha(x) [\alpha_0 - \alpha(x) \sin^2 \theta_0]}$$

$$\times \exp \left\{ i\beta \left[\pm \int_{x_0}^{x} E^{1/2} dx - k_0 \cos \theta_0 y + X_0 \right] \right\},$$ (5.72)

$$\dot{E}(x) = (n\pi)^2 [\alpha(x) \alpha_0]^{-1} [\alpha_0 - \alpha(x) \sin^2 \theta_0].$$ (5.73)

Eq. (5.72) shows that as $\alpha(x) \to 0$ the wave amplitude grows, and for $\alpha = 0$ it goes to infinity, as well as E. Thus the point $\alpha = 0$ is a caustic point of the given wave field. $N^2 \to \omega^2$ when $\alpha \to 0$, so the value of N decreases over the distance x. Therefore the amplitude of waves (or a wave group) of the given frequency ω grows while waves propagate towards decreasing N. This should support local maxima of internal waves' spectra in the vicinity of the Brunt–Väisälä frequency measured at the given point (Fig. 5.5).

Eqs. (5.72), (5.73) show that the wave's amplitude goes to infinity when $\alpha_0 = \alpha(x) \sin^2 \theta_0$, i.e., at the turning point. The latter equation is satisfied when N increases over distance. Thus a sufficient increase of the wave's amplitude is possible for both decrease and increase of N, and so an instability of the internal waves may appear. The latter may be illustrated by the calculation of the Richardson number in the wave

$$Ri = N^2 \left\{ \frac{\partial \mathbf{U}}{\partial z} \right\}^{-2} = N^2 \omega^2 l^2 [A_0^2 (\Phi')^2 |\nabla S|^2]^{-1} + \mathcal{O}[(i\beta)^{-1}].$$ (5.74)

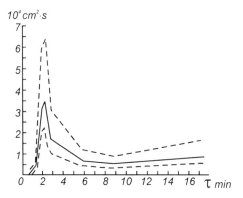

Figure 5.5. The energy spectrum of internal waves in the Brunt–Väisälä period range [179] (dashed lines correspond to the 90% ensurance).

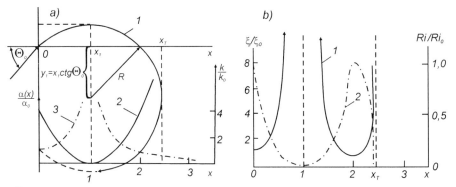

Figure 5.6. The ray picture for α varying in the lines of (5.75) ($x_1 = 1$, $\theta_0 = 45°$, $\alpha_0 = 4$): a : 1 — the ray trajectory, 2 — $\alpha(x)/\alpha_0$ vs. x, 3 — k/k_0 vs. x; b : 1 — $\hat{\xi}/\hat{\xi}_0$ vs. x, 2 — Ri/Ri_0 vs. x.

The ratio of the Richardson number in the wave to its value for the incident wave when $N^2 = N^2(x)$ is

$$\frac{Ri}{Ri_0} = \frac{[\alpha(x) + 1]\,\alpha(x)\,[\alpha_0 - \alpha(x)\sin^2\theta_0]^{1/2}}{[\alpha_0 + 1]\cos\theta_0\,\alpha_0^{3/2}} \tag{5.75}$$

Eqs. (5.75) shows that while approaching the caustic ($N^2 \to \omega^2$) and the turning point ($\alpha_0 = \alpha(x)\sin^2\theta_0$) Ri decreases considerably and may be less than $Ri^* = \frac{1}{4}$, which causes instability of the internal waves .

Consider, for example, the following case of variation of $N^2(x)$

$$N^2(x) = (N_0^2 - \omega^2)\left[\frac{x_1 - x}{x_1}\right]^2 + \omega^2,$$

which corresponds to $\alpha(x) = \alpha_0 [(x_1 - x)/x_1]^2$. Substitute this expression for α into Eq. (5.73) and then into (5.74). After integration we have the expression for the ray shape when $x_0 = 0$:

$$\hat{x}^2 + \hat{y}^2 = R^2,$$

where

$$\hat{x} = (x_1 - x), \quad \hat{y} = y + x_1 \tan \theta_0, \quad R = \frac{x_1}{\sin \theta_0}.$$

This expression represents a circle of radius R with its center at the point $\{x_1, y_1 = -x_1 \tan \theta_0\}$. The behaviour of $\alpha(x)$ and the ray trajectory, as well as variations of $\hat{\xi}/\hat{\xi}_0$ and Ri/Ri_0 versus x calculated by Eqs. (5.72), (5.75), are shown schematically in Fig. 5.6. It shows that $\hat{\xi}/\hat{\xi}_0 \to 0$, $Ri/Ri_0 \to 0$ for $x \to x_1$, and the ray inclination goes to zero, i.e., the point $x = x_1$ is a caustic; $\hat{\xi}/\hat{\xi}_0 \to \infty$, $Ri/Ri_0 \to 0$, $\theta \to \pi/2$ for $x \to x_1 [1 + 1/\sin \theta_0]$, i.e., this point is the turning point.

If the ocean density is distributed as $\rho_0 = \rho_1 = \text{const}$ for $0 \leq z \leq h(x)$ and $\rho_0 = \rho_2 = \text{const}$ for $\infty > z \geq h(x)$, where $h(x)$ is the pycnocline depth depending on the horizontal coordinate x, then internal waves amplitudes increase for decreasing h (the pycnocline lifts to the free surface) and decreases otherwise.

The ray approximation is invalid in the vicinity of caustics and turning points. More accurate methods should be used in these cases. The solution in the vicinity of the turning point was worked out in [318] for the particular case $N = N(x)$ and plane waves. The Maslov method allowing the solution of the problem in the vicinity of caustics was used by Voronovich [380]. Estimations for the wave field in caustics (where $\omega \to N(z, x)$) were given in this paper which showed that internal waves' amplitudes grow several folds while approaching these points.

3. INTERNAL WAVES PROPAGATION IN A FLOW WITH HORIZONTAL VELOCITY GRADIENT

Oceanic flow parameters usually have horizontal variations. As a rule, a characteristic scale of horizontal velocity variations of averaged ocean flows is considerably greater than the characteristic internal waves' length. In this case flows may be considered to be varied slowly, and the ray methods may be used for the investigation of internal waves. But sometimes the horizontal velocity gradient may be considerable (in the jet flows like the Gulf Stream, the Kuroshio, the Cromwell–Lomonosov flow, the Tareyev flow). In such a case the ray method is invalid and more accurate methods should be used. The problem of internal waves' propagation on the background flow with a horizontal velocity discontinuity modeling the jet flow is considered in this section. A general formalism of the ray method applied to internal waves propagating in the horizontally inhomogeneous ocean is also stated. It allows us, in particular, to study an effect of the horizontally inhomogeneous mean flow on internal waves.

Assume that the mean flow is directed along the x-axis and depends on the horizontal coordinate y only, i.e., $\mathbf{U} = \{U(y), 0, 0\}$, and $\rho_0 = \rho_0(z)$. Then small amplitude internal waves are described by Eqs. (2.46)–(2.49) for $V \equiv W \equiv 0$. Consider elementary wave solutions of the form

$$\{\mathbf{u}, p, \rho\} \sim \{\hat{\mathbf{u}}, \hat{p}, \hat{\rho}\} \exp \{ik(x - c_f t)\}, \qquad (5.76)$$

where k is the real wave number and c_f is the phase velocity which may be complex (in such a case $c_f = c_r + ic_i$). When the phase velocity is complex the internal waves modes will be stable only when $kc_i \leq 0$. Substituting Eq. (5.76) into (2.46)–(2.49), using for simplification the Boussinesq approximation and assuming $f \equiv 0$ we have

$$\rho_0 \left\{ ik(U - c_f)\hat{u} + \left(\frac{dU}{dy}\right)\hat{v} \right\} = -ik\hat{p};$$

$$ik\rho_0(U - c_f)\hat{v} = -\left(\frac{\partial \hat{p}}{\partial y}\right);$$

$$ik\rho_0(U - c_f)\hat{w} = -\left(\frac{\partial \hat{p}}{\partial z}\right) - g\hat{\rho}; \qquad (5.77)$$

$$ik\hat{u} + \left(\frac{\partial \hat{v}}{\partial y}\right) + \left(\frac{\partial \hat{w}}{\partial z}\right) = 0;$$

$$ik(U - c_f)\hat{\rho} + \left(\frac{d\rho_0}{dz}\right)\hat{w} = 0.$$

Introduce the new functions

$$P = \frac{\hat{p}}{\rho_0}; \qquad F = \frac{\hat{v}}{U - c_f}. \qquad (5.78)$$

Then Eqs. (5.77) may be reduced to the single equation for P:

$$\left[(U - c_f)^2 \frac{\partial}{\partial y} \left\{ (U - c_f)^{-2} \frac{\partial P}{\partial y} \right\} + \frac{\partial}{\partial z} \left\{ B \frac{\partial P}{\partial z} \right\} \right] - k^2 P = 0, \quad (5.79)$$

where

$$\frac{\partial P}{\partial y} = -ik (U - c_f)^2 F;$$

$$B(y, z, k, c_f) = \left\{ 1 - \frac{N^2(z)}{k^2 [U(y) - c_f]^2} \right\}^{-1}. \quad (5.80)$$

The limiting conditions for P should be required for Eq. (5.79) when $|y| \to \infty$. Under corresponding boundary conditions Eq. (5.79) is an eigenvalue problem for the eigenfunctions P and velocity c_f when k is fixed. This problem is more complicated than the ordinary problem of hydrodynamic stability (studied, e.g., in Section 4.1) because it is three-dimensional and is described by the partial differential equation (5.79).

Following Barcilon and Drasin [16] consider fluid with $N^2 = $ const. Then assuming

$$\frac{P}{\rho_0} = \widehat{P}(y) \exp \{ ik (x + \lambda z - c_f t) \}, \quad \text{i.e.,} \quad P = \widehat{P}(y) e^{ik\lambda z}, \quad (5.81)$$

from Eq. (5.79) we obtain

$$(U - c_f)^2 \frac{d}{dy} \left[(U - c_f)^{-2} \frac{d\widehat{P}}{dy} \right] - k^2 \left[\frac{\alpha^2 (U - c_f)^2 - N^2}{k^2 (U - c_f)^2 - N^2} \right] \widehat{P} = 0, \quad (5.82)$$

where $\alpha = k (1 + \lambda^2)^{1/2}$ and $k\lambda$ is the real vertical wave number. Assume as a model for $U(y)$

$$U(y) = \begin{cases} V & \text{at } y > 0 \\ -V & \text{at } y < 0. \end{cases} \quad (5.83)$$

Then Eq. (5.82) yields

$$\frac{d^2 \widehat{P}}{dy^2} = k^2 \left\{ [(1 + \lambda^2)(1 \mp \sigma)^2 - \varepsilon][(1 \mp \sigma)^2 - \varepsilon]^{-1} \right\} \widehat{P}, \quad (5.84)$$

where $\sigma = c_f / V$ and $\varepsilon = N_0^2 / k^2 V^2$ is the Richardson number based on the internal wave length. Solution (5.84) may be tried in the form (see [16])

$$\widehat{P} = A_\pm \exp(\alpha l_\pm y) + D_\pm \exp(-\alpha l_\pm y), \qquad (5.85)$$

where $A_\pm$, $D_\pm$ are some constants, and

$$l_\pm = \{[(1+\lambda^2)(1\mp\sigma)^2 - \varepsilon]/[(1\mp\sigma)^2 - \varepsilon]\}^{1/2}. \qquad (5.86)$$

The limiting condition for $\widehat{P}$ for $y \to \pm\infty$ yield

$$\widehat{P} = \begin{cases} D_+ \exp(-kl_+ y), & y > 0; \\ A_- \exp(kl_- y), & y < 0. \end{cases} \qquad (5.87)$$

For $y = 0$ the continuity conditions for the pressure $\widehat{P}$ and the velocity component $\widehat{v}$, normal to the surface $y = \mathrm{const}$, should be satisfied. From these conditions it follows that $D_+ = A_-$ and that the equation for eigenvalues is

$$l_+(1-\sigma)^2 + l_-(1+\sigma)^2 = 0,$$

which, using (5.86), may be rewritten in the following form

$$2[\varepsilon^2 + (1+\lambda^2)(\sigma^2-1)^2](\sigma^2+1) - (1+\lambda^2)\varepsilon(\sigma^2-1)^2$$

$$- \varepsilon(3 + 10\sigma^2 + 3\sigma^4) = 0. \qquad (5.88)$$

Eq. (5.88) is cubic with respect to σ^2, i.e., there are two triplets of roots corresponding to $\pm\sigma$. The roots of Eq. (5.88) depend on λ and ε values. The case of non-stratified fluid with a shear flow corresponds to $\varepsilon = 0$ and $\lambda = 0$. Then $\sigma^2 = -1$ $(c_f = \pm iV)$ and the Kelvin–Helmholtz instability occurs. Thus waves may be stable ($c_i = 0$) only when ε and λ are large enough. The condition $\sigma = 0$, i.e.,

$$(4 + \lambda^2)\varepsilon = 2(\varepsilon^2 + 1 + 1\lambda^2),$$

is a stability criterion. This leads to the following instability condition for internal waves

$$0 \le \lambda^2 \le \frac{2(\varepsilon-1)^2}{\varepsilon-2}. \qquad (5.89)$$

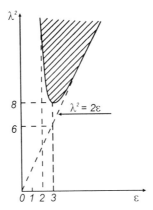

Figure 5.7. A neutral curve, stability and instability regions (shaded) for internal waves in a horizontally inhomogeneous flow [16].

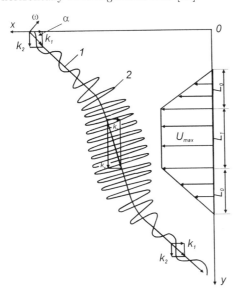

Figure 5.8. A qualitative picture of distribution of the mean velocity U and internal wave deformation in the flow: 1 — direction of the wave propagation; 2 — amplitude variations along the ray.

An instability region corresponding to Eq. (5.89) is shown in Fig. 5.7. Unstable modes will always exist for any vales of ε. So the horizontally inhomogeneous flow (5.83) is always unstable. When λ is fixed, stable modes may become unstable while ε is growing.

Consideration of more realistic models of $N(z)$ and $U(y)$ leads to considerable mathematical difficulties and belongs to the future. But there are some numerical calculations, as well as some results [16], generalizing the model presented above. In the work by Ivanov and Morozov [142] the equation similar to (5.82), with the $U(y)$ variation law shown in Fig. 5.8, was solved numerically. A progressive wave was given running at the angle $\alpha = \tan k_1 / k_2$, where k_1 and k_2 are x and y projections of the wave vector respectively. As mentioned above, the coefficients of Eq. (5.82) have singularities when $U = c_f$ and $-k_1^2 [U - c_f] = N^2$. The numerical experiment was limited to the consideration of negative values of k_1 and to Doppler frequencies $-k_1 (U - c_f)$ lower than the Brunt–Väisälä frequency. The qualitative behaviour of internal waves for $k_1 x - \omega t = \pi/2$ on the flow with the horizontal velocity gradient is shown in Fig. 5.8. The wave length decreases and the amplitude grows in the range L_0. The angle of rotation of the phase vector also increases. Internal wave parameters are constant and equal to those at the end of L_0 in the interval L_1 where the mean velocity is constant. The wavelength and the amplitude grow again and the mean velocity decreases in the range L_0'. Thus the wave has initial parameters while leaving the interval L_0'.

If the characteristic length of internal waves is much less than the characteristic horizontal scale of the variations of flow velocity, then the ray method may be used for the problem solution. In [124], [382] a general problem of the propagation of short internal waves in the horizontally inhomogeneous ocean in the ray approximation was considered. Bottom unevenness, and horizontal inhomogeneity of the velocity and density fields were taken into consideration simultaneously. Moreover, a weak instability of these fields was also considered (it was assumed that the characteristic time scale of ρ_0 and $\mathbf{U}$ variation is considerably greater than the characteristic period of internal waves). While studying internal waves in the horizontally inhomogeneous flow (e.g., to solve Eq. (5.79)) the solution method from Sections 5.1, 5.2 may be used. But here we expound, following the work [380], a more general problem where some inhomogeneities of ρ_0, $\mathbf{U}$ and of the bottom relief are taken into account.

We shall consider a complete nonlinear hydrodynamic combined equations (2.13)–(2.15) in the Boussinesq approximation for $f = 0$ with the boundary conditions (1.39), (1.41), (1.42) for $p_a = 0$. All of the hydrodynamic fields are presented in the form

$$\hat{\varphi}\left(\mathbf{x}, z, t\right) = \varphi^{(0)}\left(\mathbf{X}, z, \tau\right) + \delta\varphi\left(\mathbf{x}, z, t\right); \quad \mathbf{X} = \varepsilon\mathbf{x}; \quad \tau = \varepsilon t, \quad (5.90)$$

where $\hat{\varphi}$ is one of the hydrodynamic functions, $\varphi^{(0)}$ is the averaged 'background' field, φ a disturbance propagating on the background, ε a small parameter characterizing the slowness of horizontal and time variations of basic movement, δ a small amplitude parameter. Moreover, we assume H variations, i.e., $H = H\left(\mathbf{X}\right)$, to be slow also. Substituting Eq. (5.90) into (2.13), (2.14), (2.15 a), taking into consideration that $w^{(0)} \sim \varepsilon\,|\,u^{(0)}\,|,\ \varepsilon\,|\,v^{(0)}\,|$ (this follows from the continuity equation) and extracting values corresponding to the basic movement yield

$$\frac{\partial\mathbf{u}^{(0)}}{\partial t} + \left(\mathbf{u}^{(0)}\nabla\right)\mathbf{u}^{(0)} = -\nabla p;$$

$$g\rho_0 = -\frac{\partial p^{(0)}}{\partial z}; \quad \nabla\mathbf{u}^{(0)} = 0; \quad \frac{\partial\rho^{(0)}}{\partial t} + \mathbf{u}^{(0)}\nabla\rho^{(0)} = 0. \quad (5.91)$$

The boundary conditions for Eqs. (5.91) may be formed when adding the index 0 to all of the parameters.

A solution for disturbances φ will be tried in the form

$$\varphi = \left[\varphi^{(1)}\left(\mathbf{X}, z, \tau\right) + \varepsilon\varphi^{(2)}\left(\mathbf{X}, z, \tau\right) + \ldots\right]\exp\left\{\frac{i}{\varepsilon}S\left(\mathbf{X}, \tau\right)\right\}, \quad (5.92)$$

where S is the phase determining the local wave numbers $k_i = \partial S / \partial X_i$ $(i = 1,\,2)$ and the frequency $\omega = -\partial S / \partial\tau$. Substituting (5.92) into the disturbance equation and equating values of the order δ in Eq. (5.90) we can obtain equations and boundary conditions for $w^{(1)}$, the vertical velocity of the first order disturbances (the superscript (1) will be omitted further):

$$w'' + \left(\frac{k^2 N^2}{\omega_d^2} - \frac{\omega_d''}{\omega_d} - k^2\right)w = 0; \quad (5.93)$$

$$\left(\frac{w}{\omega_d}\right)' = \frac{gk^2}{\omega_d^3}w \quad \text{at} \quad z = \zeta^{(0)}; \quad w = 0 \quad \text{at} \quad z = -H\left(\mathbf{X}\right), \quad (5.94)$$

where $\omega_d = \omega - \mathbf{k}\mathbf{u}^{(0)}$ is the Doppler frequency depending on z when $\mathbf{u}^{(0)}$ is the function of z, and

$$N^2 = \frac{g}{\rho^{(0)}} \frac{\partial \rho^{(0)}}{\partial z},$$

a prime meaning derivation with respect to z. The boundary value problem (5.93), (5.94) yields a number of eigenvalues (dispersion relations) for different normal modes

$$\omega = f(\mathbf{k}, \mathbf{X}, \tau) \qquad (5.95)$$

and of eigenfunctions $w = w(z, \mathbf{X}, \tau)$ depending parametrically on $\mathbf{X}$ and τ. All the remaining parameters characterizing internal wave are expressed in terms of w by:

$$\mathbf{u} = \frac{i\mathbf{k}}{k^2} \omega_d \left(\frac{w}{\omega_d}\right)' - \frac{iw}{\omega_d} \frac{\partial \mathbf{u}^{(0)}}{\partial z};$$

$$p = \frac{i\omega_d^2}{k^2} \left(\frac{w}{\omega_d}\right)'; \quad \rho = -i\frac{w}{\omega_d}\frac{\partial \rho^{(0)}}{\partial z}; \quad \xi = \frac{iw}{\omega_d}. \qquad (5.96)$$

Extracting, further, the terms of order $\delta\varepsilon$ in the basic equations and the boundary conditions, after rather tedious derivations yields the equation and the boundary conditions for $w^{(2)}$:

$$[w^{(2)}]'' + \left(\frac{k^2 N^2}{\omega_d^2} - \frac{\omega_d''}{\omega_d} - k^2\right) w^{(2)} = F;$$

$$\omega_d \frac{\partial w^{(2)}}{\partial z} - \left(\frac{gk^2}{\omega_d} + \omega_d'\right) w^{(2)} = G \quad \text{at} \quad z = \zeta^{(0)};$$

$$w^{(2)} = -u_j \frac{\partial H}{\partial x_j} = Q \quad \text{at} \quad z = -H(\mathbf{X}), \qquad (5.97)$$

where F and G are functions written in terms of $\varphi^{(0)}$ and $\varphi^{(1)}$ (an explicit form of these functions is given in [382]). For solubility of the boundary value problem (5.97) the functions F, G and Q are required to be orthogonal to the eigenfunctions of the corresponding boundary value problem. This leads to the solubility condition:

$$\int_{-H}^{\zeta^{(0)}} F \frac{iw}{k^2} \, dz - \left[\frac{i}{k^2} \frac{w}{w_d} G \right]\Big|_{z=\zeta^{(0)}} - \left[\frac{i}{k^2} Q w' \right]\Big|_{z=-H} = 0. \qquad (5.98)$$

Taking into account the direct form of the functions F, G, Q, condition (5.98) after some transformations may be reduced to the form of the conservation law for the adiabatic invariant $\mathcal{F}$:

$$\frac{\partial \mathcal{F}}{\partial \tau} + \nabla_h \left(\mathbf{c}_g \mathcal{F} \right) = 0, \qquad (5.99)$$

where

$$\mathcal{F} = \int_{-H}^{\zeta^{(0)}} \left(\frac{N^2}{w_d^3} - \frac{w_d''}{2w_d^2 k^2} \right) w^2 \, dz + \left[\left(\frac{g}{w_d^3} + \frac{w_d'}{2w_d^2 k^2} \right) w^2 \right]\Big|_{z=\zeta^{(0)}}, \qquad (5.100)$$

$$\mathbf{c}_g \mathcal{F} = \int_{-H}^{\zeta^{(0)}} \left\{ \mathbf{u}^{(0)} \left[\frac{N^2}{w_d^3} - \frac{w_d''}{2w_d^2 k^2} \right] + \frac{1}{2w_d k^2} \frac{\partial^2 \mathbf{u}^{(0)}}{\partial z^2} + \frac{\mathbf{k}}{k^2} \left(\frac{N^2}{w_d^2} - 1 \right) \right\}$$

$$\times w^2 \, dz + \left\{ \left[\mathbf{u}^{(0)} \left(\frac{g}{w_d^3} + \frac{w_d'}{2w_d^2 k^2} \right) - \frac{1}{2w_d k^2} \frac{\partial \mathbf{u}^{(0)}}{\partial z} + \frac{g\mathbf{k}}{w_d^2 k^2} \right] w^2 \right\}\Big|_{z=\zeta^{(0)}} \qquad (5.101)$$

From properties of the boundary value problem (5.93) it may be shown that the ratio of Eq. (5.100) to (5.101) is the group velocity $\mathbf{c}_g = \partial f / \partial \mathbf{k}$. Thus the conservation law for $\mathcal{F}$ is a companion of the conservation law for the wave action (4.59). Therefore $\mathcal{F}$ may also be called the wave action. In the absence of mean flow ($\mathbf{u}^{(0)} = 0$) $\mathcal{F} = E / \omega$, where E is the mean energy of the wave. When the Richardson numbers of the mean flow are high (slow variation with depth of the horizontal velocity) we have for internal waves

$$\omega'' \ll \frac{N^2}{w_d^2} k^2 \quad \text{and} \quad \mathcal{F} = \int_{-H}^{\zeta^{(0)}} \frac{E}{w_d} \, dz.$$

Thus $\mathcal{F}$ is the vertically integrated wave action contained in Eq. (4.59).

It is important that the conservation law for the wave action of dynamic fields (5.99) will be invalid for arbitrary basic hydrodynamic fields, i.e., those not satisfying the hydrodynamical equations (5.91), which were described in Section 5.2 for internal waves propagating in the ocean with an arbitrary field of undisturbed density $\rho_0 = \rho_0 (z, \mathbf{x})$. Eqs. (5.91) with the according boundary conditions were used while deriving (5.99).

For a complete solution of the problem it is necessary to obtain the values of S and $\mathbf{c}_g$. Taking into consideration definitions of the wave number and the frequency ($\mathbf{k} = \nabla_h S$, $\omega = -\partial S / \partial \tau$) we derive from (5.95) the Hamilton–Jacobi equations for the wave phase

$$
\frac{\partial S}{\partial \tau} + f\,(\,\nabla_h\,S, \mathbf{X}, \tau) = 0\,.
\tag{5.102}
$$

Eqs. (5.99) and (5.102) completely define the wave behaviour in the ray approximation. The characteristics of Eq. (5.102) are solutions of the following combined equations:

$$
\frac{dX_i}{d\tau} = \frac{\partial f}{\partial k_i} = c_{g\,i}\,(k_j, X_j, \tau)\,; \qquad \frac{dk_i}{d\tau} = \frac{\partial F}{\partial k_i}\,; \qquad \frac{d\omega}{d\tau} = -\frac{\partial f}{\partial \tau}\,.
\tag{5.103}
$$

The value of S along the characteristics is calculated by

$$
S = S_0 + \int_{\tau_0}^{\tau} \left[\,k_i c_{g\,i} - \omega\,\right] d\tau'\,,
\tag{5.104}
$$

where S_0 is the initial value of S. The characteristics of Eq. (5.99) coincide with those for Eq. (5.103). We may show further (see [382]), that the following equality is valid along the characteristics

$$
|\,\mathbf{c}_g \mathcal{F}\,|\,dl = \mathrm{const}\,,
\tag{5.105}
$$

where dl is the distance between two infinitely close characteristics' projections on the coordinate subspace (x_1, x_2). From it (5.103) follows that $\mathbf{c}_g \perp dl$, so the relation (5.103) states the constancy law of the wave action flow along the ray tube.

The theory presented allows us to unify consideration of internal waves propagation in the ocean with slow time and weak horizontal inhomogeneities of the mean state. Some exact applications of this theory have been given by Voronovich [380]. Following this work we consider the influence of a plane parallel flow with a horizontal velocity gradient on internal waves in the ray approximation.

Assume

$$\mathbf{u}^{(0)}(z, \mathbf{X}, \tau) \equiv \{U(Y), 0, 0\} \quad \text{and} \quad \rho^{(0)} = \rho_0(z),$$

here $\mathbf{X} = \{X, Y\}$. Then from (5.103) it follows that $\omega = \text{const}$ along the characteristics, because the ocean parameters are independent of time, and $k_1 = \text{const}$ (k_1 is the X-axis projection of vector $\mathbf{k}$) because the ocean parameters are independent of X. In the case considered a number of equidistant curves (Fig. 5.9) serve as the characteristics, whence the conservation of $c_g^{(y)}$ follows (the projection of c_g to Y-axis)

$$c_g^{(y)} \mathcal{F} = \text{const}. \tag{5.106}$$

Consider further the simplest case in which $N^2 = \text{const}$. Eqs. (5.106) and (5.101) yield

$$\frac{k_2}{k^2}\left[\frac{N^2}{(\omega - k_1 u)^2} - 1\right] w^2 = \text{const}. \tag{5.107}$$

While deriving (5.107) we used for simplification the 'rigid lid' approximation filtering surface waves. Then all terms in (5.106), (5.107) calculated for $z = \zeta^{(0)}$ go to zero because $\zeta^{(0)} \equiv 0$ when $z = \zeta^{(0)} = 0$ and $w = 0$. From (5.107) there follows

$$\frac{w}{w_0} = R\left(\frac{\cos^2 \alpha}{R - \sin^2 \alpha}\right)^{1/4}, \tag{5.108}$$

where

$$R = \left(\frac{N^2}{\omega^2} - 1\right) \Big/ \left[\frac{N^2}{(\omega - k_1 u)^2} - 1\right]; \quad k_1 = \frac{n\pi}{H}\left(\frac{N^2}{\omega^2} - 1\right)\sin \alpha. \tag{5.109}$$

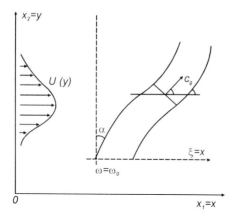

Figure 5.9. A profile of a flow with a horizontal velocity gradient and the general picture of the characteristics' projections on the coordinate subspace in case $U^{(0)} = [U(y), 0]$, $\rho^{(0)} = \rho^{(0)}(z)$.

Here $H = $ const is the ocean's depth, n the mode number, ω, w_0 the initial frequency and amplitude of the internal wave in the absence of flow, α is the angle included between the wave vector and the Y-axis. From Eqs. (5.108) and (5.109) it follows that when $Uk_1 > 0$ we have at the point where $R = \sin^2 \alpha$: $k_2 \to 0$, $c_g^{(y)} \to 0$ and $w \to \infty$ while $|U|$ grows versus the Y coordinate, i.e., this point (where $R = \sin^2 \alpha$) is the turning point of the ray. When $Uk_1 < 0$ while $|U|$ grows for $\omega - k_1 U \to N^2$ we have $k_2 \to \infty$, $w \to \infty$, and further in this case the internal wave reaches the critical level (where $\omega - k_1 U = N^2$) during a infinite time $\tau \sim |U - U^*|^{-1/2}$. In the first case the critical level is reached during a finite time, but the ray methods are invalid because $k_2 \to 0$. Thus the internal wave amplitude grows while the wave propagating both in the forwards and backwards directions of the flow growing versus the Y coordinate.

The conservation law of the adiabatic invariant similar to (5.99) was stated by Garrett [90] for the mean flow

$$\mathbf{U}(\mathbf{X}, z, t) = \{U(\mathbf{X}, z, t), \ V(\mathbf{X}, z, t), \ W(z, t)\}$$

under the condition of its slow variation versus all coordinates (including z) and an infinitely deep ocean. Internal waves in a spatially inhomogeneous flow in the ray approximation under the same assumptions were analyzed in general by Jones [146]. He showed that in such a flow the wave numbers and the group velocities of wave packets may increase over time under appropriate conditions.

A thorough study of the characteristic features of internal waves' behaviour resulting from interaction between internal waves and non-uniform large scale oceanic flows in the ray approximation was performed by Badulin, Shrira [9] and Tsimring [10]. They studied the case of a wave packet being 'trapped' by a stratified shear flow of arbitrary geometry based on a set of Hamiltonian ray equations, with the Hamiltonian being determined from the Taylor–Goldstein boundary value problem. It was shown that when a wave packet approaches the trapping layer the wave number and amplitude grow with time while the wave's motion tends to concentrate at a certain depth. Experimental evidence of the trapping of internal wave in a strong horizontally inhomogeneous Lomonosov current [11] proved that under certain conditions the trapping definitely occurs and produces strong effects like wave breaking or stimulates the generation of small scale turbulence at the depth of maximal stability (the Brunt–Väisälä frequency).

Notes

1 Scattering of gravity waves by large scale near surface turbulence and random internal waves is considered by Bunimovich and Zhmur [43], Raevsky [307] on the basis of closed equations for the average field and average intensity of gravity waves.

2 From now on we omit the symbol $\sim$ considering dimensionless variables unless stated otherwise.

Chapter 6

BASIC SOURCES OF INTERNAL WAVES GENERATION IN THE OCEAN

There are different sources which induce internal waves in the ocean: they may be caused by oscillations of atmospheric pressure, wind, regions of the ocean's bottom (during underwater earthquakes), by tidal forces etc.. Nonlinear mechanisms of internal waves generation, e.g., resonant interaction of surface waves for a flow over large bottom obstacles, are quite important. The generation of internal waves by oscillations of wind and atmospheric pressure is thoroughly studied in this chapter, the main attention being paid to resonant effects. The generation of internal waves by local initial disturbances is also considered in this chapter. Such disturbances may be induced by both atmospheric sources and different deformations of density and velocity fields. A brief overview of the other sources of the generation of internal waves is given in the conclusions of the chapter. Specific questions of wave generation by underwater earthquakes, tidal flows, and bottom obstacles are presented in [49], [179] and therefore are not considered here.

1. INDUCTION OF INTERNAL WAVES BY FLUCTUATIONS OF ATMOSPHERIC PRESSURE

The generation of internal waves in a stratified ocean by atmospheric pressure fluctuations has been studied in numerous works (e.g., [49]). But, as a rule, only stationary internal waves appearing in the fluid

153

as a result of periodic surface pressure variations are considered. This does not allow the evaluation of the efficiency of generation when the frequency of the disturbing force is equal to an eigenfrequency of the stratified fluid (resonance). But in this particular case a very considerable energy transfer from the atmospheric pressure field to internal waves may occur. Such a resonant theory taking into account the random character of the atmospheric pressure fluctuations (which is the most realistic) was worked out in [205].

Assume that the random atmospheric pressure distribution $P(\mathbf{x}, t)$ is given on the surface of a stratified rotating ocean of constant depth at a time $t > 0$ ($P(\mathbf{x}, t) = (1/\rho_0) p_a(\mathbf{x}, t)$ cm$^2 \times$ s^{-2} is the atmospheric pressure normalized by the mean fluid density ρ_0). Then the linearized equation for w (the vertical velocity component in the internal wave) and the boundary conditions will have the form (2.42), (2.43), (2.44). The initial conditions for Eq. (2.42) are

$$w\,|_{t=0} = \frac{\partial w}{\partial t}\bigg|_{t=0} = 0\,. \tag{6.1}$$

The initial condition (6.1) relates to the state of rest of the fluid for $t \leq 0$. Assume $P(\mathbf{x}, t)$ to be a homogeneous random function of $\mathbf{x}$ with a zero assembly average $\langle P(\mathbf{x}, t) \rangle \equiv 0$. Then p and w may be presented in the form of Fourier integrals[1]

$$P(\mathbf{x}, t) = \int e^{i\mathbf{kx}}\, \Pi(\mathbf{k}, t)\, d\mathbf{k}\,; \quad w(\mathbf{x}, t, z) = \int e^{i\mathbf{kx}}\, \varphi(\mathbf{k}, t, z)\, d\mathbf{k}\,. \tag{6.2}$$

Here Π and φ are Fourier images of the pressure and of the vertical velocity respectively; limits of integrals (6.2) are infinite. Substituting (6.2) into (2.42)–(2.44), after the Laplacian transformation with respect to time and taking into account (6.1) yield the following boundary value problem for the vertical coordinate z:

$$\bar{\varphi}'' + \frac{N^2(z)}{g}\,\bar{\varphi}' - \frac{p^2 + N^2(z)}{p^2 + f^2}\,k^2\bar{\varphi} = 0\,; \tag{6.3}$$

$$\left\{\bar{\varphi}' - \frac{k^2 g}{p^2 + f^2}\,\bar{\varphi}\right\}\bigg|_{z=0} = k^2 p^2\,\frac{\bar{\Pi}}{p^2 + f^2}\,; \quad \bar{\varphi}|_{z=H} = 0\,. \tag{6.4}$$

Here $\bar{\Pi}$ and $\bar{\varphi}$ are the Laplacian images of the functions Π and φ, and p is a parameter of the Laplacian transformation, a prime meaning derivation with respect to z (the z-axis is directed downwards from the ocean's surface).

The homogeneous boundary value problem (6.3), (6.4) was analyzed in details in Section 3.1. Consider the inhomogeneous problem (6.3), (6.4) based on the results of Section 3.1. Assume $\psi(k, p, z)$ to be a solution of Eq. (6.3) such that $\psi(k, p, H) = 0$. Then we shall try a solution of combined equations (6.3), (6.4) in the form $\bar{\varphi}(\mathbf{k}, p, z) = C(\mathbf{k}, p) \times \psi(k, p, z)$, where $C(\mathbf{k}, p)$ indicates the effect of the disturbing force. Satisfying the first boundary condition (6.4) for $\bar{\varphi}$, we rewrite the problem solution in the form

$$\bar{\varphi}(\mathbf{k}, p, z) = \frac{k^2 \, p \, \bar{\Pi}(\mathbf{k}, p)}{(p^2 + f^2) \, \mathcal{L}_0(p, \psi)} \, \psi(k, p, z), \qquad (6.5)$$

where

$$\mathcal{L}_0(p, \psi) = \left\{ \psi'(k, p, z) - \frac{k^2 g}{p^2 + f^2} \, \psi(k, p, z) \right\} \bigg|_{z=0}. \qquad (6.6)$$

Using the inverse Laplace transform of $\varphi(\mathbf{k}, t, z)$ we obtain

$$\varphi(\mathbf{k}, t, z) = \frac{1}{2\pi i} \int_{\sigma - i\infty}^{\sigma + i\infty} \left\{ \frac{k^2 p \, \bar{\Pi}(\mathbf{k}, p) \, e^{pt}}{(p^2 + f^2) \, \mathcal{L}_0(p, \psi)} \, \psi(k, p, z) \right\} dp, \qquad (6.7)$$

where $p = \sigma + i\omega$. Owing to the properties of the boundary value problem (6.3), (6.4), discussed in Section 3.1, one may state that all of the singularities of the integrand of (6.7) are unessential singular points which coincide with the roots of the function $\mathcal{L}_0(p, \psi)$, i.e., are the eigenvalues of the homogeneous boundary value problem (6.3), (6.4). Here the unessential singular points of (6.7) are gathered in segments $[\pm if, \pm iN]$, symmetrical with respect to zero, of the imaginary axis with the points of accumulation $\pm if$ (we consider for simplification only the practical interesting case $N > f$). The points $\pm iN$ are regular and the points $\pm if$ do not contribute to the integral (6.7). Then deforming the integration path as shown in Fig. 6.1 the value of the integral (6.7)

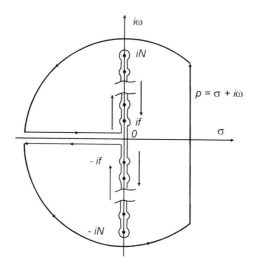

Figure 6.1.   An integration path on the complex plane $p = \sigma + i\omega$.

is defined by a sum of residues over all of the unessential singular points mentioned, i.e.,

$$\varphi\left(\mathbf{k}, t, z\right) = \sum_{n} \operatorname*{res}_{p_n} \left\{ \frac{k^2 p\, \psi\left(k, p, z\right)}{\left(p^2 + f^2\right) \mathcal{L}_0\left(p, \psi\right)} \int_0^t e^{p\,(t-\tau)} \left[\Pi\left(\mathbf{k}, \tau\right)\right] d\tau \right\},$$

$$(6.8)$$

where p_n are the zeros of the function $\mathcal{L}_0\left(p, \psi\right)$. As the functions ψ and $\mathcal{L}_0\left(p, \psi\right)$ depend only on p^2 and the roots $p_n = \pm i\omega_n$ are imaginary and symmetrical with respect to the real axis, (6.8) yields

$$\varphi\left(\mathbf{k}, t, z\right) = \sum_{n=1}^{\infty} \chi_n\left(z\right) \int_0^t \left\{\cos\left[\omega_n\left(t - \tau\right)\right] \Pi\left(\mathbf{k}, \tau\right)\right\} d\tau; \qquad (6.9)$$

$$\chi_n\left(k, z\right) = k^2 \psi\left(i\omega_n, z\right) / 2\left(\omega_n^2 - f^2\right) \mathcal{L}_0'\left(i\omega_n, \psi\right). \qquad (6.10)$$

In Eq. (6.10) $\mathcal{L}_0'\left(i\omega_n, \psi\right) = d\mathcal{L}_0 / dp^2$. Based on the results of Section 3.1 we can show that the series (6.10) in terms of eigenmodes is convergent. An evident form of the function $\chi_n\left(k, z\right)$ can be derived when the Brunt–Väisälä frequency distributions are given.

We shall calculate statistical characteristics of internal waves using Eq. (6.9) which is valid for an arbitrary ocean stratification. Introduce the correlation function $\Omega\left(\mathbf{k}, t_0, t\right)$ for the atmospheric pressure in the space of wave numbers $\mathbf{k}$ with time shift t

$$\Omega\left(\mathbf{k}, t_0, t\right) \delta\left(\mathbf{k} - \mathbf{k}'\right) = \left\langle \Pi\left(\mathbf{k}, t_0\right) \Pi^*\left(\mathbf{k}', t_0 + t\right) \right\rangle . \qquad (6.11)$$

Here angle brackets indicate the average of the theoretical chance, an asterisk means the complex conjugate value. If the pressure field is statistically stationary then $\Omega\left(\mathbf{k}, t_0, t\right)$ is independent of the initial time t_0. Introducing now into the space of wave numbers the single-time spectrum of the internal waves' vertical velocities in the form

$$S\left(\mathbf{k}, t, z\right) \delta\left(\mathbf{k} - \mathbf{k}'\right) = \left\langle \varphi\left(\mathbf{k}, t, z\right) \varphi^*\left(\mathbf{k}', t, z\right) \right\rangle \qquad (6.12)$$

and using (6.9), we obtain the relation between $\Omega\left(\mathbf{k}, t_0, t\right)$ and $S\left(\mathbf{k}, t, z\right)$:

$$S\left(\mathbf{k}, t, z\right) = \sum_{n=1}^{\infty} \sum_{m=1}^{\infty} \chi_n\left(k, z\right) \chi_m\left(k, z\right) I_{nm} , \qquad (6.13)$$

where

$$I_{nm} = \int_0^t \int_0^t \left\{ \Omega\left(\mathbf{k}, \tau', \tau' - \tau''\right) \cos \omega_n\left(t - \tau'\right) \cos \omega_m\left(t - \tau''\right) \right\} d\tau' \, d\tau'' . \qquad (6.14)$$

If the turbulent pressure field is statistically stationary, then the correlation function for the atmospheric pressure fluctuations $\Omega\left(\mathbf{k}, \tau', \tau' - \tau''\right)$ with the time shift $\tau' - \tau'' = \tau$ is independent of the initial time τ'. In this case, for $n \neq m$ all of the integrals I_{nm} will be bounded oscillating functions of time t. When $n = m$

$$I_{nn} = \frac{1}{2} \int_0^t \int_0^t \Omega\left(\mathbf{k}, \tau' - \tau''\right) \left\{ \cos \omega_n\left(\tau' - \tau''\right) + \cos \omega_n\left[2t - \left(\tau' + \tau''\right)\right] \right\} d\tau' d\tau'' . \qquad (6.15)$$

Replacing the variables $\tau' - \tau'' = \tau$ and $\tau' + \tau'' = \hat{\tau}$, (6.15) yields

$$I_{nn} = \frac{1}{2} \left\{ \int_0^t (t - \tau) \Omega(\mathbf{k}, \tau) \cos \omega_n \tau d\tau + \int_{-t}^0 (t - \tau) \Omega(\mathbf{k}, -\tau) \cos \omega_n \tau d\tau \right.$$

$$+ \int_0^t \Omega(\mathbf{k}, \tau) \left[\sin \omega_n \tau - \sin \omega_n (2t - \tau) \right] d\tau \Bigg\}. \qquad (6.16)$$

From Eq. (6.16) follows that for $\omega_n t \gg 1$ the first two integrals dominate in (6.16):

$$I_{nn} \approx \frac{1}{2} \int_0^t (t - \tau) \Omega(\mathbf{k}, \tau) \cos \omega_n \tau d\tau + \frac{1}{2} \int_{-t}^0 (t - \tau) \Omega(\mathbf{k}, -\tau) \cos \omega_n \tau d\tau.$$

$$(6.17)$$

Eq. (6.17) shows that for $t \to \infty$ the value of I_{nn} grows linearly over time proportionally to the Fourier cosine transformation $\Omega(\mathbf{k}, \tau)$ with respect to time. Defining further the space–time spectral density of pressure pulsations

$$E(\mathbf{k}, \omega) = (2\pi)^{-1} \int_0^\infty \Omega(\mathbf{k}, \tau) \cos \omega \tau d\tau, \qquad (6.18)$$

from (6.17), (6.18) for $t \to \infty$ yield

$$I_{nn} \approx \pi t \left\{ E(\mathbf{k}, \omega)|_{\omega = \omega_n(k)} + E(\mathbf{k}, \omega)|_{\omega = -\omega_n(k)} \right\}. \qquad (6.19)$$

Then using (6.13) for a long time t we derive[2]

$$S(\mathbf{k}, t, z) \approx \pi t \sum_{n=1}^\infty \left[\chi_n(k, z) \right]^2 \left\{ E(\mathbf{k}, \omega)|_{\omega = \omega_n(k)} \right.$$

$$\left. + E(\mathbf{k}, \omega)|_{\omega = -\omega_n(k)} \right\}. \qquad (6.20)$$

Some important general conclusions follow from Eq. (6.20):

1. Intense internal waves are generated by the atmospheric pressure fluctuations (a linear time growth of the internal waves' vertical velocity spectrum) only when there is a set of frequencies and wave numbers in the space–time atmospheric pressure spectrum according to dispersion relations of free internal wave modes. This corresponds in fact to a 'cut out' of resonant lines, relating to the dispersion equation $\omega = \pm \omega_n(k)$, from the function $E(\mathbf{k}, \omega)$.

2. In the resonant case different internal wave modes do not correlate for high t, which corresponds asymptotically to the presence of diagonal terms in (6.13). In other words, the internal waves' spectrum for high t may be presented as the superposition of spectra for uncorrelated normal modes.

The abovementioned properties are, seemingly, common for resonant multi-frequency statistical systems.

Unfortunately, an experimental evaluation of Eq. (6.20) is difficult because of the absence of the experimental evidence of the space–time spectrum of the atmospheric pressure pulsations above the ocean surface. But because of the same order of the atmospheric and oceanic Brunt–Väisälä frequency we may expect a considerable contribution to the atmospheric pressure spectrum from the atmospheric internal gravity waves with frequencies and wave numbers equivalent to those of the ocean. Under these conditions it is quite probable that the resonant conditions are satisfied.

To estimate the growth of velocity and possible amplitudes of oceanic internal waves it is necessary to model the distribution of $N(z)$ and to calculate the function $\chi_n(k, z)$. Consider further the two simplest models of the depth distribution of the Brunt–Väisälä frequency describing two boundary characteristic cases of ocean stratification. The first model corresponds to a sharp density jump at an arbitrary depth, and the ocean is homogeneous outside this region (the two-layer ocean model). The second model corresponds to linear ocean stratification, and the variation with depth of the mean density is exponential.

The first model. Assume the mean density $\rho_0(z)$ to be distributed in an infinitely deep ocean in accordance with

$$\rho_0(z) = \begin{cases} \rho_1, & 0 \le z < 0, \\ \rho_2, & h \le z < \infty. \end{cases} \tag{6.21}$$

Here h is the depth of the density jump, ρ_1 and ρ_2 are the density of the upper and the lower layers, respectively. Then the expression for $N^2(z)$ has the form

$$N^2 = g \ln\left(\frac{\rho_2}{\rho_1}\right) \delta(z - h) \approx g \frac{\Delta\rho}{\bar{\rho}_0} \delta(z - h),$$

where

$$\Delta\rho = \rho_2 - \rho_1 ; \quad \bar{\rho}_0 = \frac{1}{2}\left(\rho_1 + \rho_2\right).$$

Solving the boundary value problem (6.3), (6.4)[3], assuming for simplification $f = 0$ (high frequency internal waves) and using the inverse Laplace transform yields[4]:

for $0 \le z < h$

$$\chi_i\left(k, z\right) = \frac{\left\{k\left[kg\left(\Delta\rho/\bar{\rho}_0\right)\left(\operatorname{cth} kh \ \operatorname{sh} kz - \operatorname{ch} kz\right)\right] - \omega_i^2\left(1 + \operatorname{cth} kh\right)\left(\operatorname{sh} kz - \operatorname{ch} kz\right)\right\}}{\left\{kg\left[\operatorname{cth} kh \left(1 + \Delta\rho/\bar{\rho}_0\right)\right] - 2\omega_i^2\left(1 + \operatorname{cth} kh\right)\right\}}, \quad (6.22)$$

for $h \le z < \infty$

$$\chi_i\left(k, z\right) = \frac{k\omega_i^2 \ \exp\left\{-k\left(z - h\right)\right\}}{kg\left[\operatorname{cth} kh\left(1 + \Delta\rho/\bar{\rho}_0\right) + 1\right] - 2\omega_i^2\left(1 + \operatorname{cth} kh\right)}, \quad (6.23)$$

where $i = 1, 2$ and the frequencies ω_1^2, ω_2^2 satisfy the following dispersion relation:

$$\omega_{1,2}^2 = \frac{kg}{2\left(1 + \operatorname{cth} kh\right)}\left\{\left[\operatorname{cth} kh\left(1 + \frac{\Delta\rho}{\bar{\rho}_0}\right) + 1\right]\right.$$

$$\left. \pm \left[\left[\operatorname{cth} kh\left(1 + \frac{\Delta\rho}{\bar{\rho}_0}\right) + 1\right]^2 - 4\left(\frac{\Delta\rho}{\bar{\rho}_0}\right)\left(1 + \operatorname{cth} kh\right)\right]^{1/2}\right\}.$$

$$(6.24)$$

The frequency ω_2 (the minus sign in the square root in (6.24)) refers to an internal wave at the interface h. Retaining only the terms of order $\mathcal{O}\left(\Delta\rho/\bar{\rho}\right)$ in (6.24), we have the following expression for ω_2^2:

$$\omega_2^2 \approx kg\left(\frac{\Delta\rho}{\bar{\rho}_0}\right)\left[1 + \operatorname{cth} kh\right]^{-1} + \mathcal{O}\left[\left(\frac{\Delta\rho}{\bar{\rho}_0}\right)^2\right]. \quad (6.25)$$

The frequency ω_1 (a plus sign at the square root in (6.24)) refers to a surface wave. In this case the expression (6.24) for ω_1^2 may be written in the following form

$$\omega_1^2 \approx kg\left[1 + \mathcal{O}\left(\frac{\Delta\rho}{\bar{\rho}_0}\right)\right],\qquad(6.26)$$

which corresponds to the dispersion relation for deep water surface waves with an accuracy to $\mathcal{O}\left(\frac{\Delta\rho}{\bar{\rho}}\right)$. (When $kh \gg 1$ the dispersion relation (6.26) transforms strictly to the relation for surface waves $\omega_1^2 = gk$).

The second model. The mean density $\rho_0(z)$ is distributed over the ocean's depth H following the exponential law $\rho_0(z) = \bar{\rho}_0 \exp(\mathfrak{x}z)$. Then $N^2 = g\mathfrak{x} = N_0^2 = \text{const}$.

The solution of the boundary value problem (6.3), (6.4) for $N_0^2 = \text{const}$ after the Laplace transform yields

$$\chi_i(k,z)$$

$$= \frac{2\,\xi_n k^2 H^2\,(N_0^2 - f^2)\,\sin\left[\xi_n\,(1 - z/H)\right]}{(\xi_n^2 + k^2 H^2)\{(N_0^2 - f^2)H\left[\xi_n \sin\xi_n - \cos\xi_n\right] + g\cos\xi_n[\xi_n^2 + k^2 H^2]\}},$$
$$(6.27)$$

where ξ_n are roots of the equation

$$\tan = \frac{\xi\,(N_0^2 - f^2)\,H}{(k^2 H^2 + \xi^2)\,g},\qquad(6.28)$$

and the dispersion relation is given by

$$\omega_n = \left[\frac{(k^2 H^2 N_0^2 + f^2\xi_n^2)}{(k^2 H^2 + \xi_n^2)}\right]^{1/2}.\qquad(6.29)$$

For $N_0^2 - f^2 > k^2 Hg$, all of the roots of Eq. (6.28) are real, symmetrical with respect to $\xi = 0$, and satisfy the inequality

$$(n+1)\pi > |\xi_n| > \left(n + \frac{1}{2}\right)\pi,\ (n = 1,\ 2,\ \dots).$$

For $kH \gg 1$ and $|\xi_n| < kH$ for $n \gg 1$, we have $|\xi_n| \to \left(n + \frac{1}{2}\right)\pi$; for $|\xi_n| > kH$ and $n \to \infty$, we have $|\xi_n| \to n\pi$. For $kH \ll 1$, $|\xi_n| \to n\pi$

rather rapidly. For $N_0^2 - f^2 \leq k^2 Hg$ two imaginary roots $\pm i\alpha$ appear in the simultaneous equations for $\{\xi_n\}$ studied above instead of the roots $\pm \xi_0$. These roots relate to the surface wave. In this case $\chi_n(k, z)$ for the surface wave is described by expression (6.27) when replacing ξ_0 by $i\alpha$. The condition for the existence of a surface wave (which is frequently satisfied in the ocean for waves of the moderate length) is obviously well behaved: the phase velocity $c_f = \omega_\alpha / k$ of the surface waves should not exceed the long wave phase velocity $\sqrt{gH}$. This follows from Eq. (6.24):

$$ c_f = \frac{\omega_\alpha}{k} < \frac{N_0}{k} \leq \sqrt{gH}. $$

When Eq. (6.23) is satisfied, $\alpha = i\xi_0$ and the surface wave mentioned disappears, turning into an internal wave.

To evaluate the growth of internal waves, we assume for simplification that the atmospheric pressure fluctuations are induced by the harmonic wave

$$ P(\mathbf{x}, t) = P_0 \cos(\mathbf{kx} - \mathbf{kU}t); \quad P_0 \equiv 0 \text{ at } t \leq 0, \qquad (6.30) $$

where $\mathbf{U}$ is the phase velocity of the atmospheric pressure wave. Then $w(\mathbf{x}, t, z) \sim e^{i\mathbf{kx}}$ and Eq. (6.9) for $\mathbf{kU} = \omega_n$ (the resonant case) under the condition $\Delta \omega_n t \gg 1$ yields the linear amplitude growth over time

$$ w(\mathbf{x}, t, z) \sim \tfrac{1}{2} \chi_n(k, z) P_0 t \cos(\mathbf{kx} - \omega_n t). \qquad (6.31) $$

As $w = \partial \xi / \partial t$ in linear internal waves, where ξ is a displacement of a level, then Eq. (6.31) yields the following formula for $\xi(\mathbf{x}, t, z)$ (for $\Delta \omega_n t \gg 1$)

$$ \left. \begin{aligned} \xi(\mathbf{x}, t, z) &\sim A_n(k, z)\, t P_0 \sin(\mathbf{kx} - \omega_n t); \\ A_n &= \frac{\chi_n}{2\omega_n}. \end{aligned} \right\} \qquad (6.32) $$

Fig. 6.2 (curve 4) shows the dependence of the value $h^{1/2} A \left[(\Delta\rho/\bar\rho) g\right]^{-1/2}$ over kh at the depth $z = h$ for the first stratification model (6.21). This value has a maximum when $kh \sim 1$ and then decreases rapidly. Thus, the sharpest internal waves growth occurs at the wave numbers $k \sim h^{-1}$,

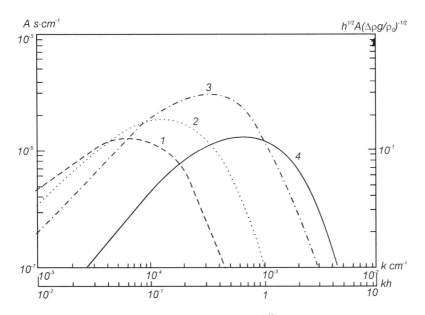

Figure 6.2.   The dependence of $h^{1/2} A \left[\frac{\Delta \rho}{\bar{\rho}} g \right]^{-1/2}$ on kh (4), and A vs. k for different h (1, 2, 3) in the two-layer model: $1-h = 10^4$ cm; $2-h = 0.5 \times 10^4$ cm; $3-0.2 \times 10^4$ cm.

i.e., at wavelengths of the order depth h of the density jump. This is also obvious from the curves presenting the dependence of A over k for different h (Fig. 6.2, *1–3*). The following values of characteristic parameters were given when calculating $A(k)$: $\Delta \rho / \bar{\rho}_0 \sim 10^{-3}$; $g = 10^3$ cm $\times$ s^{-2}; $h = 100, 50, 20$ m, respectively. The values $A(k)$ also have maxima, which shift to the longer waves range and decrease as h increases. Thus the internal waves growth is more intense for lower h.

The term $\varepsilon = (N_0^2 - f^2) H/g$ which is a multiplier in the r.h.s. of Eq. (6.28) is quite small in natural conditions. So for the characteristic values $N_0^2 \approx 10^{-4}$ s^{-2}, $H \approx 10^3$ m, $g = 10^3$ cm $\times$ s^{-2} yield $\varepsilon \approx 10^{-2}$. In this case approximate roots of Eq. (6.28) are

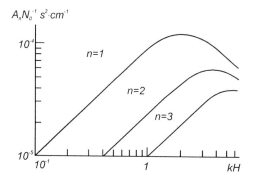

Figure 6.3.   The dependence of A_n on kH for different modes ($N = \text{const} = N_0$).

$$\xi_n \approx n\pi \left\{ 1 + \frac{(N_0^2 - f^2) H}{[\, k^2 H^2 + (n\pi)^2 \,] g} + \mathcal{O}\left(\varepsilon^2\right) \right\} . \qquad (6.33)$$

Then using (6.27) and neglecting in (6.33) the terms of order $\mathcal{O}\left(\varepsilon\right)$ and higher, we obtain the following expression for $|\chi_n\left(k, z\right)|$:

$$|\chi_n\left(k, \; z\right)| \approx \left| \frac{2n\pi k^2 H^2 \left(N_0^2 - f^2\right) \sin\left[\, n\pi \left(1 - z/H\right)\right]}{[\,(n\pi)^2 + k^2 H^2\,]\left\{ \left(N_0^2 - f^2\right) H - g\left[\,(n\pi)^2 + k^2 H^2\,\right]\right\}} \right| . \qquad (6.34)$$

Then taking into account[5] that $N^2 \gg f^2$ and the relative smallness of the first term contained in figure brackets in the denominator of Eq. (6.31) in comparison with the second term, we have

$$|A_n| = \left| \frac{\chi_n}{2\omega_n} \right| \approx \left| \frac{n\pi k H N_0 \sin\left[\, n\pi \left(1 - z/H\right)\right]}{g\left[\,(n\pi)^2 + k^2 H^2\,\right]^{3/2}} \right| . \qquad (6.35)$$

Fig. 6.3 shows the dependence $A_n N_0^{-1}$ over kH for the first three modes for $z = H/2h$. Similarly to the first stratification model A_n has a maximum, which shifts to shorter waves and decreases as the mode number increases.

For rather small values of kH ($kH \leq 5$) A_n decreases as n increases, and for large kH such a decrease is slower (Fig. 6.3). The following asymptotic formula follows from Eq. (6.35):

$$|A_n| \sim \frac{kHN_0}{g\,(n\pi)^2} \sim n^{-2} \quad \text{for} \quad \frac{k^2 H^2}{(2\pi)^2} = \left(\frac{H}{L}\right)^2 \ll \left(\frac{n}{2}\right)^2 , \qquad (6.36)$$

where $L = 2\pi / k$ is the internal wave's length. It follows from formula (6.36) that to generate rather long internal waves of high modes ($L \gg 2H/n$) a greater atmospheric pressure energy is necessary than for the induction of lower modes. In contrast, for very short waves A_n decreases slowly versus the mode number and the generation of higher modes is possible.

To evaluate the time t_M required for considerable growth of the amplitude ξ_M (e.g., $\xi_M \approx 10$ m) assume that the maximum atmospheric pressure change is of 5 to 10 mbar, i.e., $P_0 = (0.5\text{–}1) \times 10^4$ cm$^2 \times$ s^{-2}. Then using (6.32) we have:

$$t_M \approx [\, A_n\, (k,\ z_M)\, (5-10)\,]^{-1}, \qquad (6.37)$$

where z_M is the depth of the maximum amplitude of the internal waves. Then for the first stratification model $h = 50$ m, $\Delta\rho/\bar\rho_0 = 10^3$, $k = 10^{-2}$ m^{-1} yield $t_M \approx (0.63\text{–}1.26) \times 10^5$ s $\approx (17.5\text{–}35)$ h. For $\xi_M = 10$ m and $k = 10^{-2}$ m^{-1} the value of ξk characterizing the wave's nonlinearity power is about 10^{-1}, i.e., within the linear theory's validity. The value of t_M for the second stratification model is of the same order, but for longer waves. For example, $N_0^2 = 10^{-4}$ s^{-1}, $H = 1$ km, $n = 1$, $k = 2$ km^{-1}, $\xi_M = 10$ m yield $t_M \approx (0.84\text{–}1.68) \times 10^5$ s $\approx (23\text{–}46)$ h.

Thus under favorable conditions internal waves generated by atmospheric pressure fluctuations may reach quite high amplitudes over a day. Under unfavorable conditions when the time of their growth increases considerably the suggested mechanism may be sufficient at initial periods of their generation, and then another mechanisms may start, e.g., wave–mean flow interaction [252] or energy transfer to internal waves as a result of their nonlinear interaction with surface waves [34].

2. GENERATION OF INTERNAL WAVES BY WIND FIELD

Some observations (e.g., [7], [57], [67], [71], [72], [106], [131], [303], [304], [340], [341]) shows a close connection between the wind strength in the near water atmospheric layer and internal waves rising in the ocean. When the wind field is spatially inhomogeneous at the lower border of the oceanic Eckman boundary layer the flow velocity component appears, which may induce internal waves in the lower stratified water layers. We consider the lower boundary of the Eckman layer as the lower border of the upper quasi-homogeneous ocean layer. Then the following model may be appropriate for studying the generation of internal waves by wind fluctuations: the density is constant in the upper (Eckman) layer of the thickness h, but the turbulent viscosity is considerable; viscosity in the lower layer is neglected, but the fluid is stratified, i.e., the lower layer's density is a function of depth. A similar model was studied by Krauss [180] and by Miropol'sky [251] in more detail taking into account the arbitrary character of the wind velocity's fluctuations and resonant effects[6].

Consider the free surface of the ocean at rest and rotating starting from the time instant $t = 0$ under frictional stress owed to wind—the vector $\tau(\mathbf{x}, t) = \{\tau_1, \tau_2\}$, where $\mathbf{x} = \{x_1, x_2\}$, and components of the wind frictional stress normalized by the water density in the upper Eckman layer $\bar{\rho} = \text{const}$ having the dimension $\text{cm}^2 \times \text{s}^{-2}$. In the Eckman layer of constant thickness h and density $\bar{\rho}$ the equations of motion and continuity are:

$$\frac{\partial U}{\partial t} + fV = -\frac{\partial}{\partial z}\overline{u'w'}; \qquad \frac{\partial V}{\partial t} - fU = -\frac{\partial}{\partial z}\overline{v'w'}; \;;$$

$$\frac{\partial U}{\partial x_1} + \frac{\partial V}{\partial x_2} + \frac{\partial W}{\partial z} = 0, \tag{6.38}$$

where U, V, W are x_1, x_2 and y components of the drifting flow respectively, $\overline{u'w'}$ and $\overline{v'w'}$ are respective Reynolds stresses (in the right handed coordinate system $\{x_1, x_2, z\}$; the z-axis is directed from the free surface $z = 0$ to the ocean's bottom $z = H$). In Eqs. (6.38) we neglect the free surface inclination, i.e., only the drifting velocity component is taken into account.

Below the Eckman boundary layer ($z > h$) the linearized equation for the vertical velocity component w in internal waves has in the Boussinesq approximation the form (see (2.42)):

$$\frac{\partial^2}{\partial t^2}[\Delta w] + f^2 \frac{\partial^2 w}{\partial z^2} + N^2(z)\Delta_h w = 0. \tag{6.39}$$

As boundary conditions for Eqs. (6.38), (6.39) we assume

$$\text{at} \quad z = 0 \quad -\overline{u'w'} = \tau_1; \quad -\overline{v'w'} = \tau_2; \tag{6.40}$$

$$\text{at} \quad z = h \quad \overline{u'w'} = \overline{v'w'} = 0 \, ; \quad w = W\,(z = h) = W_h \, ; \qquad (6.41)$$

$$\text{at} \quad z = h \quad w = 0 \, . \qquad (6.42)$$

The boundary conditions (6.40), (6.42) are obvious and the boundary conditions (6.41) show that the level $z = h$ is the interface between the viscous and inviscid fluid. That is why the vertical velocities for $z = h$ are continuous and the horizontal velocity components may have a discontinuity. We assume further that the fluid is at rest for $t \leq 0$, i.e.,

$$w\,|_{t=0} = \frac{\partial w}{\partial t}\bigg|_{t=0} = 0 \, . \qquad (6.43)$$

We integrate Eqs. (6.38) over z between 0 and h (taking into account (6.40), (6.41)) and solve the obtained equations with respect to W_h:

$$W_h\,(\mathbf{x}, t) = -\int\limits_{0}^{t} \left\{ \left[\frac{\partial \tau_1\,(t')}{\partial x_1} + \frac{\partial \tau_2\,(t')}{\partial x_2} \right] \cos f\,(t - t') \right.$$

$$\left. + \left[\frac{\partial \tau_1\,(t')}{\partial x_2} - \frac{\partial \tau_2\,(t')}{\partial x_1} \right] \sin f\,(t - t') \right\} dt' \, . \qquad (6.44)$$

Thus the problem reduces to the solution of Eq. (6.39) with the boundary conditions (6.41), (6.42), where W_h is given by (6.44). If $\tau_1\,(\mathbf{x}, t)$ and $\tau_2\,(\mathbf{x}, t)$ are homogeneous random functions with a zero assembly average, then τ_1, τ_2, w and W_h may be presented in the form of Fourier integrals

$$\{\tau_1\,(\mathbf{x}, t), \tau_2\,(\mathbf{x}, t), w\,(\mathbf{x}, t, z), W_h\,(\mathbf{x}, t)\}$$

$$= \int e^{i\mathbf{k}\mathbf{x}} \{\hat{\tau}_1, \hat{\tau}_2, \hat{w}, W_h\}\, d\mathbf{k} \, , \qquad (6.45)$$

where $\hat{\varphi}$ are Fourier components of according functions; integrals (6.45) have infinite limits. Substituting (6.45) into (6.39), (6.41), (6.42) and assuming

$$\hat{w} = \Phi + \frac{H - z}{H - h} W_h$$

yields the inhomogeneous equation with homogeneous boundary conditions

$$\frac{\partial^2}{\partial t^2}\left[\Phi'' - k^2\Phi\right] + f^2\Phi'' - N^2(z)k^2\Phi$$

$$= k^2\left(\frac{H-z}{H-h}\right)\left\{\frac{\partial^2}{\partial t^2}\hat{W}_h + N^2(z)\hat{W}_h\right\}; \qquad (6.46)$$

$$\Phi\big|_{z=h} = \Phi\big|_{z=H} = 0; \quad \Phi\big|_{t=0} = \frac{\partial\Phi}{\partial t}\bigg|_{t=0} = 0. \qquad (6.47)$$

Here a prime means derivative with respect to z. We consider the homogeneous equation (6.46) with the solution assumed in the form

$$\Phi(z,\,t) = \varphi(z)\exp\{-i\omega t\}.$$

This yields for $\varphi(z)$:

$$\varphi'' + k^2\frac{N^2(z) - \omega^2}{\omega^2 - f^2}\varphi = 0; \quad \varphi\big|_{z=h} = \varphi\big|_{z=H} = 0. \qquad (6.48)$$

The boundary value problem (6.48) yields a number of eigenvalues $\omega_n = \omega_n(k, N^2, f^2)$ determining dispersion relations and a number of eigenfunctions $\varphi_n(k, z)$, (see Section 3.1). So the system $\{\varphi_n\}$ is orthogonal and complete when $N(z)$ is nonzero. Under this assumption we expand the functions of z in the r.h.s. of Eq. (6.46) in normalized functions:

$$\eta = \frac{H-z}{H-h} = \sum_{n=1}^{\infty}\alpha_n\varphi_n(\eta); \quad \frac{H-z}{H-h}N^2(z) = \sum_{n=1}^{\infty}\beta_n\varphi_n(\eta);$$

$$\alpha_n = \int_0^1\eta\,\varphi_n(\eta)\,d\eta; \quad \beta_n = \int_0^1 N^2(\eta)\,\eta\,\varphi_n(\eta)\,d\eta. \qquad (6.49)$$

Here $\eta = (H-z)(H-h)^{-1}$. Then presenting $\Phi(z,t)$ in the form

$$\Phi(\eta,t) = \sum_{n=1}^{\infty}\chi_n(t)\,\varphi_n(\eta), \qquad (6.50)$$

substituting Eqs. (6.49), (6.50) into (6.46), (6.47) and using Eq. (6.50) and the orthogonality condition (3.16) yields

$$\frac{\partial^2\chi_n}{\partial t^2} + \omega_n^2\chi_n = -\left[\frac{\omega_n^2 - f^2}{N^2 - f^2}\right]\left\{\alpha_n\frac{\partial^2}{\partial t^2}\hat{W}_h + \beta_n\hat{W}_h\right\};$$

$$\chi_n \big|_{t=0} = \frac{\partial \chi_n}{\partial t} \bigg|_{t=0} = 0. \qquad (6.51)$$

Solving Eq. (6.51), using Eq. (6.50) and replacing Φ by $\hat{W}$ we obtain

$$\hat{w}\,(\mathbf{k}, t, z)$$

$$= \sum_{n=1}^{\infty} \left\{ \left[\frac{(\omega_n^2 - f^2)(\alpha_n \omega_n^2 - \beta_n)}{(N^2 - f^2)\,\omega_n} \right] \varphi_n(\eta) \int_0^t \hat{W}_h(\mathbf{k}, t')\sin\omega_n(t - t')dt' \right\}$$

$$+ \left[\eta - \sum_{n=1}^{\infty} \alpha_n \varphi_n(\eta) \frac{(\omega_n^2 - f^2)}{(N^2 - f^2)} \right] \hat{W}_h\,(\mathbf{k}, t).$$

$$(6.52)$$

Using (6.45) we have from (6.44)

$$\hat{W}_h\,(\mathbf{k}, t) = -i \int_0^t \left\{ Q_1\,(t')\cos f\,(t - t') + Q_2\,(t')\sin f\,(t - t') \right\} dt', \quad (6.53)$$

where

$$Q_1 = \{ k_1 \hat{\tau}_1 + k_2 \hat{\tau}_2 \}; \quad Q_2 = \{ k_2 \hat{\tau}_1 - k_1 \hat{\tau}_2 \}.$$

Substituting (6.53) into (6.52) and using the following formula[7]

$$J = \int_0^t \hat{W}_h\,(\mathbf{k}, t')\sin\omega_n\,(t - t')\,dt' = \frac{\omega_n}{\omega_n^2 - f^2}\,\hat{W}_h$$

$$+ i \left\{ \int_0^t \left[\frac{\omega_n Q_1(t')}{\omega_n^2 - f^2}\cos\omega_n\,(t - t') + \frac{f Q_2(t')}{\omega_n^2 - f^2}\sin\omega_n(t - t') \right] dt' \right\},$$

$$(6.54)$$

we have the following expression for $\hat{w}\,(\mathbf{k}, t, z)$

$$\hat{w}\,(\mathbf{k}, t, z)$$

$$= i \sum_{n=1}^{\infty} A_n \left\{ \int_0^t Q_1\,(t')\cos\omega_n\,(t - t') + \frac{f}{\omega_n} Q_2\,(t')\sin\omega_n\,(t - t')\,dt' \right\}$$

$$= i \sum_{n=1}^{\infty} A_n \int_0^t \{ B_n^{(1)} \hat{\tau}_1\,(\mathbf{k}, t')\sin[\omega_n\,(t - t') + \ae_n^{(1)}]$$

$$+\{B_n^{(2)}\hat{\tau}_2\,(\mathbf{k},\ t')\,\sin\left[\,\omega_n\,(t-t')+\mathit{æ}_n^{(2)}\,\right]\}\,dt'\,,\qquad\qquad(6.55)$$

where

$$A_n=\frac{\alpha_n\omega_n^2-\beta_n}{(N^2-f^2)}\,\varphi_n\,(\eta)\,;\qquad B_n^{(1)}=\frac{1}{\omega_n}\,\sqrt{f^2k_2^2+\omega_n^2k_1^2}\,;$$

$$B_n^{(2)}=\frac{1}{\omega_n}\,\sqrt{f^2k_1^2+\omega_n^2k_2^2}\,;\qquad\qquad(6.56)$$

$$\mathit{æ}_n^{(1)}=\arctan\left[\frac{k_1\omega_n}{k_2f}\right]\,;\qquad\mathit{æ}_n^{(2)}=\arctan\left[-\frac{k_2\omega_n}{k_1f}\right]\,.$$

Expression (6.55) shows that in spite of modulation of the disturbing force $\widehat{W}_h$ by the inertial frequency f (see (6.53)), oscillations $\hat{w}$ occur at internal waves' eigenfrequencies ω_n with a phase shift relative to the frictional stress caused by wind which is determined by correlation between longitudinal and transverse scales of the wind field's irregularity and by the ratio ω_n/f. Oscillations of horizontal components of the velocity of internal waves, expressions of which may be easily derived from (6.55), have another nature. These oscillations represent a superposition of the inertial frequency and ω_n. Thus inertial oscillations generated by the wind in the boundary layer induce in the lower water layers quasi-horizontal waves of frequency f as well as three-dimensional internal waves oscillating at their eigenfrequencies under the disturbing force action.

Now we derive some of the statistical characteristics of internal waves using expression (6.55) valid for arbitrary ocean stratification. We introduce a spectral tensor for the frictional stress caused by wind in the space of wave numbers $\mathbf{k}$ with time shift t

$$\Omega_{ij}\,(\mathbf{k},t_0,t)\,\delta\,(\mathbf{k}-\mathbf{k}')=\langle\hat{\tau}_1\,(\mathbf{k},t_0)\,\hat{\tau}_2^*\,(\mathbf{k},t_0+t)\rangle\,;\quad i,j=1,\,2.\quad(6.57)$$

For $i=j$ components of the tensor Ω_{ij} represent spectral functions of each component of the frictional stress which are independent of the initial time instant t_0 under the assumption of a stationary wind field, and for which $\Omega_{ii}\,(\mathbf{k},t)=\Omega_{ii}\,(\mathbf{k},-t)$. For $i\neq j$ the components of Ω_{ij} represent the cross spectral function of the components of the frictional stress caused by wind, which are complex in general. When components of the frictional stress caused by wind are stationary too, the components of Ω_{ij} are also independent of t_0 and $\Omega_{ij}(\mathbf{k}, t) = \Omega_{ij}(\mathbf{k}, -t)$. Introducing now a single-time spectrum of vertical velocities of internal waves in the space of wave numbers $S(\mathbf{k}, t, z)$ in the form

$$S(\mathbf{k}, t, z) \, \delta(\mathbf{k} - \mathbf{k}') = \langle \hat{w}(\mathbf{k}, t, z) \, \hat{w}^*(\mathbf{k}', t, z) \rangle \qquad (6.58)$$

and using expression (6.55) we obtain the relation connecting $S(\mathbf{k}, t, z)$ and $\Omega_{ij}(\mathbf{k}, t, z)$:

$$S(\mathbf{k}, t, z) = \sum_{n=1}^{\infty} \sum_{m=1}^{\infty} \left\{ A_n A_m \sum_{i,j}^{2} B_n^{(i)} B_n^{(j)} J_{nm}^{(i,j)} \right\}, \qquad (6.59)$$

where

$$J_{nm}^{(i,j)} = \int_0^t \int_0^t \Omega_{ij}(\mathbf{k}, t', t' - t'') \sin\left[\omega_n(t - t') + æ_n^{(i)}\right]$$

$$\times \sin\left[\omega_m(t - t') + æ_m^{(j)}\right] dt' \, dt''. \qquad (6.60)$$

If the field of the frictional stress caused by wind is statistically stationary and its components are stationary as well, then for $n \neq m$ all integrals $J_{nm}^{(i,j)}$ will be bounded oscillating functions of time t. For $n = m$

$$J_{nn}^{(i,j)} = \frac{1}{2} \int_0^t \int_0^t \Omega_{ij}(\mathbf{k}, t' - t'')$$

$$\times \left\{ \cos\left[\omega_n(t' - t'') - (æ_n^{(i)} - æ_n^{(j)})\right] \right.$$

$$\left. - \cos\left[2\omega_n t - \omega_n(t' + t'') + (æ_n^{(i)} + æ_n^{(j)})\right] \right\} dt' \, dt''. \qquad (6.61)$$

Replacing $t' - t'' = \xi$; $t' + t'' = \xi'$ and integrating once we obtain the following expression for $J_{nn}^{(i,j)}$ valid for $\Delta\omega_n t = |\omega_n - \omega_{n+1}| t \sim \omega_n t \gg 1$:

$$J_{nn}^{(i,i)} \approx \int_0^t [t - \xi] \, \Omega_{ii}(\mathbf{k}, \xi) \cos \omega_n \xi \, d\xi + \int_{-t}^0 [t - \xi] \, \Omega_{ii}(\mathbf{k}_1 - \xi) \cos \omega_n \xi \, d\xi.$$

$$(6.62)$$

For nondiagonal terms in $J_{nn}^{(i,j)}$, taking into account properties of the tensor Ω_{ij} for $\Delta\omega_n t \sim \omega_n t \gg 1$ yields:

$$J_{nn}^{(1,2)} + J_{nn}^{(2,1)} \approx \cos\left[\,\ae_n^{(1)} - \ae_n^{(2)}\,\right] \int_0^t [\,t - \xi\,]\,[\,\Omega_{12}\,(\xi) + \Omega_{21}\,(\xi)\,]\,\cos\,\omega_n\xi\,d\xi\,.$$

$$(6.63)$$

Defining the space–time density of the frictional stress components $E_{ii}\,(\mathbf{k},\omega)$ and the space–time co-spectrum of the components $P_{12}\,(\mathbf{k},\omega)$ by

$$E_{ii}\,(\mathbf{k},\omega) = \frac{1}{2\pi} \int_0^\infty \Omega_{ii}\,(\mathbf{k},t)\,\cos\,\omega t\,dt\,;$$

$$P_{12}\,(\mathbf{k},\omega) = \frac{1}{2\pi} \int_0^\infty [\,\Omega_{12}\,(t) + \Omega_{21}\,(t)\,]\,\cos\,\omega t\,dt \qquad (6.64)$$

yields from (6.59), for $t \to \infty$,[8]

$$S\,(\mathbf{k},t,z) \approx 2\pi t \sum_{n=1}^\infty A_n^2\,\{\,[\,B_n^{(1)}\,]^2\,E_{11}\,(\mathbf{k},\omega) + [\,B_n^{(2)}\,]\,E_{22}\,(\mathbf{k},\omega)$$

$$+ \cos\left[\,\ae_n^{(1)} - \ae_n^{(2)}\,\right][\,B_n^{(1)}\,B_n^{(2)}\,]\,P_{12}\,(\mathbf{k},\,\omega)\,\}|_{\omega=\pm\,\omega_n\,(k)}. \qquad (6.65)$$

It follows from Eq. (6.65) that resonant (linear in time t) growth of the spectrum of internal waves occurs only when frequencies and wave numbers in space–time spectra of the wind frictional stress components correspond to dispersion relations of free internal wave modes. Furthermore, the spectral density of the vertical velocities of internal wave is formed both by auto-spectra of the frictional stress components and by their co-spectrum with weight functions depending on the correlation of longitudinal and transverse scales of the wind field's irregularity taken for a fixed wave number and ω_n/f ratio. Furthermore, different internal wave modes do not correlate with each other for large enough t.

Owing to the absence of experimental data about the space–time spectrum of fluctuations of the frictional stress owed to wind a general experimental evaluation of Eq. (6.65) is difficult. But the frequency spectrum of the wind velocity (the wind stress τ has a square dependence upon the wind velocity U: $\tau \sim U^2$) has its maximum in the range of the mesome-teorological minimum at periods of about ten hours, which corresponds to the time range of internal wave frequencies, [220]. So to estimate the velocity of the growth of internal waves we assume that fluctuations of the frictional stress caused by wind have a single harmonics with a period of about ten hours, i.e.,

$$\tau_1(\mathbf{x}, t) = \tau(\mathbf{x}, t) \cos \vartheta; \quad \tau_2(\mathbf{x}, t) = \tau(\mathbf{x}, t) \sin \vartheta, \tag{6.66}$$

where

$$\tau(\mathbf{x}, t) = J_m\{\tau_0 \exp[i(\mathbf{kx} - \sigma t)]\} = \tau_0 \sin(\mathbf{kx} - \sigma t). \tag{6.67}$$

Here ϑ is the angle included between the vector of the frictional stress owed to wind and the x_1-axis; σ is the frequency, and $\tau_0 = $ const the amplitude of oscillations of the frictional stress caused by wind. Substituting Eq. (6.66), (6.67) into (6.65) and taking into account that $w(\mathbf{x}, t, z) \sim \exp\{i\mathbf{kx}\}$ we obtain for $\sigma \approx \omega_n$ (resonance) and $\omega_n t \approx \sigma t \gg 1$:

$$w(\mathbf{x}, t, z) \approx t\,\frac{A_n \tau_0 k}{2}\,B_n \sin[\mathbf{kx} - \omega_n t + \mathscr{x}_n], \tag{6.68}$$

where

$$B_n = \left[\cos^2 \tilde{\vartheta} + \frac{f^2}{\omega_n^2}\sin^2 \tilde{\vartheta}\right]^{1/2}; \quad \mathscr{x}_n = \arctan\left\{\frac{\omega_n}{f}\operatorname{ctg}\tilde{\vartheta}.\right\} \tag{6.69}$$

Here $\tilde{\vartheta} = \vartheta - \vartheta'$ is the angle included between the vector of the frictional stress caused by wind and the direction of wave's propagation ($k_1 = k\cos\vartheta'$; $k_2 = k\sin\vartheta'$; $k = |\mathbf{k}|$). Using $w = \partial\xi/\partial t$, which is valid for the linear theory ($\xi = 0$ for $t \leq 0$), and (6.68) yields the expression for a level displacement in an internal wave

$$\xi \approx t\,\tau_0 \mathcal{P}_n \cos[\mathbf{kx} - \omega_n t + \mathscr{x}_n]; \quad \mathcal{P}_n = \frac{A_n B_n k}{2\omega_n}. \tag{6.70}$$

To calculate $\mathcal{P}_n$ it is necessary to give the distribution model of $N^2(z)$. Assume that $N^2 = $ const, i.e., the density ρ_0 has exponential distribution below the quasi-homogeneous layer ($z \geq h$), which is close to the natural density distribution. The solution of the boundary value problem (6.48) gives, for $N^2 = \text{const}$, the normalized eigenfunctions and the dispersion relation:

$$\varphi_n(\eta) = \sqrt{2}\sin n\pi\eta; \quad \omega_n^2 = \left[k^2\tilde{H}^2 N^2 + (n\pi)^2 f^2\right]\left[k^2\tilde{H}^2 + (n\pi)^2\right]^{-1},$$

(6.71)

where $\tilde{H} = H - h$. Calculating α_n and $\beta_n = \alpha_n\,[N^2]^{-1}$ we obtain the expression for the amplification coefficient $\mathcal{P}_n$:

$$\mathcal{P}_n(k) = (-1)^n\,\sqrt{2}n\pi$$

$$\times\left\{\frac{[k^2\tilde{H}^2 + (n\pi)^2]\,f^2\,\sin^2\tilde{\vartheta} + [k^2\tilde{H}^2 N^2 + (n\pi)^2\,f^2]\cos^2\tilde{\vartheta}}{[k^2\tilde{H}^2 N^2 + (n\pi)^2][k^2\tilde{H}^2 + (n\pi)^2]}\right\}\sin n\pi\eta;$$

$$\mathcal{P}_n(\omega_n) = (-1)^n\,\frac{\sqrt{2}\,[(N^2 - \omega_n^2)(\omega_n^2 - f^2)]^{1/2}}{\tilde{H}\,(N^2 - f^2)\,\omega_n^2}$$

$$\times[f^2\sin^2\tilde{\vartheta} + \omega_n^2\cos^2\tilde{\vartheta}]^{1/2}\sin n\pi\eta. \quad (6.72)$$

These dimensionless variables will be convenient

$$\tilde{\mathcal{P}} = |\mathcal{P}_n|\tilde{H}N; \quad \tilde{k} = \frac{k\tilde{H}}{n\pi}; \quad \tilde{\omega} = \frac{\omega}{N};$$

$$\tilde{f} = \frac{f}{N}; \quad \tilde{\eta} = n\pi\eta. \quad (6.73)$$

Fig. 6.4 shows the dependence $\tilde{\mathcal{P}}(\tilde{\omega})$ for different values of $\tilde{\vartheta}$ ($\tilde{\vartheta} = 0$; $\pi/4$; $\pi/2$) and $\tilde{f}$ ($\tilde{f} = 10^{-2}$; 10^{-1}, which for the characteristic value $f \approx 10^{-4}$ s^{-1} for moderate latitudes corresponds to the values $N = 10^{-2}$; 10^{-3} s^{-1}). The amplification coefficient $\tilde{\mathcal{P}}$ is initially growing versus $\tilde{\omega}$ (from zero for $\tilde{\omega} = \tilde{f}$) and reaches its maximum value $\tilde{\mathcal{P}}_{\max} = \sqrt{2}\,(1+\tilde{f})^{-1}$ in the point $\tilde{\omega} = \tilde{f}^{1/2}$ (for $\tilde{\vartheta} = 0$), and then is decreasing up to zero for $\tilde{\omega} = 1$, i.e., for $\omega = N$. When $\tilde{f}$ grows (i.e., N decreases for constant f) the maximum shifts to the right along the frequency axis and its absolute value slightly decreases. When the angle $\tilde{\vartheta}$ between wind and wave directions grows $\tilde{\mathcal{P}}(\tilde{\omega})$ decreases from its maximum (for

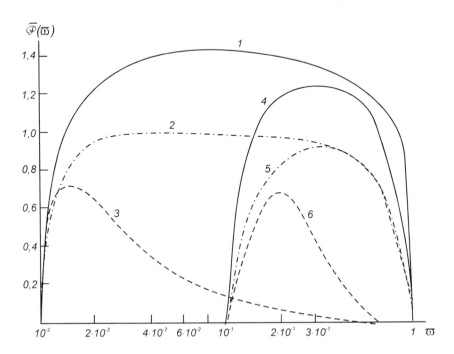

Figure 6.4. The dependence of $\widetilde{\mathcal{P}}$ on $\tilde{\omega}$: $1, 2, 3$ — at $\tilde{f} = 10^{-2}$, $\tilde{\vartheta} = (0; \pi/4; \pi/2)$; $4, 5, 6$ — at $\tilde{f} = 10^{-1}$, $\tilde{\vartheta} = (0; \pi/4; \pi/2)$.

$\tilde{\vartheta} = 0$) to the minimum (for $\tilde{\vartheta} = \pi/2$), and furthermore $\mathcal{P}_{\max}$ is shifted to the low frequency region. Thus the most intense generation of internal waves may occur when fluctuations of the frictional stress owed to wind have frequencies close to the value $\omega \approx N^{1/2} f^{1/2}$ and the frictional stress wave has the same direction as the mean wind. As we consider the resonant case ($\sigma \approx \omega_n$) the wave number modulus of the disturbing force should satisfy the equality $k \approx n\pi f^{1/2} N^{-1/2} \widetilde{H}^{-1}$ following from the dispersion relation (6.71). Under the resonance condition taking place at other frequencies and for other $\tilde{\vartheta}$ values the amplification coefficient $\widetilde{\mathcal{P}}$ decreases.

Variations of $\widetilde{\mathcal{P}}$ versus $\tilde{k}$ qualitatively coincide with behaviour of the function $\widetilde{\mathcal{P}}(\tilde{\omega})$. The function $\widetilde{\mathcal{P}}(\tilde{k})$ grows from zero for $\tilde{k} = 0$ up to its maximum for $\tilde{k} = \tilde{f}^{1/2}$ and then asymptotically goes to zero when $\tilde{k} \to \infty$. When ϑ is growing the function $\widehat{\mathcal{P}}(\tilde{k})$ is decreasing. When the mode number n grows the function $\mathcal{P}(k)$ (the dimensional coefficient of wave amplification) decreases, and furthermore for $n \gg 1$ the function $\mathcal{P}(k) \sim n^{-1}$. The variable æ_n characterizing the phase shift between the wave of the frictional stress owed to wind and the internal wave depends on ω only when $\tilde{\vartheta} \neq 0, \pi/2$ ($\text{æ}_n = \pi/2$, 0 for $\tilde{\vartheta} = 0, \pi/2$) and is steadily increasing versus growing ω/f.

To estimate the time t_M required for considerable growth of internal waves amplitude ξ_M (e.g., $\xi_M = 10$ m) we assume that resonance occurs at the frequency $\omega \approx 1.8 \times 10^{-4}$ s^{-1}. This frequency corresponds to the period of about 10 h. The atmospheric pressure spectrum has its maximum at the same period [220]. Therefore the maximum in the τ spectrum also appears here. As the characteristic ocean parameters we assume, further, the following values of variables contained in (6.70): $\tilde{H} = H - h \approx 3 \times 10^5$ cm, $N = 10^{-3}$ s^{-1}; $f = 10^{-4}$ s^{-1}, $\tau_0 = 1$ cm$^2 \times$ s^{-2} (this value of τ_0 corresponds to a wind velocity $U \approx 10$ m/s) and $\tilde{\vartheta} = 0$. Then (6.70) yields that for $\tilde{\eta} = \pi/2$ (the depth where the internal waves' amplitude is maximum) the value of $\xi_M = 10$ m will be reached during the time $t_M \approx 2.6 \times 10^5$ s, i.e., during about three days or about ten wave periods (t_M is estimated by the formula $t_M \approx \xi_M \tilde{H} N \times [\tau_0 \mathcal{P}]^{-1}$ following from (6.70)). From this estimation it follows that the generation of internal waves by the wind's fluctuation may be quite intense. The estimation presented above is somewhat approximate because it depends on the variation of numerous parameters. Moreover, the values of t_M may vary for other profiles of $N(z)$. But when the ocean is stratified in a rather narrow layer close to the interface $z = h$ the time of generation of internal waves may be only less than the one estimated[9].

3.   GENERATION OF INTERNAL WAVES BY LOCALIZED DISTURBANCES

Internal waves in the ocean may also be generated by localized initial disturbances. Besides atmospheric sources, such initial disturbances are deformations of the density field, turbulent patches induced by different sources, etc.. So we study propagation of internal waves induced by an initial impulse in the rotating ocean of finite depth. This linear problem may be described under the Boussinesq approximation by a single equation for the vertical velocity w in a wave:

$$\frac{\partial^2}{\partial t^2}[\Delta w] + f^2 \frac{\partial^2 w}{\partial z^2} + N^2(z)\,\Delta_h w = 0\,. \qquad (6.74)$$

We take the bottom 'no penetration' conditions and zero vertical velocity at the free surface (the 'rigid lid' condition) as the boundary conditions for Eq. (6.74), i.e.,

$$w\,|_{z=0} = w\,|_{z=-H} = 0\,. \qquad (6.75)$$

The Cauchy problem for Eq. (6.74) is formulated as

$$w\,|_{t=0} = F_0\,(\mathbf{x}, z)\,; \qquad \left.\frac{\partial w}{\partial t}\right|_{t=0} = F_1\,(\mathbf{x}, z)\,. \qquad (6.76)$$

The solution of the problem (6.74)–(6.76) was obtained by Miropol'sky [253] by the Fourier transform. We follow this example in our further considerations.

The fundamental solutions of Eq. (6.74) with the boundary condition (6.75) have the form of plane waves

$$w\,(\mathbf{x}, z, t) = \Phi\,(\mathbf{k}, z)\,\exp\{i\,[\mathbf{k}\mathbf{x} - \omega\,(k)\,t]\}\,, \qquad (6.77)$$

where $\Phi\,(\mathbf{k}, z)$ and the dispersion law $\omega = \omega\,(k)$ are obtained from the Sturm–Liouville boundary value problem

$$\Phi'' + k^2 \frac{N^2(z) - \omega^2}{\omega^2 - f^2}\,\Phi = 0\,; \qquad \Phi|_{z=0} = \Phi|_{z=-H} = 0\,. \qquad (6.78)$$

The boundary value problem (6.78) for $f^2 < \omega^2 < N^2$ (the usual range for internal waves) yields a denumerable set of eigenfunctions $\{\varphi_n\,(z, k)\}$ and its eigenvalues define a denumerable set of dispersion relations $\omega = \pm\omega_n\,(k)$. The general solution of (6.74) may be written in the form

$$\Phi\,(k, z) = \sum_{n=1}^{\infty}\{A_n\,(\mathbf{k}) + B_n\,(\mathbf{k})\}\,\varphi_n\,(z, k)\,, \qquad (6.79)$$

where the coefficients $A_n(\mathbf{k})$ and $B_n(\mathbf{k})$ are defined with the conditions (6.76) and will be calculated below. The general solution of Eq. (6.74) may be written in the form

$$w(\mathbf{x}, z, t) = \sum_{n=1}^{\infty} \left\{ \int\!\!\int_{-\infty}^{\infty} A_n(\mathbf{k})\, \varphi_n(z, k)\, \exp\{i\,[\mathbf{kx} - \omega_n(k)\, t]\}\, d\mathbf{k} \right.$$

$$\left. + \int\!\!\int_{-\infty}^{\infty} B_n(\mathbf{k})\, \varphi_n(z, k)\, \exp\{i\,[\mathbf{kx} + \omega_n(k)\, t]\}\, d\mathbf{k} \right\}.$$

(6.80)

As the function $\omega_n(k)$, is even with respect to k the waves propagating in right and left directions are contained in both terms of Eq. (6.80). Assuming $t = 0$ in Eq. (6.80) and in its time derivative and using the initial conditions (6.76) yield

$$F_0(\mathbf{x}, z) = \sum_{n=1}^{\infty} \left\{ \int\!\!\int_{-\infty}^{\infty} [A_n(\mathbf{k}) + B_n(\mathbf{k})]\, e^{i\mathbf{kx}}\, \varphi_n(z, k)\, d\mathbf{k} \right\};$$ (6.81)

$$F_1(\mathbf{x}, z) = -i \sum_{n=1}^{\infty} \left\{ \int\!\!\int_{-\infty}^{\infty} [A_n(\mathbf{k}) - B_n(\mathbf{k})]\, \omega_n(k)\, e^{i\mathbf{kx}}\, \varphi_n(z, k)\, d\mathbf{k} \right\}.$$

(6.82)

Formulae for the Fourier transform inversion yield

$$\frac{1}{4\pi^2} \int\!\!\int_{-\infty}^{\infty} F_0(\mathbf{x}, z)\, e^{-i\mathbf{kx}}\, d\mathbf{x} = \Phi_0(\mathbf{k}, z)$$

$$= \sum_{n=1}^{\infty} [A_n(\mathbf{k}) + B_n(\mathbf{k})]\, \varphi_n(z, k)\,;$$ (6.83)

$$\frac{1}{4\pi^2} \int\!\!\int_{-\infty}^{\infty} F_1(\mathbf{x}, z)\, e^{-i\mathbf{kx}}\, d\mathbf{x} = \Phi_1(\mathbf{k}, z)$$

$$= \sum_{n=1}^{\infty} [A_n(\mathbf{k}) - B_n(\mathbf{k})]\, \omega_n(k)\, \varphi_n(z, k).$$

(6.84)

Hence using the orthogonality condition (3.16),

$$\int_{-H}^{0} [N^2(z) - f^2]\varphi_n(z,k)\varphi_m(z,k)\,dz = \delta_{nm},$$

where $\varphi_{m,n}$ are normalized eigenfunctions [see (3.16) and (3.17)], we obtain for $A_n(\mathbf{k})$ and $B_n(\mathbf{k})$:

$$A_n(\mathbf{k}) = \frac{1}{2}\left[\overline{\Phi}_0 + \frac{i\overline{\Phi}_1}{\omega_n}\right]; \qquad B_n(\mathbf{k}) = \frac{1}{2}\left[\overline{\Phi}_0 - \frac{i\overline{\Phi}_1}{\omega_n}\right], \tag{6.85}$$

where

$$\overline{\Phi}_0 = \int_{-H}^{0} \Phi_0(\mathbf{k},z)[N^2(z) - f^2]\varphi_n\,dz;$$

$$\overline{\Phi}_1 = \int_{-H}^{0} \Phi_1(\mathbf{k},z)[N^2(z) - f^2]\varphi_n\,dz. \tag{6.86}$$

Thus the coefficients $A_n(\mathbf{k})$ and $B_n(\mathbf{k})$ are defined by the expansion of Fourier components of the initial disturbance in simultaneous functions $\{\varphi_n(z,k)\}$. As the function $F_0(\mathbf{x},z)$ and $F_1(\mathbf{x},z)$ are real, then $\Phi_0(-\mathbf{k},z) = \Phi_0^*(\mathbf{k},z)$, $\Phi_1(-\mathbf{k},z) = \Phi_1^*(\mathbf{k},z)$. Then due to evenness of the function $\omega = \omega_n(k)$ we have

$$A_n(-\mathbf{k}) = A_n^*(-\mathbf{k}); \qquad B_n(-\mathbf{k}) = B_n^*(-\mathbf{k}) \tag{6.87}$$

and the solution (6.80) is real.

Thus the integral version of (6.80) describes propagation of an arbitrary initial disturbance in stably stratified rotating fluid. To describe transformation of the disturbance during its propagation it is necessary to integrate expression (6.80). Unfortunately an exact method is absent and approximate methods are usually used.

Following [253] we consider the typical integral

$$w(\mathbf{x},z,t) = \sum_{n=1}^{\infty} \int\int_{-\infty}^{\infty} \exp\{i[\mathbf{kx} - \omega_n(k)t]A_n(\mathbf{k})\varphi_n(z,k)\,d\mathbf{k}\} \tag{6.88}$$

in one-dimensional and two-dimensional axisymmetric cases. The form
(6.88) is an exact solution of the Cauchy problem for $B_n \equiv 0$, i.e.,
$\Phi_0 = i\omega_n\Phi_1$. This corresponds to the 'self-consistent' initial conditions
when the single arbitrary function $F_0\,(\mathbf{x}, z)$ is given for $t = 0$ and the
function $\left.\dfrac{\partial w}{\partial t}\right|_{t=0}$ is obtained by time derivation of the general solution
(6.80). For plane (one-dimensional) waves $x_2 = 0$, $k_2 = 0$. Then (6.88)
will have the form

$$w\,(\mathbf{x}, z, t) = \sum_{n=1}^{\infty} \int_{-\infty}^{\infty} \exp\left\{\,i\,[\,kx - \omega_n\,(k)\,t\,]\,\right\} A_n\,(k)\,\varphi_n\,(z, k)\,dk\,, \quad (6.89)$$

where $x = x_1$, $k = k_1$. In the axisymmetric case it is convenient to operate with cylindrical coordinates in the physical $\{x_1, x_2\} \to \{r, \theta\}$ and
wave $\{k_1, k_2\} \to \{k, \vartheta\}$ spaces. If the initial disturbance F_0 is indepen-
dent of θ, i.e., $w|_{t=0} = F_0\,(r, z)$ and $A_n\,(\mathbf{k}) = A_n\,(k)$, then Eq. (6.88) in
cylindrical coordinates and taking into account the relation

$$\int_0^{\infty} \exp\left\{\,ikr\,\cos\,(\theta - \vartheta)\,\right\} d\vartheta = 2\pi\,J_0\,(kr)$$

yields

$$w\,(r, z, t) = 2\pi \sum_{n=1}^{\infty} \int_0^{\infty} A_n\,(k)\,\varphi_n\,(z, k)\,k\,J_0\,(kr)\,\exp\left\{\,-i\omega_n\,(k)\,t\,\right\} dk\,,$$

$$(6.90)$$

where the $A_n\,(k)$ are expressed through $\hat{A}_n\,(r)$ coefficients of $w\,(r, z, 0)$
expansion in simultaneous functions $\{\varphi_n\,(z, k)\}$ ($w\,(r, z, 0) = \sum_{n=1}^{\infty}\,[\,\tilde{A}_n\,(r)\times$
$\times\varphi_n\,(z)\,)\,]$ by the Fourier–Bessel formula

$$A_n\,(k) = \frac{1}{2\pi} \int_0^{\infty} r\,\hat{A}_n\,(r)\,J_0\,(kr)\,dr\,. \quad (6.91)$$

In Eqs. (6.90), (6.91) $J_0\,(x)$ is the first-genus zero-order Bessel function.
For large r the asymptotical form of $J_0\,(kr)$ may be used

$$\left| J_0\,(kr) - \sqrt{2}\,(\pi kr)^{-1/2}\,\cos\left(kr - \frac{\pi}{4}\right)\right| < 0\,(\,[\,kr\,]^{-3/2})\,.$$

Then (6.90) becomes

$$w\left(r, z, t\right)$$

$$= \sqrt{\tfrac{\pi}{r}} \sum_{n=1}^{\infty} \int_{0}^{\infty} A_n\left(k\right) \varphi_n\left(z, k\right) k^{1/2} \tag{6.92}$$

$$\times \left\{ \exp\left[i\left(kr - \omega_n t - \tfrac{\pi}{4}\right)\right] + \exp\left[-i\left(kr + \omega_n t - \tfrac{\pi}{4}\right)\right] \right\} dk .$$

To calculate the integrals (6.89), (6.92) the stationary phase method is appropriate. Stationary points are defined by

$$\frac{\partial}{\partial k} \left\{ kx - \omega_n\left(k\right) t \right\} = x - c_{gn}\left(k\right) t = 0 , \tag{6.93}$$

where $c_{gn} = \partial \omega_n / \partial k$ is the group velocity. Then for large t, integrals in the form (6.90), (6.92) will have the asymptotic form:

$$w\left(x, z, t\right) = \mathrm{Re} \left\{ \sum_{j} \sum_{n=1}^{\infty} \sqrt{2}\, \varphi_n\left(z, k\right) A_n \left[\pi t \left| \frac{\partial c_{gn}}{\partial k} \right| \right]^{-1/2} \right.$$

$$\left. \times \exp\left[i\left(kx - \omega_n t - \frac{\pi}{4}\mathrm{sign}\frac{\partial c_{gn}}{\partial k} \right) \right] \right\} \Bigg|_{k=æ_j} \tag{6.94}$$

where $æ_j$ are positive roots of Eq. (6.93). Only the integrand exponent $\exp\{i\left(kr - \omega_n t\right)\}$ plays a decisive role in the integral (6.92). This follows from (6.93) and automatically satisfies the radiation condition.

The asymptotic expression (6.94) yields when taking into consideration the first and the third expansion terms of the index at the exponent in Eqs. (6.90), (6.92) in the form

$$kx - \omega_n\left(k\right) t$$

$$= \sum_{j} \left\{ æ_j x - \omega_n\left(æ_j\right) t + \left(k - æ_j\right)\left[x - t c_{gn}\left(æ_j\right)\right] \right. \tag{6.95}$$

$$\left. - \tfrac{1}{2}\left(k - æ_j\right)^2 \frac{d c_{gn}\left(æ_j\right)}{dk} t - \tfrac{1}{6}\left(k - æ_j\right)^3 \frac{d^2 c_{gn}}{dk^2}\bigg|_{k=æ_j} t + \ldots \right\}$$

In the vicinity of stationary values of the group velocity, i.e., when $dc_{gn}/dk \to 0$, Eq. (6.94) is invalid and the next term in the expansion (6.95) should be taken into account. This leads to the expression describing the Airy wave (see, e.g., [362], [289])

$$w(x, z, t) \approx \sum_{n=1}^{\infty} \varphi_n(z) A_n(\text{æ}_0) \frac{2\pi}{3 \, | \, 2b_n \, |^{1/3}} E(v_n) \cos[\text{æ}_0 x - \omega_{0n} t];$$

$$v_n = 2 \, (3)^{-3/2} \, | \, a_n \, |^{3/2} \, | \, b_n \, |^{-1/2};$$

$$a_n = x - c_{gn}(\text{æ}_0) \, t; \qquad b_n = \frac{1}{6} \, t \, \frac{d^2 c_{gn}}{dk^2} \bigg|_{k=\text{æ}_0}, \qquad (6.96)$$

$E(v_n)$ is the Airy function expressed through the first- and second-genus fractional-order Bessel function:

$$E(v) = \begin{cases} v^{1/3} \, [\, J_{1/3}(v) + J_{-1/3}(v)\,] & ab > 0, \\ \\ v^{1/3} \, [\, I_{1/3}(v) - I_{-1/3}(v)\,] & ab < 0. \end{cases} \qquad (6.97)$$

The values æ_0 and ω_0 correspond to that value of $c_{gn}(k) = c_{0n}$ for which $dc_{gn}/dk = 0$, but x and t may vary within some ranges. The validity condition for Eqs. (6.96) was studied, e.g., in [289], where it was shown that Eqs. (6.96) are valid until the difference $t - t_0$ (where $t_0 = (\text{æ}_0/\omega_0) \, x$) becomes very large. Since Eq. (6.96) becomes invalid Eq. (6.94) is quite satisfactory. In the rotation absence ($f \equiv 0$) for small k (long waves) we have the expansion $\omega_n(k) \approx c_f^0 \, k - dk^3 + \mathcal{O}(k^4)$, where $d < 0$ [see Eqs. (3.39], (3.41)]. This dispersion relation also leads to the Airy wave description. Indeed, after substitution of this expansion into the integral (6.89) integration [391, Chapter 13] yields

$$w(x, z, t) \approx 2\pi \sum_{n=1}^{\infty} (3dt)^{-1/3} E\left[\frac{x}{(3dt)^{1/3}}\right] A_n(0) \, \varphi_n(z, 0). \qquad (6.98)$$

Thus in the rotation absence the leading wave front (long waves are propagating at the highest phase velocity) is described by the Airy function (6.98).

For particular calculations by Eqs. (6.94)–(6.96) it is necessary to set a model of the distribution $N^2(z)$ in Eq. (6.77). Assume, further, that

$N^2(z) = $ const (the mean density ρ_0 has an exponential depth distribution), which approximately models the ocean density distribution and allows us to finish calculations in a simple way.

For $N^2 = $ const the solution of the boundary value problem (6.77) yields the following normalized eigenfunctions and dispersion relations

$$\varphi_n(z) = \sqrt{2}\, \sin \frac{n\pi z}{H}\,;$$

$$\omega_n^2 = [\,k^2 H^2 N^2 + (n\pi)^2 f^2\,][\,k^2 H^2 + (n\pi)^2\,]^{-1}. \qquad (6.99)$$

It is convenient to introduce the following dimensionless variables

$$\tilde{x} = xn\pi H^{-1}\,; \quad \tilde{r} = rn\pi H^{-1}\,; \quad \tilde{t} = tN\,; \quad \tilde{k} = kH(n\pi)^{-1}\,; \quad \tilde{\omega} = \omega_n N^{-1}\,;$$

$$\tilde{f} = fN^{-1}\,; \quad \tilde{c} = (n\pi)(HN)^{-1} c_{gn}\,; \quad \tilde{w} = w(HN)^{-1}\,;$$

$$\tilde{A} = A_n(n\pi)/(HN)^2\,; \quad \eta = n\pi z H^{-1}\,;$$

$$\tilde{a} = a_n H^{-1}(n\pi)\,; \quad \tilde{b} = b_n H^{-3}(n\pi)^3\,; \quad \tilde{v} = v_n\,. \qquad (6.100)$$

Then Eqs. (6.89), (6.92), (6.99) will be rewritten in the form

$$w(x,\eta,t) = \sin \eta \int_{-\infty}^{\infty} \sqrt{2}\, \exp\{\,i\,[\,kx - \omega(k)\,t\,]\,\}\, A(k)\, dk\,; \qquad (6.101)$$

$$w(r,\eta,t) = \sin \eta \sqrt{\frac{4\pi}{r}} \int_{-\infty}^{\infty} A(k)\, k^{1/2}\, \exp\left\{\,i\left[\,kr - \omega(k)\,t - \frac{\pi}{4}\right]\right\} dk\,; \qquad (6.102)$$

$$\varphi(\eta) = \sqrt{2}\, \sin \eta\,; \quad \omega^2 = (k^2 + f^2)\,(k^2 + 1)^{-1}. \qquad (6.103)$$

In (6.101)–(6.103) a sign of the sum over n is omitted and Eqs. (6.101)–(6.103) refer to each of the normal modes. We omit, furthermore, the sign '$\sim$' implying dimensionless variables always, unless otherwise stated.

Let turn to investigation of integrals (6.101), (6.102). Stationary points of both integrals are defined by the equation $c = \partial\omega/\partial k = \xi$, where $\xi = xt^{-1}$ in the one-dimensional case and $\xi = rt^{-1}$ in the axisymmetric case. The equation for stationary points which takes into account the dispersion relation (6.103) will be

$$c = k (1 - f^2) [k^2 + 1]^{-3/2} [k^2 + f^2]^{-1/2} = \xi. \qquad (6.104)$$

Examine Eq. (6.102). In Fig. 6.5 the function $c(k)$ is shown for fixed values[10] of f ($f = 10^{-2}$). From Fig. 6.5 it is obvious that $c(k)$ is always positive, and furthermore for $k \to \infty$ $c \to 0$, and for $k = k_0 = f^{1/2} 3^{-1/4} [1 - f (2\sqrt{3})^{-1} + \mathcal{O}(f^2)]$ the function $c(k)$ has the maximum

$$c_0 \approx 1 - \sqrt{3}f + \mathcal{O}(f^2) < 1.$$

For $k = k_0$ the frequency ω is expressed as

$$\omega(k_0) = \omega_0 = f^{1/2} 3^{-1/4} [1 + f(2\sqrt{3})^{-1} + \mathcal{O}(f^2)].$$

Values of k_0 are obtained from the equation $\partial c/\partial k = 0$, where $\partial c/\partial k$ has the form

$$\frac{\partial c}{\partial k} = (1 - f^2) [f^2 - 3k^4 - 2k^2 f^2] [k^2 + 1]^{-5/2} [k^2 + f^2]^{-3/2}. \quad (6.105)$$

The group velocity maximum (and the Airy wave existence) is caused by fluid rotation. As follows from (6.105) in the rotation absence ($f = 0$) $\partial c/\partial k \neq 0$ for $k \neq 0$ and the Airy wave does not appear. But, as shown above, for $k \approx 0$, i.e., in the vicinity of the leading wave front, the wave is described by the Airy function with Eq. (6.98).

For $\xi < c_0$ Eq. (6.104) has two roots $æ_1(\xi)$ and $æ_2(\xi)$ (we define $æ_1(\xi) \leq æ_2(\xi)$, and for $\xi \to c_0$ both roots go to a single root $æ_1 \to æ_2 \to k_0$. It is difficult to define the roots of Eq. (6.104) in the general form because a fourth-order equation must be solved with respect to k. But we may use the following circumstance: to obtain the root $æ_1$, i.e., when $k < k_0$, we may neglect the value of k^2 compared with unity (it corresponds to the hydrostatic approximation $\omega \ll N$), and when obtaining the second root $æ_2$ we may neglect the effect of the Coriolis force ($f^2 \ll k^2 \sim 1$ or, which is the same, dimensional values are governed by the inequality

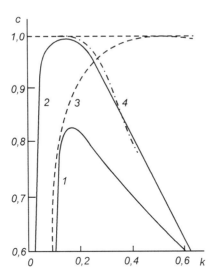

Figure 6.5. The dimensionless group velocity c vs. k: $1 - f = 10^{-1}$; $2 - f = 10^{-2}$; $3 - f = 10^{-2}$, $k \ll 1$; $4 - f = 10^{-2}$, $k \leq 1$.

$f^2 \ll \omega^2$). In Fig. 6.5 the function $c(k)$ is approximated quite well by the curves *4* and *3* for $k < k_0$ and $k > k_0$ accordingly. The crossover point of these curves is always above c_0 for any f. That is why to obtain æ_1 and æ_2 for $\xi < c_0$ we may use approximate formulae following from Eq. (6.104) for $k^2 \ll 1$ and $f \ll k^2 \sim 1$. The accuracy of derivation of æ_1 and æ_2 grows for decreasing ξ and for $\xi \approx c/2$ ($f = 10^{-2}$) is less than 1%.

Using formulae for approximate derivation of the roots of Eq. (6.104) we obtain the following expressions for variables contained in Eq. (6.94) and defining integrals (6.101), (6.102) for $\xi < c_0$:

$$\text{æ}_1 = f\xi \left(1 - \xi^2\right)^{-1/2}; \quad \omega\left(\text{æ}_1\right) = f\left(1 - \xi^2\right)^{-1/2};$$

$$\left.\frac{\partial c}{\partial k}\right|_{k=\text{æ}_1} = f^{-1}\left(1 - \xi^2\right)^{3/2}; \quad \text{æ}_2 = \xi^{-1/3}\left(1 - \xi^{2/3}\right)^{1/2};$$

$$\omega\left(\text{æ}_2\right) = \left(1 - \xi^{2/3}\right)^{1/2}; \quad \left.\frac{\partial c}{\partial k}\right|_{k=\text{æ}_2} = -3\xi^{4/3}\left(1 - \xi^{2/3}\right)^{1/2}; \quad (6.106)$$

Taking into consideration Eqs. (6.106) we have for Eqs. (6.101), (6.102) the following asymptotic expressions:

$$w(x, \eta, t) =$$

$$2\sqrt{\pi} \sin \eta \left\{ \frac{f^{1/2} A(\text{æ}_1)}{(1-\xi^2)^{3/4} t^{1/2}} \exp\left\{ i\left[f(\xi^2-1)^{1/2} t - \frac{\pi}{4}\right]\right\} \right.$$

$$\left. + \frac{A(\text{æ}_2)}{3^{1/2}\xi^{2/3}[1-\xi^{2/3}]^{1/4} t^{1/2}} \exp\left\{ i\left[-(1-\xi^{2/3})^{3/2} t + \frac{\pi}{4}\right]\right\} \right\} ;$$

$$\text{(6.107)}$$

$$w(r, \eta, t) = 2\sqrt{\pi} \sin \eta \left\{ \frac{f A(\text{æ}_1)}{(1-\xi^2) t} \exp\left\{ i\left[f(\xi^2-1)^{1/2} t - \frac{\pi}{4}\right]\right\} \right.$$

$$\left. + \frac{A(\text{æ}_2) t^{1/3}}{3^{1/2} r^{5/6}} \exp\left\{ i\left[-(1-\xi^{2/3})^{3/2} t + \frac{\pi}{4}\right]\right\} \right\}. \qquad \text{(6.108)}$$

Expressions (6.107), (6.108) represent a superposition of frequency modulated waves, one of which has a frequency close to the inertial frequency, and another to the Brunt–Väisälä frequency. Variation of the frequency and the wave number of these waves versus x, r and t is obvious from Eqs. (6.106) and is the same for both plane and axisymmetric cases.

Consider an example when the initial disturbance is the δ-impulse. Then in a plane case $w(x, \eta, 0) = \delta(x)\psi(\eta)$ ($\delta(x)$ is the delta function), $A(k) = (2\pi)^{-1}$, and in the axisymmetric case $w(x, \eta, 0) = \delta(x)\psi(\eta)/r$, $A(k) = (2\pi)^{-1}$. Expressions (6.107), (6.108) for $\xi \ll 1$ (i.e., for $t \gg x$) have the form:

$$w(x, \eta, t) = \pi^{-1/2} \sin \eta \left\{ f^{1/2} t^{-1/2} \cos\left[ft + \frac{\pi}{4}\right] \right. \qquad \text{(6.109)}$$

$$\left. + 3^{-1/2} x^{-2/3} t^{1/6} \cos\left[t - \frac{\pi}{4}\right] \right\} ;$$

$$w(r, \eta, t) = \pi^{-1/2} \sin \eta \left\{ f t^{-1} \cos\left[ft + \frac{\pi}{4}\right] \right. \qquad \text{(6.110)}$$

$$\left. + 3^{-1/2} r^{-5/6} t^{1/3} \cos\left[t - \frac{\pi}{4}\right] \right\}.$$

Eqs. (6.109), (6.110) yield for $\xi \ll 1$ the initial δ-impulse breaks up into two waves: one of them has the inertial frequency and decays in time, another one has a frequency of about N and grows in time. In the plane case both growth and decay of respective waves occur in $t^{1/3}$ and $t^{1/2}$ times slower than in the axisymmetric case. While x and r growing (for fixed t and $\xi \ll 1$) amplitudes of waves of the frequency N are decreasing and amplitudes of inertial waves do not vary. Using the obvious formula

$$a \cos \alpha + b \cos \beta = \{a^2 + b^2 + 2ab \cos (\alpha - \beta)\}^{1/2} \sin \left[\frac{1}{2} (\alpha + \beta) + \varphi\right],$$

where

$$\tan \varphi = -(a + b) (a - b)^{-1} [1 + \cos (\alpha - \beta)]^{1/2} [1 - \cos (\alpha - \beta)]^{-1/2},$$

and neglecting f relative to 1 for $t \to \infty$ from (6.109), (6.110) yield

$$w (x, \eta, t) \approx \sin \eta \, \frac{t^{1/6}}{3^{1/2} x^{2/3}} \, \sin \left[\frac{t}{2} + \varphi\right]; \qquad (6.111)$$

$$w (r, \eta, t) \approx \sin \eta \, \frac{t^{1/3}}{3^{1/2} x^{5/6}} \, \sin \left[\frac{t}{2} + \varphi\right], \qquad (6.112)$$

where

$$\tan \varphi = - \left[1 + \cos \left(t - \frac{\pi}{2}\right)\right]^{1/2} \left[1 - \cos \left(t - \frac{\pi}{2}\right)\right]^{-1/2}.$$

Thus for large enough t the initial impulse degenerates into a wave package of frequency $N/2$ with modulated phase and with amplitude growing in time and decaying with distance. The existence of such wave packages (groups) with frequencies close to N has frequently been observed in the ocean [314].

Presented asymptotic Eqs. (6.107)–(6.112) are invalid in the vicinity of stationary values of the group velocity c_0. For $c \approx c_0$ Eqs. (6.96), (6.97) describing the Airy wave should be used. Calculation of variables containing in (6.96), (6.97) by means of the dispersion relation (6.103) for a plane wave yields

$$w (x, \eta, t) \approx A (k_0) \cos \left\{\frac{f^{1/2}}{3^{1/4}} \left[\left(1 - \frac{f}{2\sqrt{3}} + \mathcal{O} (f^2)\right) x\right.\right.$$

$$- \left(1 + \frac{f}{2\sqrt{3}} + \mathcal{O}\left(f^2\right)\right) t\right]\right\} \frac{2\pi E\left(v\right)}{3\left|2b\right|^{1/3}}, \qquad (6.113)$$

where

$$v = 2^{1/2}\, 3^{3/2} \left\{\left|x - \left(1 - \sqrt{3}\, f + \mathcal{O}\left(f^2\right)\right) t\right|\right\}^{3/2}$$

$$\times \left\{\left[1 - \frac{8}{\sqrt{3}} f + \mathcal{O}\left(f^2\right)\right] t\right\}^{-1/2}$$

$$\approx \frac{\sqrt{2}}{3^{3/2}}\, \frac{\left\{\left|x - t\right|\right\}^{3/2}}{t^{1/2}}; \qquad (6.114)$$

$$a = x - \left[1 - \sqrt{3}\, f + \mathcal{O}\left(f^2\right)\right] t \approx x - t;$$

$$b = -2\left[1 - \frac{8}{\sqrt{3}} f + \mathcal{O}\left(f^2\right)\right] t \approx -2t. \qquad (6.115)$$

Expressions (6.113)–(6.115) are valid for an axisymmetric wave on replacing x by r. Expression (6.113) represents an amplitude modulated wave of the carrier frequency of order $f^{1/2}\, 3^{1/4}$ and of the dimensional wave number of order $f^{1/2} N^{-1/2} H^{-1} n\pi$. The wave package defined by Eqs. (6.113) has a pronounced maximum at $v \approx 0.7$ (Fig. 6.6). Thus wave groups of large amplitudes correspond to the stationary value of the group velocity. As $E\left(v\right)$ is a function of v only, then the waves group's amplitude is proportional to $b^{-1/3}$, i.e., approximately $x^{-1/3} c_0^{1/3}$. For other frequencies the amplitude, according to Eq. (6.94), is proportional to $x^{-1/2} c_0^{1/2}$. So while the wave package propagate in a stratified rotating fluid, the wave groups of the frequencies with $dc/dk \approx 0$ prevail. And the role of such groups grows with distance from the source of disturbance.

For given $x = x_0$, $a = x_0 - c_0 t = c_0\left(t_0 - t\right)$, where t_0 is the time of arrival of the wave of the frequency ω for which the condition $dc/dk = 0$ is valid. Thus for $c_0 = c_{\max}$ (the present case) instants $t > t_0$ $\left(a < 0\right)$ correspond to a later arrival of the disturbance than the time of arrival of the wave propagating with the group velocity. The form of the function $E\left(v\right)$ depends on the sign of ab [see (6.97)]. Using expression (6.115)

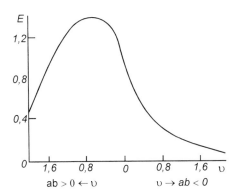

Figure 6.6. The envelope curve of the Airy wave.

yields

$$ab > 0 \quad \text{for} \quad \frac{x}{t} < c_0 \approx 1 - \sqrt{3}\,f + \mathcal{O}\left(f^2\right) > 0;$$

$$ab < 0 \quad \text{for} \quad \frac{x}{t} > c_0 \approx 1 - \sqrt{3}\,f + \mathcal{O}\left(f^2\right) > 0.$$

$$(6.116)$$

As the value b is always negative (which agrees with the sign of $\partial^2 c / \partial k^2 <$ 0) Eq. (6.113) for $ab > 0$, i.e., $x/t < c_0$, describes waves arriving later than the first appearance of the disturbance. For $ab < 0$ $(x/t > c_0)$ formula (6.113) describes a forerunner of the first appearance of the basic wave at instants $t < x_0/c_0$.

Following [362] we can evaluate an effective width of the band spectrum of the wave package (6.113). From relations (6.91), (6.92) for fixed $x = x_0$ and $v = v_M$ there follows

$$|t - t_0| = \Delta t = v_M^{2/3} 2^{-1} 3^{2/3} c_0^{-2/3} x_0^{1/3} \left| \frac{d^2 c}{d\omega^2} \right|^{1/3}$$

This expression defines the time interval (for the fixed x_0) corresponding to the values $0 \le v \le v_M$. Expanding the function $c(\omega)$ by the Taylor series with respect to $\omega = \omega_0$ and combining this expression with the previous one [362] yields

$$\Delta\omega \approx v_M^{1/3} 3^{1/3} c_0^{2/3} x_0^{-1/3} \left| \frac{d^2 c}{d\omega^2} \right|^{-1/3}, \qquad (6.117)$$

and so $\Delta\omega\,\Delta t \approx \frac{3}{2} v_M$. When considering v_M to be $v_M \approx 2$ (in this case $E(v)$ decays e times) yields $\Delta\omega\,\Delta t \approx 3$. This relation shows that an impulse of long duration has a narrow spectrum and vice versa. From the expression for $\Delta t = |t - t_0|$ (see above) follows that for large $|d^2c/d\omega^2|$ the value of $\Delta\omega$ is small and Δt is large. This means that when $c(\omega)$ has a sharp maximum the propagating impulse has the form of a long almost pure wave package. For small $|d^2c/d\omega^2|$ the wave package is short and consists of small number of oscillations. In the present case

$$\left|\frac{d^2c}{d\omega^2}\right|_{\omega=\omega_0} \approx 12\left[1 - \frac{8}{\sqrt{3}}f + \mathcal{O}\left(f^2\right)\right],$$

and that is why the internal wave package propagating at the maximum group velocity is quite wide and quasi-harmonic.

The formulae obtained relate to each of the normal internal modes, and the general solution represents their superposition. The number of modes sufficient for describing the total fluid motion is defined by the number of Fourier coefficients in expansion (6.78).

The estimations presented of the propagation of an impulse in a stratified rotating fluid may vary slightly for another stratification model. In particular, local maxima of the group velocity may appear in the short wave range for a complicated (e.g., two-maxima) distribution of the Brunt–Väisälä frequency [103]. Airy waves can then also appear in the vicinity of these maxima. A qualitative picture of propagation of internal gravity waves induced by the initial disturbance is, in general, reduced to the following: an observer standing at a fixed point $x = x_0$ far away from the coordinate of the initial disturbance will fix at initial instants a sum of the Airy waves corresponding to normal modes. Each of the modes moves at the maximal group velocity and has a frequency of order $f^{1/2}$. Then the observation point will be reached by the wave group representing the superposition of two wave packages, one of them having a frequency close to the inertial frequency and the other the frequency N. For $t \to \infty$ wave packages of almost the Brunt–Väisälä frequency will dominate in the fluid motion.

In concluding the present section we consider some particular solutions of the Cauchy problem of Eq. (6.74) for $N^2 = $ const and for an infinitely deep ocean corresponding to the following special initial conditions [255]

$$w\,|_{t=0} = \frac{Q}{R}\,; \qquad \frac{\partial w}{\partial t}\bigg|_{t=0} = 0\,, \qquad\qquad (6.118)$$

where $R = \sqrt{v^2 + z^2}$, $r = \sqrt{x^2 + y^2}$ and Q is an arbitrary constant. Such initial condition corresponds to the generation of internal waves by a 'spread initial impulse'. Eq. (6.74) has a cylindrical symmetry under the initial condition (6.118) and may be written in cylindrical coordinates

$$L_W = \left\{ \frac{\partial^2}{\partial t^2}\left[\frac{1}{r}\frac{\partial}{\partial r}\,r\,\frac{\partial}{\partial r} + \frac{\partial^2}{\partial z^2}\right] + f^2\frac{\partial^2}{\partial z^2} + N^2\frac{1}{r}\frac{\partial}{\partial r}\,r\,\frac{\partial}{\partial r} \right\} w = 0\,.$$
$$(6.119)$$

To solve this equation we use the dimensional analysis method (see [17], [255], [335]). The root w has the dimension $[w] = lT^{-1}$ and is determined by the parameters r, z, t, N, f, Q having the following dimensions: $[r] = l$, $[t] = T$, $[N] = [f] = T^{-1}$, $[Q] = l^2T^{-1}$. Then from the linearity of Eq. (6.119) and the π-theorem follows

$$w = \frac{Q}{r}\,\psi\,(\eta,\tau,\sigma)\,; \qquad \eta = \frac{r}{z}\,; \qquad \tau = Nt\,; \qquad \sigma = \frac{f}{N}\,. \qquad (6.120)$$

Substituting (6.120) into (6.119) we have

$$\frac{\partial}{\partial \eta}\left\{\left[\eta\frac{\partial}{\partial \eta}\left(\frac{1}{\eta}\psi_{\tau\tau}\right) + \eta^2\frac{\partial}{\partial \eta}\psi_{\tau\tau}\right] + \left[\eta\frac{\partial}{\partial \eta}\left(\frac{1}{\eta}\psi\right)\right] + \sigma^2\left[\eta^2\frac{\partial\psi}{\partial \eta}\right]\right\} = 0\,;$$
$$(6.121)$$

$$\psi\,|_{\tau=0} = \eta\,[1 + \eta^2]^{-1/2}\,; \qquad \frac{\partial\psi}{\partial t}\bigg|_{\tau=0} = 0\,. \qquad (6.122)$$

It is convenient to change further the unknown function and the independent variable

$$\tilde{\psi} = \psi\,(1-s)^{-1/2} = \psi\,\frac{R}{r}\,; \qquad s = (1+\eta^2)^{-1} = z^2\,R^{-2}\,; \qquad 0 \le s \le 1\,.$$
$$(6.123)$$

Then expressions (6.121) and (6.122) may be rewritten in the form

$$s^{3/2} (1 - s)^{1/2}$$

$$\times \frac{\partial}{\partial s} \left\{ s^{1/2} (1 - s) \left[\tilde{\psi}_{s\tau\tau} + (\sigma^2 - \sigma^2 s + s) \tilde{\psi}_s + \frac{1}{2} (1 - \sigma^2) \tilde{\psi} \right] \right\} = 0; \tag{6.124}$$

$$\tilde{\psi} \Big|_{\tau=0} = 1; \quad \frac{\partial \tilde{\psi}}{\partial \tau} \Big|_{\tau=0} = 0. \tag{6.125}$$

Integrating (6.124) taking into account the boundedness and regularity conditions of the solution and its derivatives at the rotation axis ($s = 1$) or at infinity ($s = 0$) yields

$$\tilde{\psi}_{s\tau\tau} + (\sigma^2 - \sigma^2 s + s) \tilde{\psi}_s + 1/2 (1 - \sigma^2) \tilde{\psi} = 0, \tag{6.126}$$

$$\tilde{\psi} \Big|_{\tau=0} = 1; \quad \frac{\partial \tilde{\psi}}{\partial \tau} \Big|_{\tau=0} = 0. \tag{6.127}$$

To solve the problem (6.126)–(6.127) for $\sigma \neq 1$ we use the Laplace transform[11]. Taking into account (6.127) we transform Eq. (6.126) according to the Laplace transform, and solving the equation obtained with respect to $\varphi(p, s)$ (the Laplace image of the function $\psi(\tau, s)$, p is the parameter of the Laplace transform) yield

$$\varphi(p, s) = C(p) \{[(1 - \sigma^2) s + \sigma^2] + p^2\}^{-1/2}, \tag{6.128}$$

where $C(p)$ is an arbitrary function. From formulae (6.127) follows that for $p \to \infty$ ($t \to 0$) $\varphi(p, s) \to 1$. Whence using expression (6.128) yields $C(p) = p$ and expression (6.128) becomes

$$\varphi(p, s) = p \{[(1 - \sigma^2) s + \sigma^2] + p^2\}^{-1/2}. \tag{6.129}$$

The inverse Laplace transform of expression (6.129) yields

$$\tilde{\psi}(\tau, s) = J_0(\xi), \tag{6.130}$$

where

$$\xi = [(1 - \sigma^2) s + \sigma^2]^{1/2};$$

$$\tau = \frac{1}{R}\sqrt{z^2 + \sigma^2 r^2 \tau} = \frac{t}{R}\sqrt{N^2 z^2 + f^2 r^2}\,; \tag{6.131}$$

$J_0(\xi)$ is the zero order Bessel function of the first kind. Using (6.120), (6.123), (6.130), (6.131) we obtain the first fundamental solution of Eq. (6.119)

$$w = \frac{Q}{R} J_0(\xi)\,. \tag{6.132}$$

Solution (6.132) satisfies the initial conditions (6.118) and Eq. (6.119) everywhere except of the point $R = 0$. For $N^2 = 0$ the solution (6.132) coincides with the solution obtained by Sobolev [335] when studying fluid in rotating vessels.

Let us analyze this solution. In spherical coordinates Eq. (6.131) becomes

$$\xi = \tau\sqrt{\cos^2\vartheta + \sigma^2\sin^2\vartheta} = t\sqrt{N^2\cos^2\vartheta + f^2\sin^2\vartheta}\,, \tag{6.133}$$

where ϑ $(0 \le \vartheta \le \pi)$ is the angle included between the radius vector **R** and the vertical z-axis. To consider time variations of the function w we study a sphere of the constant radius $R = R_0$. On this sphere at every instant w depends only on the angle ϑ and on the parameter σ [see (6.133)]. The argument of the Bessel function J_0 will vary from $\sigma\tau = ft$ (for $\vartheta = \pi/2$) to $\tau = Nt$ (for $\vartheta = 0$). While time increases, the number of waves induced by extrema of the Bessel function in the range from the equator $(\vartheta = \pi/2)$ to the pole $(\vartheta = 0)$ of the sphere will increase. The waves will appear at the equator and at the pole of the sphere, propagating towards each other and accumulating at the sphere's surface, and, moreover, longer waves will disperse into increasingly shorter waves. As $\sigma < 1$ in natural ocean conditions, waves appearing in the vicinity of the sphere's equator (inertial waves) will have larger periods, and their number at a fixed instant will be less than the number of waves appearing in the vicinity of the pole (internal gravity waves). In the absence of rotation $(f = 0)$ the waves will propagate from the pole to the equator only. In the absence of stratification the waves will propagate in the opposite direction.

Let us analyze the phase picture of generating waves. The distance between stationary phase lines $(\xi = \mathrm{const})$ is defined by the expression

$\xi = \alpha_{n+2} - \alpha_n$, where α_n is the n-th root of the expression $J_0(\alpha) = 0$. For roots of the Bessel function $J_0(\alpha)$ yields

$$\alpha_n \approx \frac{\pi}{4}(4n - 1) + \frac{1}{2\pi}(4n - 1)^{-1} + \mathcal{O}(n^{-3}),$$

then to a high accuracy

$$\alpha_{n+2} - \alpha_n \approx 2\pi n \quad \text{for} \quad n > 2.$$

Whence using expression (6.133) yields

$$\cos \vartheta = \frac{\sqrt{(2\pi n)^2 - \sigma^2 \tau^2}}{\tau (1 - \sigma^2)^{1/2}} = \frac{\sqrt{(2\pi n)^2 - f^2 t^2}}{t (N^2 - f^2)^{1/2}}, \tag{6.134}$$

where

$$2\pi n N^{-1} \le t \le 2\pi n f^{-1}.$$

For $f = 0$ and $N = 0$, respectively, we have:

$$\cos \vartheta = \frac{2\pi n}{Nt} \quad \text{for} \quad t \ge \frac{2\pi n}{N}; \tag{6.135}$$

$$\sin \vartheta = \frac{2\pi n}{Nt} \quad \text{for} \quad t \ge \frac{2\pi n}{f}. \tag{6.136}$$

Eq. (6.134) shows that the angle included between a constant phase line and the vertical grows in time. The angle ϑ varies from 0 to $\pi/2$ during the finite time $2\pi n(N - f)(Nf)^{-1}$. In the absence of rotation the angle ϑ also grows in time, but it reaches the value $\pi/2$ only when $t \to \infty$ [see (6.134)]. In the absence of stratification a constant phase line, having at the initial instant $t = 2\pi n f^{-1}$ an angle of inclination to the vertical of $\pi/2$ moves in time towards smaller angles ϑ, and for $t \to \infty$ $\vartheta \to 0$ [see (6.136)]. Eq. (6.135) was verified experimentally with good agreement of theoretical and experimental data, [104].

The analysis presented is related, strictly speaking, to a certain solution of the Cauchy problem given by Eq. (6.132). But some general properties of the Cauchy problem for inertial gravity waves mentioned in this section are qualitatively valid for other more complicated sources of generation of waves [255], [267].

4. BRIEF INFORMATION ABOUT OTHER POSSIBLE SOURCES OF GENERATION OF INTERNAL WAVES IN THE OCEAN

Generation of internal waves by fluctuations of atmospheric pressure and wind as well as by waves inducted by local disturbances have already been thoroughly studied. There are series of sources of internal waves which will be considered briefly.

Internal waves can be generated by fluctuations of the buoyancy flux at the ocean's surface. When the surface heat flow fluctuates or the surface temperature varies the pressure, fluctuations at the lower border of the Eckman boundary layer appear which induce internal waves in lower water layers. Generation of internal waves by the near surface buoyancy flux fluctuations were calculated by Magaard [227]. Calculations showed that such a mechanism of generation of internal waves is sufficient and causes growth of the amplitude of waves compared with those caused by oscillations of the friction stresses of the wind.

The survey by Thorpe [357] compared the abovementioned atmospheric sources of the generation of internal waves. Thorpe finally concluded that fluctuations of pressure, of the wind velocity and of the buoyancy flux cause the growth of velocities of internal waves of the order 1–5 m/day (generation of internal waves owed to wind friction stresses are a bit intense). The poor experimental study of spectra of atmospheric disturbances does not permit a definite determination of the main mechanism of the generation of internal waves. Different sources can seemingly dominate, depending on particular meteorological situations. Intensive energy transfer to the lower internal modes (the growth of the velocity of waves is proportional to n^{-2}, where n is the mode number) is a characteristic feature of the sources mentioned. Natural observations show that, as a rule, the lower modes dominate in the upper ocean layer.

All of the abovementioned studies related to internal waves generated in resting fluid, i.e., the influence of currents was neglected. But as horizontal currents with a vertical velocity shear are always present in the ocean, the problem of the generation of internal waves with background shear flow is of special interest. The intensity of wave generation in this case will considerably increase, and growing unsteady internal waves modes will appear. Such a hydrophysical problem has a mathematical equivalent in the hydrodynamic stability of a stratified fluid with boundary and initial conditions, a problem which is extremely difficult. A qualitative theory of steady internal waves propagation ($Ri \gg 1$) induced by random initial disturbances in the shear flow presence was developed by Townsend [363]. Townsend showed that the intensity of the generation of internal waves grows considerably owing to the action of a shear flow; certain evaluations have been made for atmospheric internal waves. In [252] a simplified two-layer model of the flow with a depth discontinuity of the velocity was considered. Waves were generated by random fluctuations of atmospheric pressure at the free ocean surface. Both steady and unsteady (in the Kelvin–Helmholtz sense) waves were studied. The spectrum of vertical displacements in internal waves for low wave numbers grows linearly versus time (steady waves) and exponentially for large wave numbers (a manifestation of unsteadiness). Steady waves grow four times as quickly as in the absence of a shear flow. During about several tens of short waves' periods the energy of unsteady waves dominates, which may cause the appearance of a local maximum in the high frequency spectrum range.

Internal waves can be generated as a result of circulation by a stratified flow around the bottom topography. When the flow velocity and the obstacle's height are small, the linear theories are valid. Such calculations for natural ocean conditions were performed by Tareyev [348] and in a more generalized form by Bell [19]. These calculations showed that induced internal waves have lengths of the order of the characteristic length of the bottom's topography. According to Bell's estimations the action of the bottom relief on the quasi-stationary flows induces internal wave resistance and the flux of momentum ~ 0.5 dyne/cm^2; this value is of about the same magnitude as the main friction stress owed to wind at the free ocean surface.

An example of the non-tidal mechanism of intensive internal wave train generation in a narrow thermocline by a near surface fresh water intrusion (a local front formed by some river water) is described in [140].

We have considered linear mechanisms of the generation of oceanic internal waves. But there is a specific nonlinear generating mechanism which has no description in the linear theory framework. This mechanism is realized in the generation of nonlinear resonant waves by the coupling of surface waves. Resonant nonlinear interaction of internal waves will be thoroughly discussed in Chapter 10. Here we briefly describe the generation of internal waves by surface waves.

When two surface waves with wave vectors $\mathbf{k}_1$ and $\mathbf{k}_2$ and frequencies ω_1 and ω_2 interact in the second order of the nonlinear perturbation theory, then internal waves synchronized with a couple of surface waves (i.e., having the horizontal wave vectors $\mathbf{k} = \mathbf{k}_1 \pm \mathbf{k}_2$ and frequency $\omega = \omega_1 \pm \omega_2$) can appear.

Frequencies of internal waves are, as a rule, lower than those of surface waves. That is why the internal wave of the frequency $\omega = \omega_1 - \omega_2$ will be synchronized. In particular, when wave numbers' moduli are almost equal (though their directions are different) the frequency difference $\omega_1 - \omega_2$ may be small and the wave numbers' difference will be comparable to $\mathbf{k}_1$ or $\mathbf{k}_2$, and the conditions $\mathbf{k}_1 - \mathbf{k}_2 = \mathbf{k}$ and $\omega_1 - \omega_2 = \omega$ may be satisfied. Owing to the resonant interaction the internal wave starts to take away the surface waves' energy. Moreover, interactive disturbances have a decay $\exp(|\mathbf{k}_1 - \mathbf{k}_2|z)$ with depth. This mechanism was analyzed by Thorpe [352] (see also [184], [278], [387], [388]) and in the most general statement (in particular, without the Boussinesq approximation) by Brekhovskikh et al. [35]. The latter work shows that the generation of internal waves by interacting surface waves is described by Eq. (7.1) of the second-order perturbation theory including surface and internal sources. For the displacement of the level in an internal wave of the m-th mode the following expression was obtained

$$z_m = \pm c_m + \sin(qx \mp \omega_m t), \quad \mathbf{q} = \mathbf{k}_2 - \mathbf{k}_1,$$

where c_m is the growth of the amplitude velocity versus time. This expression was analyzed for different stratification models; in particular, for a fixed surface waves' length the coefficient c_m for some angles included between $\mathbf{k}_1$ and $\mathbf{k}_2$ has a maximum which reduces versus increasing depth of the thermocline and shifts to more acute angles. Time estimations for internal waves show that the amplitude of an internal wave reaches surface waves once during a time of the order of several Brunt– Väisälä periods, so that the mechanism mentioned for the generation of internal waves is quite effective and can play a decisive role. The spectral method was used in [34] to calculate the interaction of surface and internal waves, which allowed one to study not only a triple-wave, but higher-order interactions. Calculations showed that the triple-wave interaction is unstable; taking into account the multi-wave interactions may lead to more effective generation of internal waves by surface waves.

The modulation mechanism of energy and momentum exchange between wind and internal waves is also a result of three-wave interaction. But, in contrast with the above mechanism, when the surface waves have random phases, here the envelope of the surface wave results from the transformation of wind waves in the field of the internal wave's surface currents [68], [183], [387]. Modulation interaction results in exponential growth or decay of internal waves depending on environmental parameters. Estimates of the effectiveness of these two mechanisms carried out by Watson [387] showed that for typical ocean conditions the latter turned out to be more effective.

Kudryavtsev [184] considered the simultaneous influence of two mechanisms described: the interaction of surface waves and the friction mechanism. The origin of the friction interaction mechanism is connected with the work done by surface stress, equal to losses of momentum of wind waves, against the long–wave orbital motions. It is important to emphasize that the modulation and friction mechanisms should exist simultaleously, as the presence of internal waves leads to variations in the wind waves' energy (momentum) losses. The effectiveness of those waves' interaction mechanisms was considered as a dependence on stratification parameters, wind velocity, and direction. The coupling of wind and internal waves was studied against the background of the drift current. The presence of turbulent friction in the upper uniform layer results in the generation of inertial oscillations. These motions receive momentum and energy which are lost in the interaction of wind and internal waves. These dynamical interactions are analysed on the basis of the integral laws of energy and momentum conservation proposed by Phillips [300].

Let us now consider sources of generation of internal waves with a tidal period. Many natural observations in different regions of the World Ocean prove that tidal internal waves exist. The energy of these waves is quite high and frequently exceeds the energy of internal waves of other periods [60], [121], [401]. The direct influence of tidal forces on the generation of internal waves in an open ocean of constant depth is quite weak [179], [349]. But when the ocean depth is not constant internal waves of tidal period may be induced [110], [13]. Rattray [309] studied the generation of tidal period internal waves by the interacting surface tide at the coastal shelf. In particular, he showed that when the shelf is broad enough these waves are almost plane and can propagate over large distances inside the ocean without considerable variation of amplitude. Rattray's theory, as well as some natural measurements [174], proved that internal tides exist in the form of standing waves in the coastal zone, but they turn into traveling waves as they go further from the coast. Rattray then switched to models based on a small perturbation approach for accounting a more realistic slope and continuous stratification [305], [310]. Cox and Sandström [54] presented a more general theory of the inducing of internal waves by a surface tide propagating over an uneven bottom in the continuously stratified ocean. It was shown that the main condition for energy transfer from the surface (barotropic) tide to internal waves is zero inequality of the horizontal component of the orbital velocity at the bottom. Using a different approach Baines [13] suggested that horizontal two-dimensionality could be accounted for by forcing his slice model with the barotropic flux normal to the shelf break. This model was compared with Rattray's model and internal tide observations in [117]. A three-dimensional numerical modeling of an internal tide was performed in [400].

However, the abovementioned theories do not explain the major incongruences: the propagation of energy along rays requiring a large number of modes (observations show that the energy is concentrated in a few modes), and the most outstanding feature of internal tidal currents is their varying modulation. Krauss [181] suggested a nonlinear generation mechanism for internal waves of near-tidal frequency by a surface (barotropic) tide passing through a baroclinic eddy field. It explains both the horizontal scale observed and the mode structure of internal tides.

One more mechanism of the generation of internal waves may act when an internal tidal wave propagates along a sloping thermocline [316]. It is similar to the generation of short internal waves as a result of nonlinear breaking of the internal tide coming to shallow waters [137], [325]. In some cases the waves have a distinct soliton character [151], [84], [127], [269], [320], [327]. A theoretical nonlinear model of internal tide transformation in the shelf zone has been developed in [128] based on the rotationally modified extended Korteweg–de Vries equation taking into account wave damping owed to turbulent bottom friction. Nonlinear internal waves emanating from the internal tide interacting with the sill were observed in the Knight Inlet [74], under the Mascarene Ridge [173], and in the Bay of Biscay [273].

Notes

1 Generally speaking, a stationary random process is described by the Fourier–Stieltjes integral $P(\mathbf{x}, t) = \int e^{i\mathbf{k}\mathbf{x}} \, d\Pi(\mathbf{k}, t)$, where $d\Pi$ is a random uncorrelated measure. But for simplification we use (here and later on) the form (6.2), i.e., we assume $d\Pi = \Pi d\mathbf{k}$, considering Π to be a generalized function.

2 An asymptotic transition from (6.13) to (6.20) requires two restrictions for a time t. The first is caused by neglect of I_{nm} in comparison with I_{nn} and by approximating I_{nn} with Eq. (6.17), and has the form $\Delta\omega_n t = |\omega_n - \omega_{n+1}| t \sim \omega_n t \gg 1$. The second restriction is caused by replacing of the finite limit in the Fourier transformation of $\Omega(\mathbf{k}, \tau)$ with respect to t to the infinite one. This is valid when $(2\pi/\theta) t \gg 1$, where $\theta(\mathbf{k})$ is an integral time correlation scale of the pulsations of the atmospheric pressure (see [299]).

3 While solving combined Eqs. (6.3), (6.4) we neglect (here and later on) the term N^2/g in Eq. (6.3), which is inessential in oceanic conditions, i.e., the Boussinesq approximation is used.

4 When $\rho_0(z)$ is described by Eq. (6.21) there is only one internal wave mode. In this connection only the term corresponding to internal waves is contained in the sum (6.20).

5 In the general case, to derive (6.35) as well as this condition we need $f^2 (n\Pi)^2 \ll k^2 H^2 N_0^2$, i.e., the Coriolis force should been taken into account for very long waves of high modes.

6 Low frequency oceanic internal waves forced by a latitude independent wind (the f-variable model) were investigated by Kundu [185], extending the hydrostatic constant-f model [187]

7 Formula (6.54) may be obtained, e.g., by the Laplace transform of J, using the Borel folding theorem twice and then turning to the original.

8 This asymptotic approximation is valid in general for $t \gg \theta(\mathbf{k})$, where $\theta(\mathbf{k})$ is the integral time scale of correlation of components of the frictional stress caused by wind.

9 A model of the propagation of a pulse of horizontal distortion of the water's surface (caused by a non-stationary wind field) from the surface to the bottom in a stratified medium is considered in [62] within the framework of the linear theory of internal waves in the long wave approximation. In the model a three-layer fluid is presented as a two-step structure of the density profile with alteration of density discontinuities and homogeneous fluid layers. The model of propagation was tested using experimental data and time distribution of the surface wind stresses during storms at the shelf and slope off the Sahara coast.

10 For moderate latitudes the dimensional Coriolis frequency in the ocean has the value $f \approx 10^{-4}$ s^{-1}. The Brunt–Väisälä frequency in the ocean is about $N \approx 10^2$–10^3 s^{-1}. That is why $\tilde{f} = fN^{-1} \approx 10^{-1}$–$10^{-2}$, and always less than 1.

11 For $\sigma = 1$ ($N^2 = f^2$) Eq. (6.126), as well as the initial equation (6.119), degenerates and the problem (6.121)–(6.122) has the solution $w = Q/R = \text{const}$.

NONLINEAR THEORY OF INTERNAL WAVES

Chapter 7

HAMILTONIAN FORMALISM FOR THE DESCRIPTION OF OCEANIC INTERNAL WAVES

Equations describing the dynamics of internal waves in incompressible stratified fluid may be written in the form of Hamiltonian equations, widely used in modern physics. The Hamiltonian formalism appears to be especially convenient for the investigation of weakly nonlinear wave fields.

A common approach to systems with a continuous number of degrees of freedom and a Hamiltonian structure, which generalize classical mechanics methods was developed by Zakharov [406] (see also [407])[1]. From a purely technical viewpoint, operation with equations written in the Hamiltonian form leads to standardization of mathematical means and to a considerable contraction of the calculations. That is why the Hamiltonian formalism will be widely used while expounding the nonlinear theory of oceanic internal waves.

The content of this chapter represents the mathematical basis for the consideration of nonlinear effects appearing while internal waves propagate. In Section 7.1 the basic equations of the dynamics of internal waves are transformed into the Hamiltonian form and some necessary information about the general Hamiltonian theory is presented. In Section 7.2 variables of the initial Hamiltonian are transformed into variables representing, in fact, wave amplitudes. Here the quadratic Hamiltonian part describing linear waves is reduced to a very simple form. Nonlinear effects are determined by higher-order terms of the Hamiltonian expansion in amplitudes powers. Coefficients of this expansion are symmetric and describe all the specifics of the physical system in the present case of internal waves. In Section 7.3 some general transformations of Hamiltonian equations are performed, simplifying their structure and describing the weakly nonlinear wave field. The resulting single equation is the basis for all further derivations.

1.   HAMILTONIAN FORM OF THE BASIC EQUATIONS OF INTERNAL WAVES THEORY

Before turning to the derivation of equations describing internal waves dynamics in an incompressible stratified fluid we shall give some information about the theory of Hamiltonian systems. Notice that the Hamiltonian formalism used in this book will stay within some algebraic means, mainly canonical transformations. We present for reference some equations necessary for further considerations.

Let the Hamiltonian $H(q_i, p_i, t)$ be given, where q_i, p_i are $i = 1, 2, \ldots, n - 2n$ independent variables called canonically conjugated coordinates and momenta. Usually H is the system's energy. Then the corresponding Hamiltonian equations have the form

$$\dot{q}_i = \frac{\partial H}{\partial p_i}; \quad \dot{p}_i = -\frac{\partial H}{\partial q_i}; \quad i = 1, 2, \ldots, n. \tag{7.1}$$

The right hand parts of these equations are coefficients at increments of independent variables in the variation of H:

$$dH = \sum_{i=1}^{n} \left(\frac{\partial H}{\partial q_i} dq_i + \frac{\partial H}{\partial p_i} dp_i \right). \tag{7.2}$$

If the Hamiltonian of the system is independent of time directly, then $H = \text{const}$ for a moving system. In this case

$$\frac{dH}{dt} = \sum_{i=1}^{n} \left(\frac{\partial H}{\partial q_i} \dot{q}_i + \frac{\partial H}{\partial p_i} \dot{p}_i \right) = 0$$

owing to Eqs. (7.1).

We replace variables

$$\tilde{q}_i = \tilde{q}_i(q_1, \ldots, q_n, p_1, \ldots, p_n);$$

$$\tilde{p}_i = \tilde{p}_i(q_1, \ldots, q_n, p_1, \ldots, p_n) \qquad (7.3)$$

and find the conditions for Eqs. (7.1) not to change form. Such transformations are called canonical. So we have

$$\dot{\tilde{q}}_i = \sum_{k=1}^{n} \left(\frac{\partial \tilde{q}_i}{\partial q_k} \frac{\partial H}{\partial p_k} - \frac{\partial \tilde{q}_i}{\partial p_k} \frac{\partial H}{\partial q_k} \right)$$

$$= \sum_{j=1}^{n} \left[\frac{\partial H}{\partial \tilde{q}_j} \sum_{k=1}^{n} \left(\frac{\partial \tilde{q}_i}{\partial q_k} \frac{\partial \tilde{q}_j}{\partial p_k} - \frac{\partial \tilde{q}_i}{\partial p_k} \frac{\partial \tilde{q}_j}{\partial q_k} \right) \right.$$

$$\left. + \frac{\partial H}{\partial \tilde{p}_j} \sum_{k=1}^{n} \left(\frac{\partial \tilde{q}_i}{\partial q_k} \frac{\partial \tilde{p}_j}{\partial p_k} - \frac{\partial \tilde{q}_i}{\partial p_k} \frac{\partial \tilde{p}_j}{\partial q_k} \right) \right].$$

Calculating p_i we obtain conditions for the transformations to be canonical, which may be written through the Poisson brackets:

$$\{\tilde{q}_i, \tilde{q}_j\} = 0; \quad \{\tilde{p}_i, \tilde{p}_j\} = 0; \qquad (7.4)$$

$$\{\tilde{q}_i, \tilde{p}_j\} = \delta_{ij}. \qquad (7.4a)$$

where, by definition, for any functions $f(q_2, \ldots, p_n)$ and $g(q_1, \ldots, p_n)$

$$\{f, g\} = \sum_{k=1}^{n} \left(\frac{\partial f}{\partial q_k} \frac{\partial g}{\partial p_k} - \frac{\partial f}{\partial p_k} \frac{\partial g}{\partial q_k} \right).$$

On the r.h.s of Eq. (7.4a) we have $c\delta_{ij}$ instead of δ_{ij}. To preserve the form of the equations it is obviously enough to redefine the Hamiltonian function by assuming $\tilde{H} = cH$. The constant c is called the valence of the transformation.

The time t may also be used in the replacement (7.3). Here canonicity conditions (7.4) strictly preserve their form, but the Hamiltonian H is transformed in a more complicated way. A transform which is inverse of the canonical transform, as well as a product of these transformations, will also be canonical. Canonical transformations form a group.

There is the following method of the canonical replacements. Consider an arbitrary function of $(2n+1)$ variables $F\left(\tilde{q}_i, p_i, t\right)$ called a generating function. Assuming

$$
q_i = -\frac{\partial F}{\partial p_i}; \quad \tilde{p}_i = -\frac{\partial F}{\partial \tilde{q}_i}; \quad \tilde{H} = H + \frac{\partial F}{\partial t} \tag{7.5}
$$

and expressing the new variables $\tilde{q}_i$, $\tilde{p}_i$ through the old ones q_i, p_i and time t by the above relations again yields the canonical transformation.

A conversion from systems with a finite number of variables to continuously dimensionalized systems causes only a change of notation. A subscript, in this case a variable continuously varying within some range (usually these are coordinates of points of the physical space $\mathbf{r}$ or of the k-space) is used. Summation is changed to integration, partial derivatives turns to so called variational derivatives, which are obtained using the formula companion of Eq. (7.2) by calculation of the variation of the Hamiltonian H. The Kronecker symbol in Eq. (7.4a) is replaced by the δ-function.

Let us show, e.g., that the Fourier transform of the variables $q_\mathbf{r}$, $p_\mathbf{r}$, where $\mathbf{r}$ is a vector in Cartesian space, into variables $q_\mathbf{k}$, $p_{-\mathbf{k}}$ by

$$
q_\mathbf{k} = (2\pi)^{-3/2} \int q_\mathbf{r}\, e^{-i\mathbf{kr}}\, d\mathbf{r}; \quad p_\mathbf{k} = (2\pi)^{-3/2} \int p_\mathbf{r}\, e^{-i\mathbf{kr}}\, d\mathbf{r}, \tag{7.6}
$$

will be the canonical transform. Indeed:

$$
\delta q_\mathbf{k} = (2\pi)^{-3/2} \int \delta q_\mathbf{r}\, e^{-i\mathbf{kr}}\, d\mathbf{r}; \quad \delta p_\mathbf{k} = (2\pi)^{-3/2} \int \delta p_\mathbf{r}\, e^{-i\mathbf{kr}}\, d\mathbf{r},
$$

so that

$$
\frac{\delta q_\mathbf{k}}{\delta q_\mathbf{r}} = (2\pi)^{-3/2}\, e^{-i\mathbf{kr}}; \quad \frac{\delta q_\mathbf{k}}{\delta p_\mathbf{r}} = 0; \quad \frac{\delta p_\mathbf{k}}{\delta q_\mathbf{r}} = 0; \quad \frac{\delta p_\mathbf{k}}{\delta p_\mathbf{r}} = (2\pi)^{-3/2}\, e^{-i\mathbf{kr}}
$$

and, e.g.,

$$
\{q_{\mathbf{k}_1}, p_{-\mathbf{k}_2}\} = \int \left(\frac{\delta q_{\mathbf{k}_1}}{\delta q_\mathbf{r}} \frac{\delta p_{-\mathbf{k}_2}}{\delta p_\mathbf{r}} - \frac{\delta q_{\mathbf{k}_1}}{\delta p_\mathbf{r}} \frac{\delta p_{-\mathbf{k}_2}}{\delta q_\mathbf{r}} \right) d\mathbf{r}
$$

$$
= (2\pi)^{-3} \int e^{-i\mathbf{k}_1\mathbf{r} + i\mathbf{k}_2\mathbf{r}}\, d\mathbf{r} = \delta\left(\mathbf{k}_1 - \mathbf{k}_2\right).
$$

Eqs. (7.4) may be verified in a similar way.

Let us now turn to the formulation of the desired equations by the Hamiltonian following the work by Voronovich [381]. We further always use the isopycnic approximation [see Eq. (2.15a)] and neglect the Earth's rotation [see Eqs. (2.13) for $f = 0$]. The Hamiltonian usually represents the system's energy, but it should be expressed through canonical variables, which are, generally speaking, nontrivial. According to [381] we proceed as follows: assume that

$$H = \int \left(\frac{(\rho_0 + \rho)\,\mathbf{v}^2}{2} + \pi\,(\rho, z) \right)\, d\mathbf{r}\,, \qquad (7.7)$$

where ρ is the density's deviation from the stable value $\rho_0\,(z)$,

$$\pi\,(\rho, z) = \pi\,(\rho_0 + \rho) - \pi\,(\rho_0) + \rho g z$$

π is the density of the potential energy, the expression for which will be derived below; $\mathbf{v}$ is the fluid velocity

$$\mathbf{v} = \frac{1}{\rho_0 + \rho}\,(\nabla\Phi + \lambda\nabla\,(\rho_0 + \rho))\,, \qquad (7.8)$$

Φ and λ are some new functions. Expression (7.8) corresponds to the Clebsch form [189] widely known in hydrodynamics. Owing to the fluid's incompressibility div $\mathbf{v} = 0$. As the Hamiltonian should be a function of the canonical variables λ and ρ we consider Φ to be the functional of λ and ρ yielding the solution of the equation div $\mathbf{v} = 0$, where $\mathbf{v}$ is given in the form (7.8). We further consider λ to be a canonical coordinate and ρ is a momentum. The radius vector $\mathbf{r}$ of a point in space is used as the subscript. Assume also that the fluid occupies some volume with fixed boundaries at rest. Let us calculate the variation of H. Integrating by parts yields

$$\delta H = \int \left[-\nabla\mathbf{v}\,\delta\Phi + \left(-\frac{\mathbf{v}^2}{2} - \nabla\left(\lambda\frac{v}{\rho} \right) + \Pi'\,(\rho_0 + \rho) + g z \right) \delta\rho \right.$$

$$\left. + \ \mathbf{v}\nabla\,(\rho_0 + \rho)\,\delta\lambda \right]\, d\mathbf{r}\,.$$

Hence it follows that the incompressibility condition may be defined by the equality $\delta H/\delta \rho = \operatorname{div} \mathbf{v} = 0$ and the canonical equations have the form

$$\dot{\lambda} = \frac{\delta H}{\delta \rho} = -\mathbf{v}\nabla\lambda + \Pi'\left(\rho_0 + \rho\right) + gz - \frac{\mathbf{v}^2}{2}; \qquad (7.9)$$

$$\dot{\rho} = -\frac{\delta H}{\delta \lambda} = -\mathbf{v}\nabla\left(\rho_0 + \rho\right). \qquad (7.9a)$$

Eq. (7.9a) is the continuity equation. The Euler equation is satisfied due to (7.9)–(7.9a). This follows from the directly verified identity:

$$\frac{\partial}{\partial t}\left(\left(\rho_0 + \rho\right)\mathbf{v}\right) + \left(\mathbf{v}\nabla\right)\left(\left(\rho_0 + \rho\right)\mathbf{v}\right) + \left(\rho_0 + \rho\right)\nabla\left(gz\right) + \nabla p$$

$$= \lambda\nabla\left(\frac{\partial \rho}{\partial t} + \left(\mathbf{v}\nabla\right)\left(\rho_0 + \rho\right)\right)$$

$$+ \nabla\left(\rho_0 + \rho\right)\left(\frac{\partial \lambda}{\partial t} + \left(\mathbf{v}\nabla\right)\lambda - \frac{\partial}{\partial \rho}\left(\Pi\left(\rho_0 + \rho\right) + gz\left(\rho_0 + \rho\right)\right) + \frac{\mathbf{v}^2}{2}\right)$$

$$+ \nabla\left(\frac{\partial \Phi}{\partial t} + \left(\mathbf{v}\nabla\right)\Phi + \Pi\left(\rho_0 + \rho\right) + gz\left(\rho_0 + \rho\right) - \frac{\left(\rho_0 + \rho\right)\mathbf{v}^2}{2} + p\right).$$

The expression (7.8) for the velocity is taken into account while deriving the latter equation. Simultaneously we obtain the expression for the pressure companion of the Cauchy–Lagrange integral:

$$p = \frac{\left(\rho_0 + \rho\right)\mathbf{v}^2}{2} - \frac{\partial \Phi}{\partial t} - \left(\mathbf{v}\nabla\right)\Phi - \Pi\left(\rho_0 + \rho\right) - \left(\rho_0 + \rho\right)gz + \text{const}. \quad (7.10)$$

In the previous discussion we have assumed gz to be a potential depending arbitrarily on coordinates, but later we consider the homogeneous gravity field.

We define the function $\Pi\left(R\right)$, which still may be arbitrary, by

$$\Pi\left(R\right) = -g\int_0^R \eta\left(\xi\right)d\xi,$$

where $\eta\left(\xi\right)$ is the function inverse to $\rho_0\left(z\right)$, i.e., $\eta\left(\rho_0\right) = z$. Then the potential energy density may be written in a form showing that it is equal to the work necessary for replacing a fluid particle of the density $(\rho_0 + \rho)$ form its equilibrium level at the level $\eta \, (\rho_0 + \rho)$ to the point in space considered

$$\pi \, (\rho, z) = \Pi \, (\rho_0 + \rho) - \Pi \, (\rho_0) + \rho g z = g \int\limits_{\eta \, (\rho_0 + \rho)}^{z} (\rho_0 + \rho - \rho_0 \, (\xi)) \, d\xi \, .$$

Now it is finally clear that the Hamiltonian in this case coincides with the energy of the medium. It is easy to obtain the following expansion of the potential energy density in powers of ρ:

$$\pi \, (\rho, z) = -\frac{g}{\rho_0'} \frac{\rho^2}{2!} + \frac{g\rho_0''}{\rho_0'^3} \frac{\rho^3}{3!} + \frac{g}{\rho_0'} \left(\frac{\rho_0''}{\rho_0'^3} \right)' \frac{\rho^4}{4!} + \dots , \qquad (7.11)$$

where a prime means partial derivation with respect to z.

The vector Euler equation (2.13)

$$\operatorname{curl} (\rho_0 + \rho) \, \mathbf{v} = [\, \nabla \lambda, \, \nabla \, (\rho_0 + \rho) \,] . \qquad (7.12)$$

If this equation may be solved with respect to λ for initially given distributions of ρ and $\mathbf{v}$, then expression (7.8) exists. Indeed, in this case

$$\operatorname{curl} (\rho_0 + \rho) \, \mathbf{v} - \lambda \nabla \, (\rho_0 + \rho) = 0,$$

and therefore we can find a function Φ such that

$$(\rho_0 + \rho) \, \mathbf{v} - \lambda \nabla \, (\rho_0 + \rho) = \nabla \, \Phi.$$

The values of λ, ρ, Φ versus time obtained from Eqs. (7.9) yield the solution of the initial hydrodynamics equations which is considered to be unique. Thus expression (7.8) will be valid. The required solubility condition of the equation is the orthogonality of the mass flux curl $(\rho_0 + \rho) \, \mathbf{v}$ of the density gradient. Let us define some class of motions which certainly solve Eq. (7.12) and thus is covered by the present consideration. In particular, assume that $\rho = 0$ for $t = 0$ and that the vector $\mathbf{v}$ has two components, e.g., v_x and v_z, independent of y. Then (7.12) is reduced to

$$\frac{\partial \lambda}{\partial t} = \frac{1}{\rho_0'} \left(\frac{\partial}{\partial x} (\rho_0 v_z) - \frac{\partial}{\partial z} (\rho_0 v_x) \right) ,$$

which is always soluble for $\rho_0' \neq 0$. For $\rho = 0$ Eq. (7.12) becomes linear with respect to λ and $\mathbf{v}$. This means that the velocity field for $t = 0$ is formed by a superposition of such plane motions. The resulting fluid motion belongs to the suggested class, which appears to be satisfactory for all of possible physical situations and is certainly valid for all of problems considered further.

We may turn to the Lagrange function L from the Hamiltonian function H:

$$L = \sum_{i=1}^{n} p_i \dot{q}_i - H\left(q_i, p_i\right).$$

In the present case the validity of the identity

$$\rho \dot{\lambda} - \frac{dH}{dv} = p + \frac{\partial}{\partial t} \left(\rho \lambda + \Phi\right) + \Pi\left(\rho_0\right) + \rho_0 g z,$$

where

$$\frac{dH}{dv} = \frac{\rho_0 + \rho}{2} v^2 + \pi\left(\rho, z\right),$$

may be verified by the relations (7.9), (7.10).

As the Lagrange function is defined with an accuracy of the increment of the arbitrary function $\frac{d}{dv}\Phi\left(q_i, t\right)$, in this case the pressure may be used as the Lagrangian density, Seliger and Whitham [323]. To pass finitely to the Lagrange formalism, which is also frequently used in the nonlinear wave theory [391], we should express the generalized momentum ρ through λ using Eq. (7.9) and rewrite ρ in the form of the λ and $\dot{\lambda}$ functional. But it is, seemingly, impossible to express ρ through λ and $\dot{\lambda}$ directly. Nevertheless, there is another opportunity to construct the Lagrange function connected with an increase of the dimensionality of the space and leading to a Lagrangian degenerate in some respect. Thus when assuming

$$L = \sum \dot{p}_i q_i + H\left(q_i, p_i\right)$$

and considering q_i and p_i to be generalized coordinates, such a Lagrangian will have the equations of motion

$$\frac{d}{dt} \frac{\partial L}{\partial \dot{q}_i} - \frac{\partial L}{\partial q_i} = 0$$

coinciding with the canonical equations (7.1). In this case such a method leads to:

$$L = \int \left(\dot\rho\lambda + \frac{(\rho_0 + \rho)\,\mathbf{v}^2}{2} + \pi\,(\rho, z) \right) d\mathbf{r} \,,$$

$\mathbf{v}$ is given by Eq. (7.8). The equations of motion (7.9) and the incompressibility condition div $\mathbf{v} = 0$ follow from the principle of least action $\delta S = 0$, where $S = \int L\, dt$, for independently varying Φ, ρ and λ. This Lagrangian is similar that presented in [323], although it does not coincide with it.

Knowledge of L appears to be useful when studying nonlinear waves, allowing the use of the averaged Lagrangian technique suggested by Whitham [8], [391]. The Boussinesq approximation (see Section 2.1) will be frequently used later on. In this case $\rho \ll \rho_0$ and $\rho_0 \approx 1$ g/cm^3, which allows us to write the Hamiltonian in the following approximate form:

$$H \approx H_{(B)} = \int \left(\frac{1}{2}\mathbf{v}^2 + \pi\,(\rho, z) \right) d\mathbf{r}\,, \tag{7.13}$$

where

$$v = \nabla\Phi + \lambda\,\nabla\,(\rho_0 + \rho) \tag{7.14}$$

(the density is now considered to be dimensionless and is measured as a part of the mean water density), and the expression for $\pi\,(\rho, z)$ has remained the same. While solving the equation div $\mathbf{v} = 0$ we derive the variable Φ,

$$\Phi = \Delta^{-1}\,\{\nabla\,(\,\lambda\,\nabla\,(\rho_0 + \rho)\,)\}\,.$$

The canonical equations now have the form

$$\dot\lambda = \frac{\delta H}{\delta\rho} = -(\,\mathbf{v}\nabla)\,\lambda + g\,(z - \eta\,(\rho_0 + \rho))\,; \tag{7.15}$$

$$\dot\rho = -\frac{\delta H}{\delta\lambda} = -(\,\mathbf{v}\nabla)\,(\rho_0 + \rho)\,. \tag{7.15a}$$

The value $z - \eta\,(\rho_0 + \rho)$ is the vertical displacement of a fluid particle. By analogy with what has been previously stated it is easy to make sure that writing the Euler equation in the Boussinesq approximation, i.e., assuming

$$\frac{\partial\mathbf{v}}{\partial t} + (\,\mathbf{v}\nabla)\,\mathbf{v} + g\rho\,\nabla z = -\nabla p\,,$$

the latter equation is satisfied as a consequence of the relations (7.15). So the system satisfies the Hamiltonian under the Boussinesq approximation.

For two-dimensional motion occurring on the plane (x, z) the stream function may be derived by the formulae

$$v_x = + \frac{\partial \psi}{\partial z}; \quad v_z = - \frac{\partial \psi}{\partial x}.$$

Then Eqs. (7.14), (7.15) will be rewritten in the form

$$\dot{\lambda} = \frac{\partial \psi}{\partial x} \frac{\partial \lambda}{\partial z} - \frac{\partial \psi}{\partial z} \frac{\partial \lambda}{\partial x} - g \left(\eta \left(\rho_0 + \rho \right) - z \right);$$

$$\dot{\rho} = \frac{\partial \psi}{\partial x} \frac{\partial}{\partial t} \left(\rho_0 + \rho \right) - \frac{\partial \psi}{\partial z} \frac{\partial}{\partial x} \left(\rho_0 + \rho \right);$$

$$\Delta \psi = \frac{\partial \lambda}{\partial z} \frac{\partial}{\partial x} \left(\rho_0 + \rho \right) - \frac{\partial \lambda}{\partial x} \frac{\partial}{\partial z} \left(\rho_0 + \rho \right). \tag{7.16}$$

Eqs. (7.16) lead to equation (2.33) for stationary internal waves. When writing the solution for all variables in the form of functions of $(x - ct, z)$ it is easy to see that

$$\rho = \rho_0 \left(z - \frac{\psi}{c} \right) - \rho_0 \left(z \right),$$

and the remaining equations immediately yield

$$\Delta \psi + \frac{\psi}{c^2} N_0^2 \left(z + \frac{\psi}{c} \right) = 0.$$

In concluding this section we present the Hamiltonian describing internal waves at a plane parallel flow.

Assume that together with the stratification there is a plane fluid flow with the velocity profile $\mathbf{u}_0$. Such a state is described by the solution of Eqs. (7.15):

$$\rho^{(0)} = 0; \quad \lambda^{(0)} = \frac{\mathbf{u}_0'}{\rho_0'} \left(\mathbf{u}_0 t - \mathbf{r} \right); \quad \Phi^{(0)} = - \frac{\mathbf{u}_0^2}{2} t + \mathbf{u}_0 \mathbf{r}$$

(the Boussinesq approximation is used here). Direct consideration of wave motions appearing at this 'basic' state background is problematical, as the latter is described by functions directly depending on coordinates and time. To avoid this difficulty we perform a canonical transformation with the following generating function

$$F\left(\rho, \tilde{\lambda}\right) = -\int \left[\tilde{\lambda}\rho + \frac{1}{2}\,\mathbf{u}_0^2\left(\rho_0 + \rho\right)t - \mathbf{u}_0\left(\rho_0 + \rho\right)\mathbf{r} \right]\, d\mathbf{r}$$

and assume that

$$\Phi = \Phi + \mathbf{u}_0\left(\rho_0 + \rho\right)\mathbf{r} - \frac{\mathbf{u}_0^2\left(\rho_0 + \rho\right)}{2}$$

Here $\tilde{\lambda}$ is a new canonical coordinate; in addition we assume that the undisturbed velocity profile $\mathbf{u}_0$ is a function of ρ_0 without loss of generality, because ρ_0 may be considered to be a single-valued function of z under the condition $\rho_0' \neq 0$. The transformation equations (7.5) are written as

$$\lambda = -\frac{\delta F}{\delta \rho}; \quad \tilde{\rho} = -\frac{\delta F}{\delta \tilde{\lambda}}; \quad \tilde{H} = H + \frac{\partial F}{\partial t}$$

and yield the relations

$$\tilde{\rho} = \rho; \quad \lambda = \tilde{\lambda} + \frac{\mathbf{u}_0'}{\rho_0'}\left(\mathbf{u}_0 t - \mathbf{r}\right);$$

$$H = \int \left[\frac{1}{2}\mathbf{v}^2 - \frac{1}{2}\mathbf{u}_0^2\left(\rho_0 + \rho\right) + \pi\left(\rho, z\right) \right]\, d\mathbf{r}. \tag{7.17}$$

Here a prime denotes derivation with respect to z, so that

$$\frac{d u_0}{d \rho_0} = \frac{u_0'}{\rho_0'}$$

Substituting expressions for λ and Φ into Eq. (7.14) yields:

$$\mathbf{v} = \mathbf{u}_0\left(\rho_0 + \rho\right) + \nabla\,\tilde{\varphi} + \tilde{\lambda}\,\nabla\left(\rho_0 + \rho\right).$$

The potential energy density is again given by (7.11) and Φ yields from the continuity equation div $\mathbf{v} = 0$

$$\Phi = \Delta^{-1}\left\{-\nabla\left(\lambda\,\nabla\left(\rho_0 + \rho\right)\right) - \nabla\,\mathbf{u}_0\left(\rho_0 + \rho\right)\right\}.$$

The Hamiltonian (7.17) represents the disturbance energy of the mean medium state since it is equal to the work of the system's conversion from the state of rest to the present state.

2.   INTRODUCTION OF ORTHOGONAL VARIABLES. HAMILTONIAN FORM OF EQUATIONS FOR A WEAKLY NONLINEAR WAVE FIELD

In this section we present the procedure of canonical transformation of the initial variations $\lambda_{\mathbf{r}}$, $\rho_{\mathbf{r}}$ to internal wave amplitudes (the subscript $\mathbf{r}$ denotes that these variables are spatial coordinate functions). This procedure is usually called a diagonalization of the Hamiltonian. To simplify formulae the Boussinesq approximation will be used, which usually appears to be sufficient. The initial Hamiltonian has the form:

$$H = \int \left[\frac{1}{2} \mathbf{v}^2 + \pi \left(\rho, z \right) \right] dV \,, \tag{7.18}$$

where

$$\mathbf{v} = \nabla \varphi + \lambda \nabla \left(\rho_0 + \rho \right) ; \tag{7.19}$$

$$\pi \left(\rho, z \right) = \Pi \left(\rho_0 + \rho \right) - \Pi \left(\rho_0 \right) + \rho g z$$

$$= -\frac{g}{\rho_0'} \frac{\rho^2}{2!} + \frac{g \rho_0''}{(\rho_0')^3} \frac{\rho^3}{3!} + \frac{g}{\rho_0'} \left(\frac{\rho_0''}{(\rho_0')^3} \right) \frac{\rho^4}{4!} + \ldots \tag{7.20}$$

Integrating by parts the term $\nabla \Phi \mathbf{v}$ in Eq. (7.18), taking into account that $\mathbf{v}_n = 0$ at rigid boundaries, and neglecting in Eq. (7.20) terms of orders higher than three (which corresponds to the weakly nonlinear approximation) yields

$$H = \frac{1}{2} \int \left[\lambda \nabla \left(\rho_0 + \rho \right) \left(\nabla \Phi + \lambda \nabla \left(\rho_0 + \rho \right) \right) - \frac{g}{\rho_0'} \rho^2 + \frac{g \rho_0''}{\rho_0'^3} \frac{\rho^3}{3} \right] dV \,. \tag{7.21}$$

Assume now that our system is horizontally homogeneous. This means that the present boundaries (which may lie, in particular, at infinity) should be horizontal. Let us turn to the Fourier transform with respect to the vector $\mathbf{r}$, which will, furthermore, always relate to horizontal coordinates:

$$f_{\mathbf{r}} \left(z \right) = \frac{1}{2\pi} \int f_{\mathbf{k}} \left(z \right) e^{i\mathbf{k}\mathbf{r}} \, d\mathbf{k} ; \quad f_{\mathbf{k}} \left(z \right) = \frac{1}{2\pi} \int f_{\mathbf{r}} \left(z \right) e^{i\mathbf{k}\mathbf{r}} \, d\mathbf{r} \,.$$

This transform, as shown in Section 7.1, is canonical; the variables $\lambda_{\mathbf{k}}$, $\rho_{\mathbf{k}}$ will be now conjugated. As the values of λ and ρ are real they should be $\lambda_{\mathbf{k}} = \bar{\lambda}_{-\mathbf{k}}$, $\rho_{\mathbf{k}} = \bar{\rho}_{-\mathbf{k}}$, where a bar means complex conjugation. Obtain first a direct expression for $\Phi_{\mathbf{k}}$ through $\lambda_{\mathbf{k}}$ and $\rho_{\mathbf{k}}$. The incompressibility condition div $\mathbf{v} = 0$ has the form

$$\frac{\partial}{\partial z} \left(\frac{\partial \Phi_{\mathbf{k}}}{\partial z} + \left[\lambda \frac{\partial}{\partial z} (\rho_0 + \rho) \right]_{\mathbf{k}} \right) - k^2 \Phi_{\mathbf{k}} = - \left[\frac{\partial}{\partial x_i} \left(\lambda \frac{\partial \rho}{\partial x_i} \right) \right]_{\mathbf{k}}. \quad (7.22)$$

Here $[\ ]_{\mathbf{k}}$ denotes the $\mathbf{k}$-th component of the Fourier transform for the corresponding expression and $\frac{\partial}{\partial x_i}$, $i = 1,\ 2$ is a derivative operator with respect to horizontal coordinates (a summation for the same subscripts is performed). Derivation of this equation with respect to z yields:

$$\frac{\partial^2}{\partial z^2} \left(\frac{\partial \Phi_{\mathbf{k}}}{\partial z} + \left[\lambda \frac{\partial}{\partial z} (\rho_0 + \rho) \right]_{\mathbf{k}} \right) - k^2 \left(\frac{\partial \Phi_{\mathbf{k}}}{\partial z} + \left[\lambda \frac{\partial}{\partial z} (\rho_0 + \rho) \right]_{\mathbf{k}} \right)$$

$$= - \frac{\partial}{\partial z} \left[\frac{\partial}{\partial x_i} \left(\lambda \frac{\partial \rho}{\partial x_i} \right) \right]_{\mathbf{k}} - k^2 \left[\lambda \frac{\partial}{\partial z} (\rho_0 + \rho) \right] \mathbf{k}. \quad (7.23)$$

Hence the following expression for the vertical velocity component yields

$$\frac{\partial \Phi_{\mathbf{k}}}{\partial z} + \left[\lambda \frac{\partial}{\partial z} (\rho_0 + \rho) \right]_{\mathbf{k}} = - \int G_k (z, z') \left\{ \frac{\partial}{\partial z} \left[\frac{\partial}{\partial x_i} \left(\lambda \frac{\partial \rho}{\partial x_i} \right) \right]_{\mathbf{k}} \right.$$

$$\left. + k^2 \left[\lambda \frac{\partial}{\partial z} (\rho_0 + \rho) \right]_{\mathbf{k}} \right\}_{z=z'} dz',$$

where $G_k (z, z')$ is the Green's function:

$$\left(\frac{\partial^2}{\partial z^2} - k^2 \right) G_k = \delta (z - z') \quad \text{and} \quad G_k (z, z') = 0\,,$$

if z belongs to the region's boundary, here $G_k (z, z') = G_k (z', z)$. If fluid boundaries are at infinity, then

$$G_k (z, z') = - \frac{1}{2k} e^{-k|z-z'|},$$

where $k = |\mathbf{k}|$. But we do not need the direct expression for the Green's function. One more differentiation of Eq. (7.23) with respect to z and using (7.22) yield

$$\Phi_{\mathbf{k}} = \frac{1}{k^2} \left[\frac{\partial}{\partial x_i} \left(\lambda \frac{\partial \rho}{\partial x_i} \right) \right]_{\mathbf{k}}$$

$$-\frac{1}{k^2}\frac{\partial}{\partial z}\int G_k(z,z')\left\{\frac{\partial}{\partial z}\left[\frac{\partial}{\partial x_i}\left(\lambda\frac{\partial\rho}{\partial x_i}\right)\right]_k\right.$$

$$\left.+k^2\left[\lambda\frac{\partial}{\partial z}(\rho_0+\rho)\right]_k\right\}_{z=z'}dz'.\qquad(7.24)$$

The solution obtained is unique with an accuracy to an additional unessential constant. Retaining in the Hamiltonian (7.21) only quadratic terms with respect to variables λ, ρ we have

$$H=-\frac{1}{2}\int\left\{k^2\rho_0'\lambda_{-k}\int G_k(\rho_0'\lambda_k)_{z=z'}dz'+\frac{g}{\rho_0'}\rho_k\rho_{-k}\right\}dz\,dk.\qquad(7.25)$$

Canonical equations for H_0 have the following form:

$$\dot\lambda_k=\frac{\delta H_0}{\delta\rho_{-k}}=-\frac{g}{\rho_0'}\rho_k;\quad\dot\rho_k=-\frac{\delta H_0}{\delta\lambda_{-k}}=\rho_0'k^2\int G_k(\rho_0'\lambda_k)_{z=z'}dz'.$$

Assuming

$$\lambda_k=\varphi_k^\nu(z)\exp(-i\omega_k^\nu t),\quad\rho_k=i\omega_k^\nu\frac{\rho_0'}{g}\varphi_k^\nu(z)\exp(-i\omega_k^\nu t),$$

where the superscript ν corresponds to different modes, the integral equation yields

$$\varphi_k^\nu=-\frac{k^2}{(\omega_k^\nu)^2}\int G_k(N_0^2\varphi_k^\nu)_{z=z'}dz',\quad\text{where}\ \ N_0^2=-g\rho_0'.$$

Applying the operator $\frac{\partial^2}{\partial z^2}-k^2$ and using the definition of the Green's function we have

$$\frac{d^2}{dz^2}\varphi_k^\nu+k^2\left(\frac{N_0^2}{(\omega_k^\nu)^2}-1\right)\varphi_k^\nu=0;\quad\varphi_k^\nu=0,\quad z\in\text{the region boundary}.$$

$$(7.26)$$

We have obtained a basic boundary value problem for internal waves in the Boussinesq approximation, which was thoroughly discussed in Section 3.1 [see formulae (3.3) and (3.4a)]. The complete eigenfunction system for this boundary value problem is normalized by the condition

$$\int N_0^2\varphi_k^\nu\bar\varphi_k^\mu\,dz=\delta(\nu-\mu).\qquad(7.27)$$

Frequencies $\omega_{\mathbf{k}}^{\nu}$ and functions $\varphi_{\mathbf{k}}^{\nu}$ depend in fact on k^2 so that $\varphi_{\mathbf{k}}^{\nu} = \varphi_{-\mathbf{k}}^{\nu} = \varphi_{k}^{\nu}$, $\omega_{\mathbf{k}}^{\nu} = \omega_{-\mathbf{k}}^{\nu} = \omega_{k}^{\nu}$ and $\omega_{\mathbf{k}}^{\nu} > 0$. If internal waves propagate in the form of a superposition of modes, which occurs when boundaries are not infinitely far apart, then φ_{k}^{ν} may be chosen to be real. If internal waves propagate infinitely in the vertical direction, then these functions will be complex. The latter relates to internal waves with inclined wave vector. Introduce new variables $a_{\mathbf{k}}^{\nu}$ and $\bar{a}_{\mathbf{k}}^{\nu}$:

$$a_{\mathbf{k}}^{\nu} = -\frac{i}{g} \sqrt{\frac{\omega_{k}^{\nu}}{2}} \int \left(N_0^2 \lambda_{\mathbf{k}}(z,t) + \frac{i\,g^2}{\omega_{k}^{\nu}} \rho_{\mathbf{k}}(z,t) \right) \varphi_{k}^{\nu}\, dz \,; \qquad (7.28)$$

$$\bar{a}_{\mathbf{k}}^{\nu} = \frac{i}{g} \sqrt{\frac{\omega_{k}^{\nu}}{2}} \int \left(N_0^2 \lambda_{\mathbf{k}}(z,t) - \frac{i\,g^2}{\omega_{k}^{\nu}} \rho_{-\mathbf{k}}(z,t) \right) \bar{\varphi}_{k}^{\nu}\, dz \,. \qquad (7.28a)$$

From (7.27) follows

$$N_0^2 \int \varphi_{k}^{\nu}(z)\, \bar{\varphi}_{k}^{\nu}(z')\, d\nu = \delta(z - z') \,. \qquad (7.29)$$

Indeed, multiplying (7.27) by $\varphi_{k}^{\nu}(z')$ and integrating over ν yields (7.29) owing to the completeness of the system φ_{k}^{ν}. Write the Poisson brackets for $a_{\mathbf{k}}^{\nu}$ and $\bar{a}_{\mathbf{k}}^{\nu}$, which are now the generalized coordinates and momenta accordingly, and make sure that the canonicity conditions for Eq. (7.4) in Section 2.4 are satisfied. From (7.28) we obtain:

$$\frac{\delta a_{\mathbf{k'}}^{\nu}}{\delta \lambda_{\mathbf{k}}(z)} = -\frac{i}{g} \sqrt{\frac{\omega_{k}^{\nu}}{2}} N_0^2 \varphi_{k}^{\nu}(z)\, \delta(\mathbf{k'} - \mathbf{k}) \,;$$

$$\frac{\delta a_{\mathbf{k'}}^{\nu}}{\delta \rho_{-\mathbf{k}}(z)} = \frac{g}{\sqrt{2\omega_{k}^{\nu}}} \varphi_{k}^{\nu}(z)\, \delta(\mathbf{k'} + \mathbf{k}) \,;$$

$$\frac{\delta \bar{a}_{\mathbf{k}}^{\nu}}{\delta \lambda_{\mathbf{k}}(z)} = \frac{i}{g} \sqrt{\frac{\omega_{k}^{\nu}}{2}} N_0^2 \bar{\varphi}_{k}^{\nu}(z)\, \delta(\mathbf{k'} + \mathbf{k}) \,;$$

$$\frac{\delta \bar{a}_{\mathbf{k'}}^{\nu}}{\delta \rho_{-\mathbf{k}}(z)} = \frac{g}{\sqrt{2\omega_{k}^{\nu}}} \bar{\varphi}_{k}^{\nu}(z)\, \delta(\mathbf{k'} - \mathbf{k}) \,.$$

Further we use the simple formula:

$$\int N_0^2 \varphi_{k}^{\nu} \varphi_{k}^{\mu}\, dz = 0 \quad \text{for} \quad \nu \neq \mu, \qquad (7.30)$$

which is obtained from Eqs. (7.26) by the standard procedure. Let us now calculate the Poisson brackets

$$\{a^\nu_{\mathbf{k'}}, a^\mu_{\mathbf{k''}}\} = \int \left(\frac{\delta a^\nu_{\mathbf{k'}}}{\delta \lambda_{\mathbf{k}}} \frac{\delta a^\mu_{\mathbf{k''}}}{\delta \rho_{-\mathbf{k}}} - \frac{\delta a^\nu_{\mathbf{k'}}}{\delta \rho_{-\mathbf{k}}} \frac{\delta a^\mu_{\mathbf{k''}}}{\delta \lambda_{\mathbf{k}}} \right) d\mathbf{k}\, dz$$

$$= \frac{1}{2} \left(\sqrt{\frac{\omega^\nu_k}{\omega^\mu_k}} - \sqrt{\frac{\omega^\mu_k}{\omega^\nu_k}} \right) \int N_0^2 \varphi^\nu_k \varphi^\mu_k \, dz\, \delta\,(k' + k'') = 0$$

taking into consideration the above equation. By analogy $\{\bar a^\nu_{\mathbf{k'}}, \bar a^\mu_{\mathbf{k''}}\} = 0$. Let us prove the latter condition

$$\{\bar a^\nu_{\mathbf{k'}}, \bar a^\mu_{\mathbf{k''}}\} = \int \left(\frac{\delta a^\nu_{\mathbf{k'}}}{\delta \lambda_{\mathbf{k}}} \frac{\delta \bar a^\mu_{\mathbf{k''}}}{\delta \rho_{-\mathbf{k}}} - \frac{\delta a^\nu_{\mathbf{k'}}}{\delta \rho_{-\mathbf{k}}} \frac{\delta \bar a^\mu_{\mathbf{k''}}}{\delta \lambda_{\mathbf{k}}} \right) d\mathbf{k}\, dz$$

$$= -\frac{i}{2} \left(\sqrt{\frac{\omega^\nu_k}{\omega^\mu_k}} + \sqrt{\frac{\omega^\mu_k}{\omega^\nu_k}} \right) \int N_0^2 \varphi^\nu_k \bar\varphi^\mu_k \, dz\, \delta\,(k' - k'')$$

$$= -i\,\delta\,(k' - k'')\,\delta\,(\nu - \mu)\,.$$

The valency of this transform is equal to $-i$. The inverse transform is described by

$$\lambda_{\mathbf{k}}(z) = i g \int \frac{1}{\sqrt{2\omega^\nu_k}} (\bar\varphi^\nu_k a^\nu_{\mathbf{k}} - \varphi^\nu_k \bar a^\nu_{-\mathbf{k}})\, d\nu\,; \qquad (7.31)$$

$$\rho_{\mathbf{k}}(z) = -\rho_0' \int \sqrt{\frac{\omega^\nu_k}{2}} (\bar\varphi^\nu_k a^\nu_{\mathbf{k}} + \varphi^\nu_k \bar a^\nu_{-\mathbf{k}})\, d\nu\,. \qquad (7.31a)$$

They may be obtained by the direct substitution of (7.31) into (7.28) and then using (7.29).

As the latter transform is canonical, equations of motion versus new variables then have the form

$$\dot a^\nu_{\mathbf{k}} = -i \frac{\delta H}{\delta \bar a^\nu_{\mathbf{k}}}\,; \qquad \dot{\bar a}^\nu_{\mathbf{k}} = i \frac{\delta H}{\delta a^\nu_{\mathbf{k}}}\,. \qquad (7.32)$$

As a consequence of the reality conditions for $\lambda_{\mathbf{k}} = \bar\lambda_{-\mathbf{k}}$, $\rho_{\mathbf{k}} = \bar\rho_{-\mathbf{k}}$ and Eqs. (7.28) the values of variables $a^\nu_{\mathbf{k}}$ and $\bar a^\nu_{\mathbf{k}}$ are complex conjugate. Thus the second of Eqs. (7.32) may be omitted. The initial simultaneous equations result in the single equation

$$\dot a^\nu_{\mathbf{k}} + i \frac{\delta H}{\delta \bar a^\nu_{\mathbf{k}}} = 0\,, \qquad (7.33)$$

and we should only calculate the Hamiltonian H, substituting the relations (7.31) into (7.21). Its quadratic part H_0 has the form

$$H_0 = \int \omega_k^\nu a_k^\nu \bar{a}_k^\nu \, d\nu \, d\mathbf{k} . \tag{7.34}$$

So the value of the energy of the linear wave field is equal to $\int \omega_k^\nu |a_k^\nu|^2 \, d\nu \, d\mathbf{k}$. In our case $\omega_k^\nu > 0$ and the energy of the internal waves is positive. The sign of the wave energy in the general case coincides with the sign of the frequency. The appearance of wave disturbances increases, as a rule, the system's energy. But the situation when the system's energy decreases is possible as well. Such waves are called 'waves with negative energy' and have negative frequencies $\omega_k^\nu < 0$. If there are such waves in a system then it should be unstable from the energy viewpoint. One of such mechanisms causing this instability will be considered in Section 10.2 for internal waves at a shear flow.

The equation of motion (7.33) for $H = H_0$ yields $\dot{a}_\mathbf{k}^\nu + i \omega_k^\nu a_\mathbf{k}^\nu = 0$, whence

$$a_\mathbf{k}^\nu = a_\mathbf{k}^\nu (0) \, e^{-i \omega_k^\nu t} . \tag{7.35}$$

It is obvious that a_k^ν are normal variables of the oscillating system formed by a stably stratified incompressible liquid in the gravity field. The relations (7.28), (7.35) and (7.31) yield a solution of the Cauchy problem for the initial linear equations. Assuming

$$a_\mathbf{k}^\nu (0) = a_0 \delta (\nu - \nu_0) \delta (\mathbf{k} - \mathbf{k}_0) ,$$

yields

$$\rho = -C \rho_0' \, \omega_{k_0}^{\nu_0} \bar{\varphi}_{k_0}^{\nu_0} + \text{c.c.} ; \quad \lambda = i \, C \, g \bar{\varphi}_{k_0}^{\nu_0} + \text{c.c.} ;$$

$$\Phi = -i \, C \, \frac{(\omega_{k_0}^{\nu_0})^2}{k_0^2} \frac{\partial}{\partial z} \, \bar{\varphi}_{k_0}^{\nu_0} + \text{c.c.} , \tag{7.36}$$

the vertical and the horizontal velocity components have the form

$$w = -i \, C \, (\omega_{k_0}^{\nu_0})^2 \bar{\varphi}_{k_0}^{\nu_0} + \text{c.c.} ; \quad \bar{\mathbf{v}} = C \, \frac{(\omega_{k_0}^{\nu_0})^2}{k_0^2} \mathbf{k}_0 \frac{\partial}{\partial z} \, \bar{\varphi}_{k_0}^{\nu_0} + \text{c.c.} , \tag{7.36a}$$

where

$$C = \frac{a_0}{2\pi} \frac{1}{\sqrt{2\omega_{k_0}^{\nu_0}}} \exp i \, (\mathbf{k}_0 \mathbf{r} - \omega_{k_0}^{\nu_0} t) .$$

Such a solution describes a monochromatic internal wave, propagating in the mode form or in the form of a plane wave (when boundaries are infinitely far).

To interpret solutions of different problems containing the values $a_{\mathbf{k}}^{\nu}$ it is useful to stress the following. Assume that there is an internal waves packet with a thin spectrum accumulated in the vicinity of $\mathbf{k} = \mathbf{k}_0$. Consider the case of propagation of the ν_0-th discrete mode and introduce the value

$$\psi\,(\mathbf{r},t) = \frac{1}{2\pi} \int a_{\mathbf{k}}^{\nu_0}\, e^{i\,\mathbf{k}\mathbf{r}}\, d\mathbf{k}.$$

Then owing to the Parseval equality

$$\int |\,\psi\,|^2\, d\mathbf{r} = \int |\,a_{\mathbf{k}}^{\nu_0}\,|^2\, d\mathbf{k}\,,$$

and to a wave packet's thickness the value $\omega_{k_0}^{\nu_0} |\,\psi\,|^2$ represents the density of the internal wave energy per unit horizontal plane (i.e., the energy density dH/dV integrated over z).

The replacement of H by H_0 corresponds to the linear approximation in which waves propagate independently. To take into consideration the interaction of waves it is necessary to save higher-order terms in the Hamiltonian. Retaining up to the third-order terms inclusive yields

$$H = \int \omega_k^{\nu} a_k^{\nu} \bar{a}_k^{\nu}\, d\mathbf{k}\, d\nu$$

$$+ \int (V_{\mathbf{k}\mathbf{k}_1\mathbf{k}_2}^{\nu\nu_1\nu_2} \bar{a}_k^{\nu} a_{k_1}^{\nu_1} a_{k_2}^{\nu_2} + \bar{V}_{\mathbf{k}\mathbf{k}_1\mathbf{k}_2}^{\nu\nu_1\nu_2} a_k^{\nu} \bar{a}_{k_1}^{\nu_1} \bar{a}_{k_2}^{\nu_2})\, \delta\,(\mathbf{k} - \mathbf{k}_1 - \mathbf{k}_2)\, \Pi\, d\mathbf{k}_i d\nu_i$$

$$+ \frac{1}{3} \int (U_{\mathbf{k}\mathbf{k}_1\mathbf{k}_2}^{\nu\nu_1\nu_2} a_k^{\nu} a_{k_1}^{\nu_1} a_{k_2}^{\nu_2} + \bar{U}_{\mathbf{k}\mathbf{k}_1\mathbf{k}_2}^{\nu\nu_1\nu_2} \bar{a}_{k_1}^{\nu_1} \bar{a}_{k_2}^{\nu_2} \bar{a}_{k_3}^{\nu_3})$$

$$\times\, \delta\,(\mathbf{k} + \mathbf{k}_1 + \mathbf{k}_2)\, \Pi\, d\mathbf{k}_i d\nu_i\,,$$

$$(7.37)$$

where a bar means the complex conjugate value. The expression (7.37) obtained has a standard Hamiltonian form taking into account cubic terms (the quadratic nonlinearity) [407]. Such a structure follows from the reality condition for H and that $\lambda_{\mathbf{k}}$ and $\rho_{\mathbf{k}}$ are expressed only through $a_{\mathbf{k}}^{\nu}$ and $\bar{a}_{-\mathbf{k}}^{\nu}$ (but not through, e.g., $a_{\mathbf{k}}^{\nu}$ and $\bar{a}_{\mathbf{k}}^{\nu}$ or $a_{-\mathbf{k}}^{\nu}$). The presence of the δ-function is caused by the horizontal homogeneity of the system. The exact Hamiltonian for the present problem contains summands of all orders of $a_{\mathbf{k}}^{\nu}$. The kinetic energy contributes the terms of orders up to four and the remaining series appears as a result of the potential energy expansion in the density orders. Thus in the propagation of internal waves the nonlinearity appears as a result of two factors: the 'dynamic' nonlinearity, corresponding to inertial terms in equations of motion and described by the kinetic energy expression in the Hamiltonian; and to the 'stratification' nonlinearity connected, obvious from Eq. (7.20), with the curvature of the undisturbed density profile.

All of the peculiarities of the system's behaviour connected with the quadratic nonlinearity are accumulated in the coefficients V and U. After substitution of expressions (7.31) into (7.21) we give without derivation the final formulae:

$$
V_{\mathbf{k}\mathbf{k}_1\mathbf{k}_2}^{\nu\nu_1\nu_2} = \frac{1}{8\pi}\left(\frac{1}{2}\,\omega_k^{\nu}\omega_{k_1}^{\nu_1}\omega_{k_2}^{\nu_2}\right)^{1/2}
$$

$$
\times \left\{ \left[\frac{\omega_k^{\nu}}{k^2}\left(\frac{\mathbf{k}\mathbf{k}_1}{\omega_{k_1}^{\nu_1}}+\frac{\mathbf{k}\mathbf{k}_2}{\omega_{k_2}^{\nu_2}}\right) + \frac{\omega_{k_1}^{\nu_1}+\omega_{k_2}^{\nu_2}-\omega_k^{\nu}}{\omega_k^{\nu}}\right]\int N_0^2\bar{\varphi}_{k_1}^{\nu_1}\bar{\varphi}_{k_2}^{\nu_2}\frac{d\varphi_k^{\nu}}{dz}\,dz\right.
$$

$$
+ \left[\frac{\omega_{k_1}^{\nu_1}}{k_1^2}\left(\frac{\mathbf{k}_1\mathbf{k}}{\omega_k^{\nu}}+\frac{\mathbf{k}_1\mathbf{k}_2}{\omega_{k_2}^{\nu_2}}\right) + \frac{\omega_k^{\nu}-\omega_{k_1}^{\nu_1}-\omega_{k_2}^{\nu_2}}{\omega_{k_1}^{\nu_1}}\right]\int N_0^2\frac{d\bar{\varphi}_{k_1}^{\nu_1}}{dz}\bar{\varphi}_{k_2}^{\nu_2}\bar{\varphi}_k^{\nu}\,dz
$$

$$
\left.+ \left[\frac{\omega_{k_2}^{\nu_2}}{k_2^2}\left(\frac{\mathbf{k}_2\mathbf{k}}{\omega_k^{\nu}}+\frac{\mathbf{k}_2\mathbf{k}_1}{\omega_{k_1}^{\nu_1}}\right) + \frac{\omega_k^{\nu}-\omega_{k_1}^{\nu_1}-\omega_{k_2}^{\nu_2}}{\omega_{k_2}^{\nu_2}}\right]\int N_0^2\bar{\varphi}_{k_1}^{\nu_1}\frac{d\bar{\varphi}_{k_2}^{\nu_2}}{dz}\varphi_k^{\nu}\,dz\right\} ;
$$

$$(7.38)$$

$$U^{\nu\nu_1\nu_2}_{\mathbf{k}\mathbf{k}_1\mathbf{k}_2} = \frac{1}{8\pi} \left(\frac{1}{2} \omega^{\nu}_{k} \omega^{\nu_1}_{k_1} \omega^{\nu_2}_{k_2} \right)^{1/2}$$

$$\times \left\{ \left[\frac{\omega^{\nu}_{k}}{k^2} \left(\frac{\mathbf{k}\mathbf{k}_1}{\omega^{\nu_1}_{k_1}} + \frac{\mathbf{k}\mathbf{k}_2}{\omega^{\nu_2}_{k_2}} \right) - \frac{\omega^{\nu_1}_{k_1} + \omega^{\nu_2}_{k_2} + \omega^{\nu}_{k}}{\omega^{\nu}_{k}} \right] \int N_0^2 \bar{\varphi}^{\nu_1}_{k_1} \bar{\varphi}^{\nu_2}_{k_2} \frac{d\bar{\varphi}^{\nu_3}_{k_3}}{dz} \, dz \right.$$

$$+ \left[\frac{\omega^{\nu_1}_{k_1}}{k_1^2} \left(\frac{\mathbf{k}_1\mathbf{k}}{\omega^{\nu}_{k}} + \frac{\mathbf{k}_1\mathbf{k}_2}{\omega^{\nu_2}_{k_2}} \right) - \frac{\omega^{\nu}_{k} + \omega^{\nu_1}_{k_1} + \omega^{\nu_2}_{k_2}}{\omega^{\nu_1}_{k_1}} \right] \int N_0^2 \frac{d\bar{\varphi}^{\nu_1}_{k_1}}{dz} \bar{\varphi}^{\nu_2}_{k_2} \varphi^{\nu}_{k} dz$$

$$+ \left. \left[\frac{\omega^{\nu_2}_{k_2}}{k_2^2} \left(\frac{\mathbf{k}_2\mathbf{k}}{\omega^{\nu}_{k}} + \frac{\mathbf{k}_2\mathbf{k}_1}{\omega^{\nu_1}_{k_1}} \right) - \frac{\omega^{\nu}_{k} + \omega^{\nu_1}_{k_1} + \omega^{\nu_2}_{k_2}}{\omega^{\nu_2}_{k_2}} \right] \int N_0^2 \bar{\varphi}^{\nu_1}_{k_1} \frac{d\bar{\varphi}^{\nu_2}_{k_2}}{dz} \varphi^{\nu}_{k} \, dz \right\} .$$

$$(7.39)$$

Consider the important particular example of the medium with a constant Brunt–Väisälä frequency $N_0 = $ const. Then Eq. (7.26) has the solution in the form of the wave running along z-axis

$$\varphi^{\nu}_{k} = \frac{1}{\sqrt{2\pi}} \frac{1}{N_0} e^{-i\nu z} . \qquad (7.40)$$

An arbitrary coefficient is chosen to satisfy the normalization condition (7.27). The dispersion relation has the form

$$\omega^{\nu}_{k} = \frac{N_0 |\mathbf{k}|}{\sqrt{k^2 + \nu^2}} . \qquad (7.41)$$

From Eqs. (7.36) it follows that the superscript ν relates in this case to the vertical wave number. Let now simplify the Hamiltonian (7.37) and the relations (7.38), (7.39), introducing the three-dimensional wave vector æ instead of the pair $(\nu, \mathbf{k})$. Substitution of Eq. (7.40) into Eq. (7.38) yields

$$V_{æ æ_1 æ_2} = \frac{1}{(2\pi)^{3/2}} \frac{1}{4N_0} \left(\frac{1}{2} \omega_æ \omega_{æ_1} \omega_{æ_2} \right)^{1/2}$$

$$\times \left\{ -\left[\frac{\omega_æ}{k^2} \left(\frac{\mathbf{k}\mathbf{k}_1}{\omega_{æ_1}} + \frac{\mathbf{k}\mathbf{k}_2}{\omega_{æ_2}} \right) + \frac{\omega_{æ_1} + \omega_{æ_2} - \omega_æ}{\omega_æ} \right] \nu \right.$$

$$+ \left. \left[\frac{\omega_{æ_1}}{k_1^2} \left(\frac{\mathbf{k}_1\mathbf{k}}{\omega_æ} + \frac{\mathbf{k}_1\mathbf{k}_2}{\omega_{æ_2}} \right) + \frac{\omega_æ - \omega_{æ_1} - \omega_{æ_2}}{\omega_{æ_1}} \right] \nu_1 \right.$$

$$+ \left[\frac{\omega_{\ae_2}}{k_2^2} \left(\frac{\mathbf{k}_2\mathbf{k}}{\omega_{\ae}} + \frac{\mathbf{k}_2\mathbf{k}_1}{\omega_{\ae_1}} \right) + \frac{\omega_{\ae} - \omega_{\ae_1} - \omega_{\ae_2}}{\omega_{\ae_2}} \right] \nu_2 \right\} . \qquad (7.42)$$

A similar relation yields for the coefficient U:

$$U_{\ae\ae_1\ae_2} = \frac{i}{(2\pi)^{3/2}} \frac{1}{4N_0} \left(\frac{1}{2} \omega_{\ae}\omega_{\ae_1}\omega_{\ae_2} \right)^{1/2}$$

$$\times \left\{ \left[\frac{\omega_{\ae}}{k^2} \left(\frac{\mathbf{k}\mathbf{k}_1}{\omega_{\ae_1}} + \frac{\mathbf{k}\mathbf{k}_2}{\omega_{\ae_1}} \right) - \frac{\omega_{\ae} + \omega_{\ae_1} + \omega_{\ae_2}}{\omega_{\ae}} \right] \nu \right.$$

$$+ \left[\frac{\omega_{\ae_1}}{k_1^2} \left(\frac{\mathbf{k}_1\mathbf{k}}{\omega_{\ae}} + \frac{\mathbf{k}_1\mathbf{k}_2}{\omega_{\ae_2}} \right) - \frac{\omega_{\ae_1} + \omega_{\ae_2} + \omega_{\ae}}{\omega_{\ae_1}} \right] \nu_1$$

$$+ \left. \left[\frac{\omega_{\ae_2}}{k_2^2} \left(\frac{\mathbf{k}_2\mathbf{k}}{\omega_{\ae}} + \frac{\mathbf{k}_2\mathbf{k}_1}{\omega_{\ae_1}} \right) - \frac{\omega_{\ae_1} + \omega_{\ae_2} + \omega_{\ae}}{\omega_{\ae_2}} \right] \nu_2 \right\} .$$

$$(7.43)$$

Here $\mathbf{k}$ and ν are respectively horizontal and vertical components of the wave vector $\ae$. In Eqs. (7.42) and (7.43) the multipliers $\delta(\nu - \nu_1 - \nu_2)$ and $\delta(\nu + \nu_1 + \nu_2)$, combined with the corresponding δ-functions in (7.37), are omitted. Omitting now the superscript ν and considering $\ae$ to be a three-dimensional vector yields the Hamiltonian (7.37) in the standard form for a spatially homogeneous medium. The coefficients V and U in the case $N_0 = $ const are given by formulae (7.42) and (7.43).

Substituting also Eq. (7.40) into Eqs. (7.31) after the Fourier transformation yields the expression for the variables λ, ρ through $a_{\ae}$:

$$\lambda = \frac{1}{(2\pi)^{3/2}} \frac{ig}{N_0} \int \frac{1}{\sqrt{2\omega_{\ae}}} a_{\ae} e^{i\ae\mathbf{R}} \, d\ae + \text{c.c.} \qquad (7.44)$$

$$\rho = -\frac{1}{(2\pi)^{3/2}} \frac{\rho_0'}{N_0} \int \sqrt{\frac{\omega_{\ae}}{2}} a_{\ae} e^{i\ae\mathbf{R}} \, d\ae + \text{c.c.} \qquad (7.45)$$

Here $\mathbf{R}$ is a three-dimensional radius-vector of a domain point.

3. SOME ADDITIONAL TRANSFORMS OF HAMILTONIAN EQUATIONS

For further considerations we shall need some transformations of the Hamiltonian (7.37) which deal only with its common structure and are valid for any system [407]. We repeat them for the sake of being methodical.

The coefficients V and U in (7.37) may be always symmetrized so that

$$V^{\nu\nu_1\nu_2}_{\mathbf{k}\mathbf{k}_1\mathbf{k}_2} = V^{\nu\nu_2\nu_1}_{\mathbf{k}\mathbf{k}_2\mathbf{k}_1}; \quad U^{\nu\nu_1\nu_2}_{\mathbf{k}\mathbf{k}_1\mathbf{k}_2} = U^{\nu\nu_2\nu_1}_{\mathbf{k}\mathbf{k}_2\mathbf{k}_1} = U^{\nu_1\nu\nu_2}_{\mathbf{k}_1\mathbf{k}\mathbf{k}_2}. \tag{7.46}$$

In particular, Eqs. (7.38)–(7.39) and (7.42)–(7.43) will always satisfy these conditions.

Let us study further simplifications available from (7.37) by means of the canonical transformations. We write the unknown transformation in the form of an integral power series:

$$
\begin{aligned}
a^{\nu}_{\mathbf{k}} = b^{\nu}_{\mathbf{k}} + \int &\left(A^{\nu\nu_1\nu_2}_{\mathbf{k}\mathbf{k}_1\mathbf{k}_2} b^{\nu_1}_{\mathbf{k}_1} b^{\nu_2}_{\mathbf{k}_2} + B^{\nu\nu_1\nu_2}_{\mathbf{k}\mathbf{k}_1\mathbf{k}_2} b^{\nu_1}_{\mathbf{k}_1} \bar{b}^{\nu_2}_{\mathbf{k}_2} \right. \\
&\left. + \; C^{\nu\nu_1\nu_2}_{\mathbf{k}\mathbf{k}_1\mathbf{k}_2} \bar{b}^{\nu_1}_{\mathbf{k}_1} \bar{b}^{\nu_2}_{\mathbf{k}_2} \right) d\mathbf{k}_1 \, d\mathbf{k}_2 \, d\nu_1 \, d\nu_2 + \dots \; ;
\end{aligned}
$$

$$
\begin{aligned}
\bar{a}^{\nu}_{\mathbf{k}} = \bar{b}^{\nu}_{\mathbf{k}} + \int &\left(\bar{A}^{\nu\nu_1\nu_2}_{\mathbf{k}\mathbf{k}_1\mathbf{k}_2} \bar{b}^{\nu_1}_{\mathbf{k}_1} \bar{b}^{\nu_2}_{\mathbf{k}_2} + \bar{B}^{\nu\nu_1\nu_2}_{\mathbf{k}\mathbf{k}_1\mathbf{k}_2} \bar{b}^{\nu_1}_{\mathbf{k}_1} b^{\nu_2}_{\mathbf{k}_2} \right. \\
&\left. + \; \bar{C}^{\nu\nu_1\nu_2}_{\mathbf{k}\mathbf{k}_1\mathbf{k}_2} b^{\nu_1}_{\mathbf{k}_1} b^{\nu_2}_{\mathbf{k}_2} \right) d\mathbf{k}_1 \, d\mathbf{k}_2 \, d\nu_1 \, d\nu_2 + \dots \; . \tag{7.47}
\end{aligned}
$$

The coefficients A and C may be considered to be symmetrized with respect to the second pair of indices. To verify the canonicity conditions it is necessary to write again the Poisson brackets [see Eqs. (7.4)]. Now, we have

$$\frac{\delta a^{\nu}_{\mathbf{k}}}{\delta b^{\mu}_{\mathbf{q}}} = \delta(\nu - \mu)\,\delta(\mathbf{k} - \mathbf{q}) + \int \left(2A^{\nu\mu\nu_2}_{\mathbf{k}\mathbf{q}\mathbf{k}_2} b^{\nu_2}_{\mathbf{k}_2} + B^{\nu\mu\nu_2}_{\mathbf{k}\mathbf{q}\mathbf{k}_2} \bar{b}^{\nu_2}_{\mathbf{k}_2} \right) d\mathbf{k}_2 \, d\nu_2 + \dots \; ;$$

$$\frac{\delta a^{\nu}_{\mathbf{k}}}{\delta \bar{b}^{\mu}_{\mathbf{q}}} = \int \left(B^{\nu\nu_1\mu}_{\mathbf{k}\mathbf{k}_1\mathbf{q}} b^{\nu_1}_{\mathbf{k}_1} + 2C^{\nu\nu_1\mu}_{\mathbf{k}\mathbf{k}_1\mathbf{q}} \bar{b}^{\nu_1}_{\mathbf{k}_1} \right) d\mathbf{k}_1 \, d\nu_1 + \dots$$

Corresponding formulae for $\bar{a}^{\nu}_{\mathbf{k}}$ are obtained by complex conjugation. With an accuracy to the first-order, values of $b^{\nu}_{\mathbf{k}}$ yield

$$\{\bar{a}^{\nu}_{\mathbf{k}}, \bar{a}^{\nu'}_{\mathbf{k}'}\} = \delta(\nu - \nu')\,\delta(\mathbf{k} - \mathbf{k}')$$

$$+ \int \left[\left(2A^{\nu\nu'\nu_2}_{\mathbf{k}\mathbf{k}'\mathbf{k}_2} + \bar{B}^{\nu'\nu\nu_2}_{\mathbf{k}'\mathbf{k}\mathbf{k}_2} \right) b^{\nu_2}_{\mathbf{k}_2} + \left(B^{\nu\nu'\nu_2}_{\mathbf{k}\mathbf{k}'\mathbf{k}_2} + 2\bar{A}^{\nu'\nu\nu_2}_{\mathbf{k}'\mathbf{k}\mathbf{k}_2} \right) \bar{b}^{\nu_2}_{\mathbf{k}_2} \right]$$

$$\times \, d\nu_2 \, d\mathbf{k}_2 + \mathcal{O}\left(b^2\right).$$

So it should be $\bar{B}^{\nu'\nu\nu_2}_{\mathbf{k}'\mathbf{k}\mathbf{k}_2} = -2A^{\nu\nu'\nu_2}_{\mathbf{k}\mathbf{k}'\mathbf{k}_2}$. Considering the expression for $\left\{a^{\nu}_{\mathbf{k}}, a^{\nu'}_{\mathbf{k}'}\right\}$ we have by analogy a symmetry of the coefficients B and C with respect to the first and the third indexes: $(B,C)^{\nu'\nu_2\nu}_{\mathbf{k}'\mathbf{k}_2\mathbf{k}} = (B,C)^{\nu\nu_2\nu'}_{\mathbf{k}\mathbf{k}_2\mathbf{k}'}$. A canonicity of the last Poisson bracket $\left\{\bar{a}^{\nu}_{\mathbf{k}}, \bar{a}^{\nu'}_{\mathbf{k}'}\right\}$ follows automatically from a canonicity of $\left\{a^{\nu}_{\mathbf{k}}, a^{\nu'}_{\mathbf{k}'}\right\}$ due to complex conjugation. The symmetry conditions obtained combined with the initial conditions may be finally expressed with the following relations

$$A^{\nu\nu_1\nu_2}_{\mathbf{k}\mathbf{k}_1\mathbf{k}_2} = A^{\nu\nu_2\nu_1}_{\mathbf{k}\mathbf{k}_2\mathbf{k}_1}\,; \qquad B^{\nu_1\nu\nu_2}_{\mathbf{k}_1\mathbf{k}\mathbf{k}_2} = -2\bar{A}^{\nu\nu_1\nu_2}_{\mathbf{k}\mathbf{k}_1\mathbf{k}_2}\,;$$

$$C^{\nu\nu_1\nu_2}_{\mathbf{k}\mathbf{k}_1\mathbf{k}_2} = C^{\nu\nu_2\nu_1}_{\mathbf{k}\mathbf{k}_2\mathbf{k}_1} = C^{\nu_1\nu\nu_2}_{\mathbf{k}_1\mathbf{k}\mathbf{k}_2}\,. \qquad (7.48)$$

A, B, C will be expressed as follows

$$A^{\nu\nu_1\nu_2}_{\mathbf{k}\mathbf{k}_1\mathbf{k}_2} = -\frac{V^{\nu\nu_1\nu_2}_{\mathbf{k}\mathbf{k}_1\mathbf{k}_2}\,\delta\left(\mathbf{k} - \mathbf{k}_1 - \mathbf{k}_2\right)}{\omega^{\nu}_{k} - \omega^{\nu_1}_{k_1} - \omega^{\nu_2}_{k_2}}\,; \qquad B^{\nu_1\nu\nu_2}_{\mathbf{k}_1\mathbf{k}\mathbf{k}_2} = -2\bar{A}^{\nu\nu_1\nu_2}_{\mathbf{k}\mathbf{k}_1\mathbf{k}_2}\,; \qquad C = 0\,;$$

$$(7.49)$$

$$A = B = 0\,; \qquad C^{\nu\nu_1\nu_2}_{\mathbf{k}\mathbf{k}_1\mathbf{k}_2} = -\frac{\bar{U}^{\nu\nu_1\nu_2}_{\mathbf{k}\mathbf{k}_1\mathbf{k}_2}}{\omega^{\nu}_{k} + \omega^{\nu_1}_{k_1} + \omega^{\nu_2}_{k_2}}\,\delta\left(\mathbf{k} + \mathbf{k}_1 + \mathbf{k}_2\right). \qquad (7.50)$$

The symmetry conditions (7.48) are obviously satisfied as a result of (7.46). When expressing coefficients in the form (7.49) and (7.50) the transformation (7.47) makes vanish, respectively, the first and the second groups of cubic terms in (7.37). Indeed, while substituting into (7.37) the expression

$$a^{\nu}_{\mathbf{k}} = b^{\nu}_{\mathbf{k}} + \int \left(\frac{-V^{\nu\nu_1\nu_2}_{\mathbf{k}\mathbf{k}_1\mathbf{k}_2}\,\delta\left(\mathbf{k} - \mathbf{k}_1 - \mathbf{k}_2\right)}{\omega^{\nu}_{k} - \omega^{\nu_1}_{k_1} - \omega^{\nu_2}_{k_2}}\, b^{\nu_1}_{\mathbf{k}_1} b^{\nu_2}_{\mathbf{k}_2} \right.$$

$$+\, 2\,\frac{\bar{V}^{\nu_1\nu\nu_2}_{\mathbf{k}_1\mathbf{k}\mathbf{k}_2}\,\delta\left(\mathbf{k}_1 - \mathbf{k} - \mathbf{k}_2\right)}{\omega^{\nu_1}_{k_1} - \omega^{\nu}_{k} - \omega^{\nu_2}_{k_2}}\, b^{\nu_1}_{\mathbf{k}_1} \bar{b}^{\nu_2}_{\mathbf{k}_2}$$

$$\left. -\, \frac{\bar{U}^{\nu\nu_1\nu_2}_{\mathbf{k}\mathbf{k}_1\mathbf{k}_2}\,\delta\left(\mathbf{k} + \mathbf{k}_1 + \mathbf{k}_2\right)}{\omega^{\nu}_{k} + \omega^{\nu_1}_{k_1} + \omega^{\nu_2}_{k_2}}\, \bar{b}^{\nu_1}_{\mathbf{k}_1} \bar{b}^{\nu_2}_{\mathbf{k}_2} \right)\, d\mathbf{k}_1 \, d\mathbf{k}_2 \, d\nu_1 \, d\nu_2\,.$$

$$(7.51)$$

the term $\int \omega_k^\nu a_k^\nu \bar{a}_k^\nu \, d\mathbf{k} \, d\nu$ induces the third-order summands, which are mutually cancelled by the cubic Hamiltonian part. As a result of this substitution the additional four-order terms also appear in (7.37) (omitted here).

The Hamiltonian in new variables b_k^ν has the form

$$H = \int \omega_k^\nu b_\mathbf{k}^\nu \bar{b}_\mathbf{k}^\nu \, d\mathbf{k} \, d\nu + \mathcal{O}\left(b^4\right),$$

and equations of motion are accordingly written as:

$$\dot{b}_\mathbf{k}^\nu + i\,\omega_k^\nu b_\mathbf{k}^\nu = \mathcal{O}\left(b^3\right).$$

If waves' amplitudes are small enough then the r.h.s. of the equality may be neglected and the equations of motion are easily integrated:

$$b_\mathbf{k}^\nu = b_\mathbf{k}^\nu\left(0\right) e^{-i\omega_k^\nu t}. \tag{7.52}$$

But the transforms (7.49) and (7.50) can not be performed in the $\mathbf{k}$-space regions, where the following relations are valid

$$\mathbf{k}_1 + \mathbf{k}_2 = \mathbf{k}\,;$$

$$\omega_{\mathbf{k}_1}^{\nu_1} + \omega_{\mathbf{k}_2}^{\nu_2} = \omega_\mathbf{k}^\nu \tag{7.53}$$

or

$$\mathbf{k}_1 + \mathbf{k}_2 + \mathbf{k}_3 = 0\,;$$

$$\omega_{\mathbf{k}_1}^{\nu_1} + \omega_{\mathbf{k}_2}^{\nu_2} + \omega_{\mathbf{k}_3}^{\nu_3} = 0\,, \tag{7.54}$$

because non-integrable peculiarities appear in (7.51). Surfaces defined by systems (7.53) and (7.54) are called resonant and the equations themselves are the synchronism conditions.

For the stable stratification in the resting medium Eqs. (7.54) have no solutions because $\omega_k^\nu > 0$. So the second group of the Hamiltonian terms in (7.37) may be neglected by means of the regular change of variables (7.47), which differs from the identical one by the second-order terms only. Satisfaction of conditions (7.53) is quite possible. This case will be considered in Section 10.1. We write the equation of motion

$$\dot{a}_\mathbf{k}^\nu + i\,\frac{\delta H}{\delta \bar{a}_k^\nu} = 0 \tag{7.55}$$

for the case, when the above described procedure is fulfilled and the terms proportional to U are absent in (7.37). It yields

$$\dot{a}_{\mathbf{k}}^{\nu} + i\omega_k^{\nu} a_{\mathbf{k}}^{\nu} + i \int V_{\mathbf{k}\mathbf{k}_1\mathbf{k}_2}^{\nu\nu_1\nu_2} a_{\mathbf{k}_1}^{\nu_1} a_{\mathbf{k}_2}^{\nu_2} \delta\left(\mathbf{k} - \mathbf{k}_1 - \mathbf{k}_2\right) d\mathbf{k}_1\, d\mathbf{k}_2\, d\nu_1\, d\nu_2$$

$$+ 2i \int \bar{V}_{\mathbf{k}_1\mathbf{k}\mathbf{k}_2}^{\nu_1\nu\nu_2} a_{\mathbf{k}_1}^{\nu_1} \bar{a}_{\mathbf{k}_2}^{\nu_2} \delta\left(\mathbf{k}_1 - \mathbf{k} - \mathbf{k}_2\right) d\mathbf{k}_1\, d\mathbf{k}_2\, d\nu_1\, d\nu_2 = 0. \quad (7.56)$$

This equation and Eqs. (7.38), (7.42) for coefficients V, describing total specifics of weakly nonlinear internal waves behaviour, are the final results of this chapter. Further we consider on the basis of Eq. (7.56) some particular cases corresponding to different physical situations:

1. *Long internal waves in a shallow water.* The internal waves length is much greater than the fluid depth. Then Eq. (7.56) becomes to the Korteveg–de Vries equation (Chapter 8).

2. *Propagation of thin spectrum internal wave packets.* A thin spectrum corresponds to pure slightly modulated wave packets. Propagation of such packets is described by the so called nonlinear Schrödinger equation. While such packets pass through a medium some mean flows appear which can produce, under certain conditions, a thin vertical structure of the oceanic hydrophysical fields (Chapter 9).

3. *Resonant triple interactions of internal waves.* Here we consider slightly nonlinear interactions of three internal waves whose parameters satisfy the synchronism relations (7.53)–(7.54). Here the instability of the respective 'dispersive' or 'explosive' internal waves takes place (Chapter 10).

4. *Description of statistical ensemble of weakly nonlinear internal waves.* On the basis of Eq. (7.56) an interaction of numerous internal waves of arbitrary amplitudes and phases is considered, which corresponds to the wide spectrum internal waves. From Eq. (7.56), by means of statistical physics, a so called kinetic equation, describing dynamic variations of the internal waves' spectrum, is obtained (Chapter 10).

Notes

1 Milder [241] constructs a Hamiltonian which requires the absence of vertical vorticity in a stratified fluid. The Hamiltonian for internal motions of a stratified fluid in terms of Eulerian coordinates is described by Henyey [126]. It admits both internal waves and motions with vertical vorticity and the couplings between them. The Hamiltonian formulation of the description of interfacial solitary waves is considered by Grimshaw and Pugjaprasetya [112]. The Hamiltonian description of hydrodynamic systems developed by Zakharov and Kuznetsov in application to weakly nonlinear waves is reviewed in [414]. The problem of small scale and small amplitude internal waves' interaction with a vigorous, large scale undulating shear flow in the Hamiltonian terms was treated by Bruhwiler and Kaper [42]. The problem of the Hamiltonian description of internal waves in the ocean was solved by Goncharov and Pavlov in [105], written entirely from the viewpoint of the Hamiltonian formalism.

Chapter 8

LONG WEAKLY NONLINEAR INTERNAL WAVES

In this chapter we study trivial nonlinear effects appearing while long internal waves propagate. There may be two situations and each of them needs its own analysis. The first is the propagation of internal waves in shallow water (e.g., in a shallow sea like the Baltic Sea) when the wavelength λ considerably exceeds the sea's depth H. In this case equations of internal waves dynamics are reduced to the Korteweg–de Vries equation (KdV) (see a review of its evolution in the Appendix). Derivation and analysis of this equation for long internal waves is given in Section 8.1. In Section 8.2 stationary solutions of the KdV equation describing long nonlinear internal waves of the stable type are analyzed. Here we also give the numerical results for particular stratification profiles. The second possible situation is propagation of long waves within a thin pycnocline (the wavelength λ is much greater than the pycnocline's thickness h). This is quite typical for natural conditions and is studied in Section 8.3. In this case internal waves are described by an equation of the KdV type but with integral dispersion. An analysis of the stationary solutions for this equation is presented as well.

1. KORTEWEG–DE VRIES EQUATION FOR LONG INTERNAL WAVES

Let us consider propagation of long internal waves when their length λ exceeds many times the water's depth H. We stop at some heuristic considerations before derivation of the equation describing these waves by the Hamiltonian formalism method.

As shown in Section 3.1 [see Eqs. (3.39), (7.40)] the dispersion relation for the n-th internal waves mode in the ocean of the finite depth (shallow water) without rotation looks like

$$\omega_n(k) = c_{f,n}^{(0)} k - d_n k^3 + \mathcal{O}(k^5),\tag{8.1}$$

where $c_{f,n}^{(0)} > 0$, $d_n > 0$ are some constants depending on the stratification profile. By the operational relations $i\omega \leftrightarrow -\partial/\partial t$, $ik \leftrightarrow \partial/\partial x$, Eq. (8.1) relates to the following differential equation for the amplitude function of the n-th mode:

$$\frac{\partial u_n}{\partial t} + c_{f,n}^{(0)} \frac{\partial u_n}{\partial x} + d_n \frac{\partial^3 u_n}{\partial x^3} = 0.$$

After substitution of the expression $u_n = a_n \exp(-i\omega_n t + ikx)$ this equation results in the dispersion relation (8.1). If we add to it the characteristic hydrodynamic nonlinear term $\gamma_n u_n \, \partial u_n/\partial x$, where γ_n is some constant, then the nonlinear nonstationary equation known as the Korteweg–de Vries equation [177] yields

$$\frac{\partial u_n}{\partial t} + (\gamma_n u_n + c_{f,n}^{(0)}) \frac{\partial u_n}{\partial x} + d_n \frac{\partial^3 u_n}{\partial x^3} = 0.\tag{8.2}$$

In the presented heuristic derivation of Eq. (8.2), widely used in some monographs on the waves dispersion theory [148], [158], [391].

This equation can be accurately derived from the original hydrodynamics equations by one of the asymptotic procedures (see, e.g., [86], [203], [291]). But we use the Hamiltonian formalism method presented in Chapter 7.

We proceed from Eq. (7.56) for the amplitudes of the normal modes $a_{\mathbf{k}}^{\nu}$

$$\dot{a}_{\mathbf{k}}^{\nu} + i\omega_k^{\nu} a_{\mathbf{k}}^{\nu} + i\int V_{\mathbf{k}\mathbf{k}_1\mathbf{k}_2}^{\nu\nu_1\nu_2} a_{\mathbf{k}_1}^{\nu_1} a_{\mathbf{k}_2}^{\nu_2} \delta(\mathbf{k} - \mathbf{k}_1 - \mathbf{k}_2)\, d\Gamma$$

$$+ 2i\int \bar{V}_{\mathbf{k}_1\mathbf{k}\mathbf{k}_2}^{\nu_1\nu\nu_2} a_{\mathbf{k}_1}^{\nu_1} \bar{a}_{\mathbf{k}_2}^{\nu_2} \delta(\mathbf{k}_1 - \mathbf{k} - \mathbf{k}_2)\, d\Gamma = 0.\tag{8.3}$$

Let the mode number ν be fixed as n. We consider the wave packet of long internal waves running in the positive x-axis direction. And therefore a_k^ν are non-zero only within the narrow limits $0 < k < \varepsilon$. For k, k_1, $k_2 \to 0$ the coefficient $V_{kk_1k_2}^{nnn}$ has the form

$$V_{kk_1k_2}^{nnn} \approx \frac{1}{8\pi}\frac{1}{\sqrt{2}} \left(\omega_k^n \, \omega_{k_1}^n \, \omega_{k_2}^n\right)^{1/2} \left(6 + (k - k_1 - k_2)\right) \int N_0^2 \varphi_0^2 \varphi_0' \, dz \,,$$

$$(8.4)$$

where $\varphi_0 = \lim\limits_{k\to 0} \varphi_k^n$. Let us now introduce the function

$$A_n\left(x,\ t\right) = \frac{1}{2\pi\sqrt{2}} \int \sqrt{\omega_k^n} a_k^n e^{ikx} \, dk \,.$$

$$(8.5)$$

From Eq. (7.45) for ρ, the vertical displacement for $\zeta = -\rho/\rho_0'$ is expressed through A as:

$$\zeta = \left(A + \bar{A}\right) \varphi_0\left(z\right).$$

$$(8.6)$$

Substituting (8.4) for the coefficient V into (8.3) and using (8.5) we rewrite (8.3) in terms of A

$$\frac{\partial A_n}{\partial t} + c_{f,n}^{(0)} \frac{\partial A_n}{\partial x} + d_n \frac{\partial^3 A_n}{\partial x^3} + \frac{\gamma_n}{2} \frac{\partial}{\partial x} \left(A_n^2 + A_n \bar{A}_n\right) = 0\,,$$

$$(8.7)$$

where

$$\gamma_n = -\int \frac{dN_0^2}{dz} \varphi_0^3 \, dz \,; \qquad d_n = \frac{\left(c_{f,n}^{(0)}\right)^3}{2} \int_{-H}^{0} \varphi_0^2 \, dz \,.$$

$$(8.8)$$

While deriving (8.7) we should take into account that $\int\limits_{-\infty}^{\infty} a_k^\nu \, dk = \int\limits_{0}^{\infty} a_k^\nu \, dk$ and the expansion of the dispersion relation (8.1).

Let us sum Eq. (8.7) with its conjugate complex and introduce the expression

$$U_n = c_0 \left(A_n + \bar{A}_n\right).$$

The resulting Eq. (8.7) has the form

$$\frac{\partial U_n}{\partial t} + c_{f,n}^{(0)} \frac{\partial U_n}{\partial x} + d_n \frac{\partial^3 U_n}{\partial x^3} + \gamma_n U_n \frac{\partial U_n}{\partial x} = 0\,.$$

$$(8.9)$$

The KdV equation (8.9) can be derived for the 3D case also. Here the free surface's influence may be taken into account, which is sometimes sufficient for long internal waves. Such an equation was derived by Leonov [203] by multi-scale expansions applied to the original hydrodynamics equations (taking into account the free surface presence and without the Boussinesq approximation). Here we give the heuristic derivation of this equation.

The dispersion relation (8.1) was obtained for 3D internal waves, for which $k = \sqrt{k_x^2 + k_y^2}$, where k_x, k_y are components of the wave vector of an internal wave placed in a horizontal plane x, y. Let consider a case when $k_y \ll k_x$, $k_x H \ll 1$, i.e., a long wave perturbation propagating along the x-axis is modulated by a longer wave perturbation along the y-axis. Expanding now the dispersion relation (8.1) and assuming that $k_y \sim k_x^2$ we obtain

$$\omega_n = c_{f,n}^{(0)} k_x - d_n k_x^3 + c_{f,n}^{(0)} \frac{k_y^2}{2k_x} + \mathcal{O}(k_x^5).$$

Considering again the latter relation to be an operational one, i.e., assuming that $i\omega \leftrightarrow -\partial/\partial t$, $ik_x \leftrightarrow \partial/\partial x$, $ik_y \leftrightarrow \partial/\partial y$, we easily derive the linear equation for the n-th mode amplitude variations $u_n(x, y, t)$

$$\frac{\partial u_n}{\partial t} + c_{f,n}^{(0)} \frac{\partial u_n}{\partial x} + d_n \frac{\partial^3 u_n}{\partial x^3} + \frac{c_{f,n}^{(0)}}{2} \int_a^x \frac{\partial^2 u_n(\xi, y, t)}{\partial y^2} d\xi = 0,$$

while adding the characteristic nonlinear term $\gamma_n u_n \, \partial u_n/\partial x$ we have [35]

$$\frac{\partial}{\partial x} \left[\frac{\partial u_n}{\partial t} + (\gamma_n u_n + c_{f,n}^{(0)}) \frac{\partial u_n}{\partial x} + d_n \frac{\partial^3 u_n}{\partial x^3} \right] + \frac{c_{f,n}^{(0)}}{2} \frac{\partial^2 u_n}{\partial y^2} = 0. \quad (8.10)$$

The equation of the (8.10) type is frequently called the Kadomtsev–Petviashvili equation and was derived for the first time for plasma waves. The coefficients of this equation obtained in [203] for surface and internal waves have the following form

$$\gamma_n = \frac{3}{\|W\|^2} \left\{ \frac{g\rho_a W^3(0)}{c_f^{(0)2}} + 2 \int_{-H}^0 \rho_0 \mu W^2 W' dz - \frac{c_f^{(0)}}{g} \int_{-H}^0 \rho_0 \mu W W'^2 dz \right\};$$

$$d_n = \frac{c_f^{(0)^3}}{2\|W\|^2} \int\limits_{-H}^{0} \rho_0 W^2 \, dz > 0; \; \|W\|^2 = g\rho_a W^2(0) + \int\limits_{-H}^{0} \rho_0 \mu W^2 \, dz \,.$$

(8.11)

The coefficients γ_n, d_n are given in terms of Section 3.1. Eqs. (8.11), under the Boussinesq and the 'rigid lid' approximations, turn into Eq. (8.8). In Eq. (8.10) u_n is the amplitude of a stream function. All the other parameters are expressed through u_n as follows:

$$u = U W'; \quad v = \frac{\partial}{\partial y} \int U \, d\xi \, W'; \quad W = -\frac{\partial U}{\partial x} W ;$$

$$p = c_f^0 \rho_0 U W'; \quad \zeta = \frac{1}{c_f^0} U W(0); \quad \rho_0 = -\frac{\rho_0'}{c_f^0} W U \,.$$

As was shown in [203], the superposition of different modes is valid for the derivation of Eq. (8.10).

The KdV equation, applied for various physics and mechanics problems, was analyzed initially by numerical methods, revealing some characteristic peculiarities of the evolution of an initial disturbance [404], [24]. And only after that was the analytical theory of the KdV solution, discovered primarily in [85], developed. The operational version of this method as well as some strict proofs is given in [196]. This method developed and applied further with respect to a number of nonlinear partial differential equations [408], [409] was called 'the method of an inverse problem of scattering theory'. This method is valid for Eq. (8.10) as well. In the next section elementary stationary solutions of Eqs. (8.9), (8.10) describing stationary long nonlinear internal waves are considered.

2. STATIONARY SOLUTIONS OF THE KORTEWEG–DE VRIES EQUATION

Consider the one-dimensional KdV equation describing plane stationary internal waves with the stream function $\psi = U(\theta) W(z)$ [see (8.1)], as well as the expressions presented above for velocity and pressure components in terms of $W(z)$, where $\theta = x - c_f t$ is the total wave phase of the considered n-th mode. Then by (8.9) the equation for the stationary wave's amplitude will have the form

$$(\gamma U + \tilde{c}) \frac{\partial U}{\partial \theta} + d \frac{d^3 U}{d\theta^3} = 0 \,,$$

where $\tilde{c} = c_f - c_f^0$ is the addition to the phase velocity of the long wave disturbance, which will be called further 'the phase velocity' in analogy with gas dynamics. Integration of the latter equation over θ yields:

$$d \frac{d^2 U}{d\theta^2} + \tilde{c}U + \frac{1}{2}\gamma U^2 = U_1 . \tag{8.12}$$

Here U_1 is some integration constant. The stream function ψ_* has a 'phase shift' while turning from the θ, z coordinates, moving at the phase velocity c_f, to the coordinates θ_*, z, moving relatively to the resting observer at the velocity c_f^*:

$$\psi_* = \psi(\theta_*, z) + (c_f - c_f^*)z , \tag{8.13}$$

The horizontal velocity u varies by the value of the phase shift $(c_f - c_f^*)$ and the boundary conditions and equations do not violate. The transform (8.13) is convenient when studying different waves in the same coordinates.

It is useful to introduce further a number of dimensionless variables:

$$\xi = \theta\lambda^{-1}; \quad \eta = zH^{-1}; \quad F = UW(c_f H)^{-1}; \quad \Omega = N_0^{-2}N^2(z);$$

$$k_0 = H\lambda^{-1}; \quad \beta = H^2 N_0^2 c_f^{-2}; \quad \delta = HN_0^2 g^{-1}; \quad \hat{\rho} = \rho_0\rho_a^{-1}. \tag{8.14}$$

Here and later the function $U(\theta)$ is considered to be dimensionless and $W(z) \rightarrow \widetilde{W}(\eta)$ and the tilde will be also omitted. In (8.14) λ and H are characteristic horizontal and vertical scales of wavy motion respectively, so that $k_0 \ll 1$, $N_0 = $ const is the maximum Brunt–Väisälä value (wherefrom $0 < \Omega < 1$), UW is the stream function with the phase shift (8.13) subtracted, ρ_a is the density value at the free ocean surface, δ is the parameter characterizing deviation from the Boussinesq approximation.

Taking into account Eq. (8.14) yields after some transformations

$$U''(\xi) + \gamma U^2 + \alpha\beta_1 U = U_2 , \tag{8.15}$$

where U_2 is a constant, and the variables α, β and γ have the forms

$$k_0^2\beta_1 \approx -\frac{2H^2 N_0^2\tilde{c}}{\left(c_f^0\right)^3}; \quad \alpha = \|W\|^2 \left(\int_{-1}^{0} \hat{\rho}W^2 \, d\eta\right)^{-1} > 0;$$

$$\gamma = \left[\frac{3}{2} \frac{\beta_0^2}{\delta^2} W^3(0) + 3 \int_{-1}^{0} \hat{\rho} \Omega W \left(\beta_0 W'W - \frac{\delta}{2} W'^2 \right) d\eta \right] \left(\int_{-1}^{0} \hat{\rho} W^2 d\eta \right)^{-1} ;$$

(8.16)

$$\|W\|^2 = \delta^{-1} W^2(0) + \int_{-1}^{0} \hat{\rho} \Omega W^2 \, d\eta > 0 .$$

It follows from Eqs. (8.16) that $\alpha > 0$ and the sign of γ is undefined. Assuming that $\gamma \neq 0$ (otherwise higher approximations should be considered) we shall suppose without any restrictions of generality that $\gamma > 0$. If $\gamma < 0$ then using the transform $\xi \to -\xi$, $U \to -U$, $U_2 \to -U_2$ we shall again have Eq. (8.15). Introduce now, instead of the amplitude function U and the parameter U_2, a new unknown function u and a parameter $\tilde{U}_0$ with the following expression:

$$U = u + \widetilde{W}_0 ; \quad U_2 = \tilde{U}_0 \left(\tilde{U}_0 + \frac{\alpha \beta_1}{\gamma} \right) ; \quad \frac{2}{3} U_0 = 2U_0 - \frac{\alpha \beta_1}{\gamma} , \quad (8.17)$$

where the latter formula is just the definition of U_0. Then Eq. (8.15) may be rewritten in the following 'canonical' form:

$$\gamma^{-1} u'' = \frac{2}{3} U_0 u - u^2 = -\frac{\partial \Pi}{\partial u} ; \quad \Pi = \frac{1}{3} u^2 (u - U_0) . \quad (8.18)$$

For qualitative description of wave motions we use now a convenient mechanical analogy [148]: the Eq. (8.18) describes the motion of a material particle with its weight γ^{-1} under the influence of the force field of a potential Π shown in Fig 8.1. The variable ξ corresponds, in these terms, to the time τ. It is obvious from Fig. 8.1 that while the energy level grows (the horizontal lines in figure) the amplitude $u_{max} - u_{min} \equiv u_m - u_*$ of the particle's oscillations inside the potential hole increases. And if low amplitude oscillations in the vicinity of the potential hole's maximum are almost symmetric, the amplitude (or the energy) of the growing oscillations becomes more and more asymmetric: the 'particle' stays for more time in the range of low u values bypassing high values (see Fig. 8.2).

In the limiting case when $u_{max} \to U_0$ the 'particle' motion becomes aperiodic: the 'particle' comes from the range $t = -\infty$, where it was

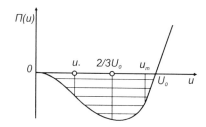

Figure 8.1.   The 'potential energy' of periodic waves.

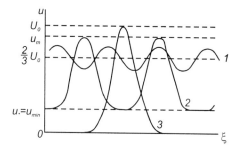

Figure 8.2.   The dependence $u(\xi)$ for Eq. (8.18): 1 — low amplitude oscillations in the vicinity of $\Pi(U)$ minimum; 2 — finite amplitude oscillations; 3 — a solitary wave.

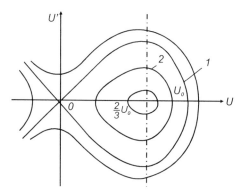

Figure 8.3.   The phase picture for Eq. (8.18): 1 — a solitary wave; 2 — periodic waves.

staying for infinitely long time at $u = 0$, to the potential hole where it reaches the value U_0 then reflects from that point and goes again to $u = 0$ for $t \to \infty$. Such a limiting aperiodic wave motion is called a solitary wave or a soliton. Thus a solitary wave represents the limiting case of a periodic wave of the highest amplitude [in terms of Eq. (8.18)] and of the infinite length.

The phase picture for Eq. (8.18) is presented in Fig. 8.3. The solitary wave relates to the separatrix *1* which has inner stable focuses *2* corresponding to nonlinear periodic wave motion.

In case of small amplitude oscillations in the vicinity of the potential hole minimum (see Fig. 8.1), when nonlinearity is small, from Eqs. (8.17), (8.18) yields

$$U = \frac{2}{3} U_0 + \widehat{U}_0 + \hat{u}_0 \exp\left(i\xi \sqrt{\frac{2}{3} \gamma U_0} \right). \qquad (8.19)$$

Here the oscillations amplitude $|\hat{u}_0| \ll 1$ and $\widehat{U}_0$ means the $\widetilde{U}_0$ value for $\hat{u}_0 \to 0$. Comparing (8.19) with the nonlinear problem solution $U = \hat{u}_0 \exp(i\xi)$ it appears that for the case considered of the infinitesimal wave $|\hat{u}_0| \ll 1$ the following equations are valid:

$$U_0 = \frac{3}{2\gamma}; \quad \hat{\beta}_1 = \frac{2}{3} \frac{\gamma}{\alpha} U_0 \approx -\frac{2H^2 N_0^2 \hat{c}}{k_0^2 c_f^{0\,3}} > 0. \qquad (8.20)$$

Here and further all of the parameters of the infinitesimal wave are marked with the upper hat and in the approximate formula for $\hat{\beta}_1$ the relation $\hat{c} \sim k_0^2$ is taken into account.

The first formula in (8.20) shows that $\hat{c}_f < c_f^0$, because for positive γ, α, U_0 the value of $\hat{c} = \hat{c}_f - c_f^0 < 0$, i.e., infinitesimal disturbances, are 'subsonic', which is a characteristic feature of a medium with negative dispersion.

In case of solitary waves' propagation in quiet water (only such a case is considered in the present chapter) from the natural physical condition of the decay of wave motion at infinity in Eqs. (8.17) yields $\widetilde{U}_0 = 0$. The exact solution of Eqs. (8.18) for $\widetilde{U}_0 = 0$ has the form

$$u = U_s = \frac{U_0}{\text{ch}^2 \left(\sqrt{\gamma U_0 / 6} \, \xi \right)} = \frac{3}{2\gamma} \, \text{ch}^{-2} \left(\xi/2 \right). \qquad (8.21)$$

In Eq. (8.21) we have used the expression for U_0 from Eq. (8.20). So the characteristic dimensionless 'width' of the soliton is equal to 2. As for its velocity of propagation, from the third equation in (8.17) for $\tilde{U}_0 = 0$ we have

$$\beta_{1_s} = -\frac{2}{3}\frac{\gamma}{\alpha}U_0 \approx -\frac{2H^2 N_0^2 c_s}{k_0^2 c_f^{0\,3}} < 0,\qquad(8.22)$$

where $c_s = c_{f_s} - c_f^0$ is an addition to the soliton phase velocity. From here it follows that $c_{f_s} > c_f^0$ and the addition c_s is equal to the modulus of the addition $|\hat{c}|$ to the phase velocity of an infinitesimal disturbance and has the opposite sign ($c_s > 0$). The latter means that the soliton is 'supersonic'. Further comparison of Eqs. (8.21) and (8.22) shows that the soliton amplitude is proportional to its propagation velocity, which is a typically nonlinear effect.

In the intermediate case of nonlinear periodic waves we have $c_f^0 - c_s < c_f < c_f^0 + c_s$. We assume further that the following condition is satisfied for such waves

$$\langle U \rangle = \frac{1}{\tilde{\lambda}}\int_0^{\tilde{\lambda}} U\left(\xi\right)d\xi = 0,\qquad(8.23)$$

where λ is the dimensionless wavelength which corresponds to the assumption of the absence of mean flow. This requirement is automatically satisfied for limiting cases of the infinitesimal wave and for the soliton. And for the nonlinear periodic waves it is in fact the definition of $\tilde{U}_0$ (a dimensionless addition to the phase wave velocity)[1]. In this case the solution (8.18) taking (8.17) into account can be expressed through elliptic functions:

$$\frac{U}{U_0} = \frac{s^2}{m^2}\,\text{cn}^2\left(\frac{\xi}{2m}\right) + \frac{\left(1 - s^2\right)K\left(s\right) - E\left(s\right)}{m^2\left(s\right)K\left(s\right)};$$

$$m\left(s\right) = \left(1 - s^2 + s^4\right)^{1/4};\qquad(8.24)$$

$$\tilde{\lambda} = 4K\left(s\right)m;\quad \frac{\tilde{U}_0}{U_0} = \frac{\left(2 - s^2\right)K\left(s\right) - 3E\left(s\right)}{m^2\left(s\right)K\left(s\right)};\quad U_0 = \frac{3}{2\gamma}.\qquad(8.25)$$

Here $\text{cn}\left(y\right)$ is the elliptic Jacobi function (that is why such waves are frequently called *cnoidal*) with the modulus s ($0 \leq s \leq 1$), $K\left(s\right)$ and $E\left(s\right)$ are accordingly first and second genus elliptic integrals. The parameter s^2 is in this case the amplitude. Eqs. (8.25) define the *nonlinear dispersion relation*, i.e., the relation between the wavelength, the wave velocity and the wave amplitude. As the function $K(s)$ grows monotonically versus s from $\pi/2$ to ∞, the first Eq. (8.25) shows that while s^2 grows the wavelength also grows, and $\tilde{\lambda} \to \infty$ for $s \to 1$.

From (8.24), (8.25) the following expansions yield for $s \to 0$:

$$U(\xi) = \frac{3s^2}{\gamma}\left(\cos\xi + \frac{s^4}{8}\cos 2\xi + \dots\right);$$

$$\frac{\tilde{\lambda}}{2\pi} = 1 + \frac{15}{64}s^4 + \dots, \qquad \frac{\tilde{U}_0}{U_0} = -1 + \frac{9}{16}s^4 + \dots$$

As the solution of Eqs. (8.24), (8.25) we have for $s \to 1$ Eq. (8.21). Herewith the distance between solitons $\tilde{\lambda} \approx 4\gamma^{-1/2}\ln\left[4(1-s^2)^{-1/2}\right]$ is defined by the asymptotics of $K(z)$.

A common physical phenomenon in this approximation is the wavelength and the wave velocity's growth versus its growing amplitude. When $\gamma < 0$ 'subsonic' solitons (troughs) will appear instead of 'supersonic' solitons (crests).

Now we study briefly, in accordance with [206], [207], some examples of the propagation of stationary weakly nonlinear internal waves in shallow water[2].

1. *Two-layer stratification model.*

$$\hat{\rho}(\eta) = 1 \ (0 > \eta > -l); \quad \hat{\rho}(\eta) = \hat{\rho}_0 > 1 \ (-l > \eta > -1).$$

In this case the Brunt–Väisälä frequency is defined by the δ-function accumulated in the point $\eta = -l$; the solution of the boundary value problem (3.3)–(3.4a) has the form

$$W(\eta) = \begin{cases} C\frac{1-l\beta_0/\delta}{1-l}(1+\eta) & (\eta > -l); \\ C(1+\beta_0\eta/\delta) & (\eta > -l), \end{cases}$$

where C is a normalizing constant, $\delta = 2(\hat{\rho}_0 - 1)(\hat{\rho}_0 + 1)^{-1}$ and the parameter β_0 (the dispersion equation solution) has the form

$$\beta_0^{0,1} = \frac{1+\delta/2}{2l(1-l)}\left[1 \pm \sqrt{1 - \frac{4\delta l(1-l)}{1+\delta/2}}\right].$$

In this case there are only two waves: the surface wave corresponding to β_0^0 and the internal wave corresponding to β_0^1. We can easily obtain expressions for the coefficients α and γ contained in the formula for nonlinear waves. But they are rather cumbrous and we simplify them by taking into account that $\delta \sim 10^2 \ll 1$.

For the surface wave:

$$\beta_0^0 = \delta\left[1 + \mathcal{O}\left(\delta\right)\right]; \quad C = \delta^{1/2};$$

$$\alpha^0 = 3\delta^{-1}\left[1 + \mathcal{O}\left(\delta\right)\right]; \quad \gamma^0 = 9/2\,\delta^{1/2}\left[1 + \mathcal{O}\left(\delta\right)\right].$$

These formulae show that the surface wave in this case is similar to that for the barotropic medium.

The internal wave:

$$\beta_0^1 = \left[l\left(1 - l\right)\right]^{-1}\left[1 + \mathcal{O}\left(\delta\right)\right]; \quad C = \delta\left(1 - l\right);$$

$$\alpha^1 = 3\left[1 + \mathcal{O}\left(\delta\right)\right]; \quad \gamma^1 = \frac{9}{2}\frac{2l - 1}{l^2\left(1 - l\right)^2}\left[1 + \mathcal{O}\left(\delta\right)\right].$$

These asymptotics are valid when $\min\{l,\ (1 - l)\} \ll \delta$. If $l > \frac{1}{2}$ then $\gamma_1 > 0$ and positive solitons (crests) can propagate in the fluid, and if $l < 1/2$ then $\gamma_1 < 0$ and only negative solitons (troughs) can appear. So the stratification profile can qualitatively change the behaviour of the nonlinear internal waves.

2. *Exponential stratification.*
In this case

$$\hat{\rho}\left(\eta\right) = e^{-\delta\eta}; \quad \Omega\left(\eta\right) = 1; \quad W_m\left(\eta\right) = c_m\sin\left[\ae_m\left(1 + \eta\right)\right]$$

and the dispersion relation (3.43) has the form

$$\ae_m\,\mathrm{tg}\,\ae_m = \frac{\delta}{1 - \frac{5}{8}\left(\delta/\ae_m\right)^2}; \quad \beta_m^0 = \ae_m^2 - \delta^2/4.$$

Again taking into account that $\delta \sim 10^2 \ll 1$ we can easily obtained parameters α, β, γ.

For the surface wave:

$$\alpha_0 = 3\delta^{-1}\left[1 + \mathcal{O}\left(\delta\right)\right]; \quad \beta_0^0 = \delta\left[1 + \mathcal{O}\left(\delta\right)\right]; \quad \gamma_0 = 9/2\,\delta^{1/2}\left[1 + \mathcal{O}\left(\delta\right)\right];$$

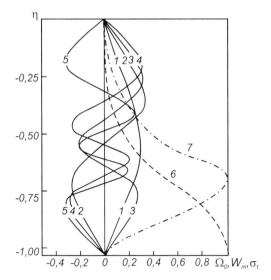

Figure 8.4. Vertical distribution of the first five eigenfunctions (1–5), the conventional density (6) [164] and the dimensionless Brunt– Väisälä frequency (7).

i.e., the formulae are similar to the two layer formulae.

For the internal wave:

$$\alpha_m = [1 + \mathcal{O}(\delta)]; \quad \beta^0_m = m^2\pi^2 + 3/4\,\delta^2 + \mathcal{O}\left(\frac{\delta^3}{m\pi}\right);$$

$$\gamma_m = \frac{\delta m\pi}{\sqrt{2}}\left\{3\,(-1)^m + 2\left[(-1)^m - 1\right] + \mathcal{O}(\delta)\right\}, \quad (m = 1,\ 2,\ 3,\ \ldots).$$

$$(8.26)$$

The formula for γ_m in (8.26) shows that solitons are positive for even mode numbers and negative for uneven modes. The expression for γ_m obtained in [207] differs from presented in (8.26) with the absence of the first term in the curly brackets. It is only because the 'rigid lid' approximation was used in [207] contrary to (8.26). All of the linear internal waves formulae are almost independent of whether the exact boundary condition or the 'rigid lid' approximation is used for the free surface.

3. *Natural stratification.*

In this case the evidence [164] obtained for the Arkona Basin of the Baltic Sea was used where $H = 47$ m, $N_0 = 5.6 \times 10^{-2}$ s^{-1}. The

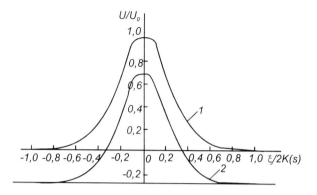

Figure 8.5. The normalized amplitude function U/U_0 vs. $\xi/2K$ (s): 1 — a solitary wave; 2 — a cnoidal wave.

Table 8.1.   Values of the parameters for the n-th mode

n	1	2	3	4	5
$10^2 \alpha_n$	2.6	1.6	1.5	1.4	1.4
$\beta_{0,n}$	19	134	306	565	843
γ_n	0.26	3.00	-5.00	10.60	9.20
$c^0_{f,n}$ (cm/s)	61	23	15	11	9
$10^2 \hat{c}_n$ (cm/s)	62	5.35	-1.63	0.70	0.38
T_n (h)	1.34	3.56	5.45	7.45	9.10
ζ_n (cm)	81.0	7.0	4.2	2.0	2.3

smoothed distributions of the dimensionless Brunt–Väisälä frequency $\Omega_0(\eta) = N^2/N_0^2$, of the conventional density $\sigma_t(\eta) = [\rho_0(\eta)-1]\cdot 10^3$ and of five first eigenfunctions[3] $W_m(\eta)$ of the main boundary value problem with the boundary conditions $W(0) = W(-1) = 0$ and the normalization $\int_{-1}^{0} W_m^2(\eta)\, d\eta = 1$ are shown in Fig. 8.4. The dimensionless wave number was chosen to be equal to $k_0 = 0,1$ and the infinitesimal wavelength $\hat{\lambda} = 2\pi H k_0^{-1} = 2.95$ km and its period $T_n = \lambda/c^0_{f,n}$. All of the wave parameters are given it Table 8.1.

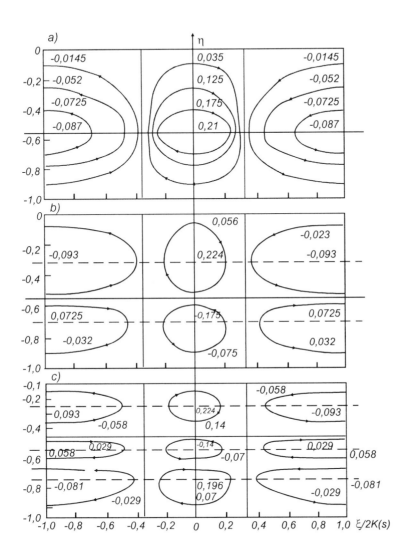

Figure 8.6.   Dimensionless streamlines $\frac{U_N(\xi)\,W_n(\eta)}{U_{n,\,0}}$ = const : a) n=1; b) n=2; c) n=3.

For further calculations of the amplitude function $U(\xi)$ we chose a cnoidal wave with the parameter $s = 0.99$ and a soliton. Here $\lambda/\hat{\lambda} = 2.14$ and $\tilde{c}_n = 0.111\,\hat{c}_n$ (λ and $\tilde{c}_n$ are the wavelength and the addition to

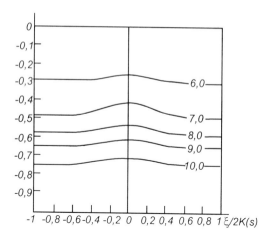

Figure 8.7.   Isolines of the conventional density σT for the first-mode solitary wave.

the 'sound velocity' for cnoidal waves, $\hat{\lambda}$ and $\hat{c}_n$ for the infinitesimal waves). It is obvious from Fig. 8.5 that the amplitude distribution for the cnoidal waves in the coordinates $U_n(\xi)/U_{n,0}$, $\xi/2K(s)$ is independent of the mode number, and for $s = 0.99$ differs insignificantly from the soliton. In Fig. 8.6 the streamlines for the first three modes of the cnoidal waves are shown[4]. To obtain the values of dimensional streamlines we should multiply their values from Fig. 8.6 by $U_{0,n}k_0^2 Hc_{f,n}^0$. Maximum displacements ζ_n for each of the solitary wave modes are also presented in Table 8.1.

In Fig. 8.7 numerical results for variations of the conventional density $\sigma_T(\eta, \xi)$ caused by propagating solitons are shown, where $\sigma_T(\eta)$ was defined by the formula

$$\sigma_T(\eta) = \sigma_t(\eta) + 10^3 k_0^2 \hat{\rho}_0(\eta) \Omega(\eta) W(\eta) U(\xi).$$

The characteristic feature of these curves is the accumulation of streamlines in the vicinity of the Brunt–Väisälä frequency and of the maximum of the amplitude functions. In this region of greater local density gradients the secondary nonlinear effects probably result in the stratified fluid mixture.

To conclude this section we remark that stationary nonlinear internal waves in the horizontally inhomogeneous ocean were studied in [293], [294].

3. PROPAGATION OF WEAKLY NONLINEAR INTERNAL WAVE PACKETS IN A THIN PYCNOCLINE

Results presented in the previous section relate to the long wave case (or to a shallow water) when $H \ll \lambda$. But the case when deep water is stratified in a thin layer (a pycnocline) and is homogeneous outside the layer is quite important for oceanology. Such a problem was considered in [21], [58], [207].

Equations describing nonlinear internal waves in a narrow pycnocline can probably be obtained with the method of the Hamiltonian formalism, but following [21], [58], [207] it seems to be more convenient to obtain these equations from the input hydrodynamics equations by methods of perturbation theory.

Following [207] we consider plane stationary nonlinear internal waves which, as was shown in Section 2.2, are described by Eqs. (2.28), (2.29).

Let the fluid be arbitrarily stratified in a layer of the thickness d. The fluid density is constant outside this layer. The pycnocline's depth is h and the ocean's depth is H. We introduce some dimensionless variables:

$$\zeta = \theta\lambda^{-1}; \quad \eta = z\tilde{d}^{-1}; \quad F = \psi\left(c_f\tilde{d}\right), \qquad (8.27)$$

where λ is the characterisitc horizontal dimension of the wave (the wavelength for periodic waves and the wave width for a soliton) and $\tilde{d}$ is the characteristic vertical scale. Eq. (2.28) in those dimensionless variables has the form

$$\varepsilon^2\frac{\partial^2 F}{\partial\zeta^2} + \frac{\partial^2 F}{\partial\eta^2} + \left\{\beta F + \frac{\delta}{2}\left[\varepsilon^2\left(\frac{\partial F}{\partial\zeta}\right)^2 + \left(\frac{\partial F}{\partial\eta}\right)^2 - 2\frac{\partial F}{\partial\eta}\right]\right\}\Omega(\eta - F) = 0,$$
$$(8.28)$$

where

$$\varepsilon = \tilde{d}\lambda^{-1}; \quad \beta = \tilde{d}^2 N_0^2 c_f^{-2}; \quad \delta = \tilde{d}N_0^2 g^{-1}; \quad \Omega = \frac{N^2}{N_0^2}. \qquad (8.29)$$

Here N_0 is the maximum Brunt–Väisälä frequency ($0 < \Omega < 1$), the parameter δ characterizes the deviation from the Boussinesq approximation. Assume that the pycnocline is so narrow that $d \ll h$, $d \ll H - h$.

While simplifying the problem we may consider that outside the pycnocline the incompressible barotropic fluid to be infinite in the vertical direction. Then we put the zero of the coordinate system ($z = 0$) in the middle of the stratified layer and consider $d/2$ to be the vertical scale $\tilde{d}$ in (8.28), (8.29) (i.e., $d/2 = \tilde{d}$, $\eta = 2z/d$ etc.). In this case distributions of $\Omega(\eta)$ and $\Omega(\eta - F)$ in (8.28) yield:

$$\Omega(\eta) = \begin{cases} 0 & (1 < \eta < \infty) \\ \Omega_0(\eta) & (-1 < \eta < 1) \\ 0 & (-\infty < \eta < -1) \, ; \end{cases}$$

$$\Omega(\eta - F) = \begin{cases} 0 & (1 + F < \eta < \infty) \\ \Omega_0(\eta - F) & (F - 1 < \eta < F + 1) \\ 0 & (-\infty < \eta < F - 1) \, . \end{cases}$$

Define now:

$$F = F^+ \quad \text{for} \quad 1 + F < \eta < \infty,$$
$$F = F^0 \quad \text{for} \quad F - 1 < \eta < F + 1,$$
$$F = F^- \quad \text{for} \quad -\infty < \eta < F - 1.$$

Then Eqs. (8.28), (8.29) yield

$$\varepsilon^2 \frac{\partial^2 F^+}{\partial \xi^2} + \frac{\partial^2 F^+}{\partial \eta^2} = 0 \, ; \quad \varepsilon^2 \frac{\partial^2 F^-}{\partial \xi^2} + \frac{\partial^2 F^-}{\partial \eta^2} = 0 \, , \tag{8.30}$$

and $F^0(\eta, \xi)$ meets Eq. (8.28) for $\Omega = \Omega_0$. Here $\varepsilon = d/2\lambda$. The boundary conditions should be set at the pycnocline boundaries as well as for $\eta \to \pm\infty$. We consider waves to decay at $\eta \to \pm\infty$. The dynamic and kinematic conditions, which require continuity of F and pressure, should be satisfied at the wavy pycnocline boundaries. It can be easily shown that these conditions also mean the continuity of the velocity field while crossing the boundary (when we assume the density ρ_0 to be continuous). Then the boundary conditions have the form (see [58], [350]):

$$F^+ = F^0 \, , \quad \frac{\partial F^+}{\partial \eta} = \frac{\partial F^0}{\partial \eta} \quad \text{at} \quad \eta = 1 + F^+ \, ;$$

$$F^- = F^0, \quad \frac{\partial F^-}{\partial \eta} = \frac{\partial F^0}{\partial \eta} \quad \text{at} \quad \eta = -1 + F^-;$$

$$F^\pm = 0, \quad \frac{\partial F^\pm}{\partial \eta} = 0 \quad \text{at} \quad \eta \to \pm\infty. \tag{8.31}$$

In such a statement the problem was solved in [207].

We assume that the parameter $\varepsilon \ll 1$ in Eqs. (8.28), (8.30). As in this assumption the fluid is infinite outside the pycnocline, and to prevent the degeneration of Eq. (8.30) we introduce the stretched vertical coordinates: $\eta_- = \varepsilon (\eta - 1)$ for the upper layer; and $\eta_+ = \varepsilon (\eta + 1)$ for the lower layer. Hereafter Eqs. (8.30) will have the form

$$\Delta^+ F^+ = 0; \quad \Delta^- F^- = 0; \quad \Delta^\pm = \frac{\partial^2}{\partial \xi^2} + \frac{\partial^2}{\partial r^{\pm 2}}. \tag{8.32}$$

Following [58] we try the solution of Eq. (8.28), (8.30) with the boundary conditions (8.31) in the form

$$F = \{F^+, \ F^0, \ F^-\} = \varepsilon F_1 + \varepsilon^2 F_2 + \dots, \quad \beta = \beta_0 + \varepsilon \beta_1 + \dots. \tag{8.33}$$

While substituting Eq. (8.33) into (8.28)–(8.30), developing F in the vicinity of $\eta = \pm 1$ and $\Omega (\eta - F^0)$ in the vicinity of η as a Taylor series, then equating terms at the same ε power we obtain the terms of order ε_0:

$$\Delta^+ F_1^+ = 0; \quad F_1^+ (0, \ \xi) = F_1^0 (1, \ \xi); \quad F_1^+ \big|_{\eta^+ \to +\infty} = 0; \tag{8.34}$$

$$\mathcal{L}\left(F_1^0\right) \equiv \frac{\partial^2 F_1^0}{\partial \eta^2} - \delta \Omega_0 (\eta) \frac{\partial F_1^0}{\partial \eta} + \beta_0 \Omega_0 (\eta) F_1^0 = 0; \quad \frac{\partial F_1^0}{\partial \eta} \bigg|_{\eta = \pm 1} = 0;$$
$$\tag{8.34a}$$

$$\Delta^- F_1^- = 0; \quad F_1^- (0, \ \xi) = F_1^0 (1, \ \xi); \quad F_1^- \big|_{\eta^- \to -\infty} = 0. \tag{8.34b}$$

and the terms of order ε^2:

$$\Delta^+ F_2^+ = 0; \quad F_2^+ (0, \ \xi) = F_2^0 (1, \ \xi); \quad F_2^+ \big|_{\eta^+ \to +\infty} = 0. \tag{8.35}$$

$$\mathcal{L}\left(F_2^0\right) = \chi\left(F_1^0\right) \equiv \beta_0 \Omega_0' F_1^{0^2} - \beta_1 \Omega_0 F_1^0 - \delta \Omega_0' F_1^0 \frac{\partial F_1^0}{\partial \eta} - \frac{\delta}{2} \Omega_0 \left(\frac{\partial F_1^0}{\partial \eta}\right)^2;$$
$$\tag{8.35a}$$

$$\left. \frac{\partial F_2^0}{\partial \eta} \right|_{\eta=\pm 1} = \chi_\pm \equiv \left. \frac{\partial F_1^\pm}{\partial \eta^\pm} \right|_{\eta^\pm=0} + \beta_0 \Omega_0 \, (\pm 1) \, F_1^{0^2} \, (\pm 1, \, \xi) \, ; \qquad (8.35b)$$

$$\Delta^- F_2^- = 0 \, ; \quad F_2^- \, (0, \, \xi) = F_2^0 \, (-1, \, \xi) \, ; \quad \left. F_2^- \right|_{\eta^- \to -\infty} = 0 \, . \qquad (8.35c)$$

Here $\delta = N_0^2 \alpha / 2g$; N_0 is the maximum Brunt–Väisälä frequency within the pycnocline.

The homogeneous boundary value problem (8.34) is factorisable, i.e., allows a solution in the form

$$F_1^0 = W \, (\eta) \, U \, (\xi) \, . \qquad (8.36)$$

Substituting Eq. (8.36) into (8.36a) we make sure that $W \, (\eta)$ is the solution of the Sturm–Liouville boundary value problem:

$$W'' - \delta \Omega_0 \, (r) \, W' + \beta_0 \Omega_0 \, (r) \, W = 0 \, , \quad W' \, (\pm 1) = 0 \, , \qquad (8.37)$$

and the amplitude function $U \, (\xi)$ is still undefined.

We stop at some properties of the boundary value problem (8.37) which may be rewritten in the 'self-adjoint' form:

$$(\hat{\rho} W')' + \beta_0 \hat{\rho} \Omega_0 \, (\eta) \, W = 0 \, ; \quad W' \, (\pm 1) = 0 \, . \qquad (8.38)$$

Here $\hat{\rho} \, (\eta) = \rho_0 \, (\eta) / \rho_a$ is the dimensionless density normalized to the mean fluid density $\rho_a = (\rho_+ + \rho_-)/2$ which is defined in every of infinite homogeneous layers (ρ_+ and ρ_- respectively). While going from (8.37) to (8.38) the definition of the Brunt–Väisälä frequency in the pycnocline was used: $\Omega_0 = -\delta^{-1} \hat{\rho}' \, (\eta) / \hat{\rho} \, (\eta)$.

The boundary value problem (8.38) is self-adjoint. If we assume that $\Omega_0 \, (\eta) > 0$ everywhere outside the pycnocline, then it has a denumerable set of positive eigenvalues $\{\beta_{0,n}\}$ and a system of according eigenfunctions $\{W_n\}$ which is complete and orthogonal:

$$(W_n, W_m) = \int_{-1}^{1} \hat{\rho} \, \Omega_0 W_m W_n \, d\eta = 0 \quad n \neq m \, ;$$

$$(W_n, W_n) = \|W_n\|^2 = \int_{-1}^{1} \hat{\rho} \, \Omega_0 W_n^2 \, d\eta > 0 \, . \qquad (8.39)$$

For large n the evaluation $\beta_{0,n} \sim n^2$ is valid. Assuming $V = \hat{\rho} W'$ the following boundary value problem yields for V:

$$[(\hat{\rho}\Omega_0)^{-1} V']' + \beta_0 \hat{\rho}^{-1} V = 0; \quad V(\pm 1) = 0, \tag{8.40}$$

which has the properties similar to the boundary value problem (8.38) (described above). In particular, the complete eigenfunction system $\{V_n\}$ satisfies the orthogonality condition:

$$(V_n, V_m) = \int_{-1}^{1} \hat{\rho}^{-1} V_n V_m \, d\eta = 0 \quad n \neq m;$$

$$(V_n, V_n) \equiv \|V_n\|^2 = \int_{-1}^{1} \hat{\rho}^{-1} V_n^2 \, d\eta > 0. \tag{8.41}$$

Consider the solution of the inhomogeneous boundary value problem

$$(\hat{\rho} f')' + \beta \hat{\rho} \, \Omega_0 f = \hat{\rho} \chi(\eta); \quad f'(\pm 1) = \chi_{\pm}, \tag{8.42}$$

where β does not coincide with any of the eigenvalues $\beta_{0,n}$ of the boundary value problem (8.38). While defining $\hat{\rho} f' = \psi$ we make sure that the function ψ should be obtained from the solution of the following boundary value problem:

$$\hat{\rho}\left[(\hat{\rho}\Omega_0)^{-1} \psi'\right]' + \beta\psi = \hat{\rho}\left(\chi\Omega_0^{-1}\right)'; \quad \psi(\pm 1) = \hat{\rho}_{\pm}\chi_{\pm}, \tag{8.43}$$

where $\hat{\rho}_{\pm} = \hat{\rho}(\pm 1)$. Consideration $\psi = \tilde{\psi} + \psi_0$, where

$$\psi_0 = \frac{\hat{\rho}_+\chi_+ - \hat{\rho}_-\chi_-}{\hat{\rho}_+ - \hat{\rho}_-} [\hat{\rho}(\eta) - \hat{\rho}_-] + \hat{\rho}_-\chi_-, \tag{8.44}$$

yields that ψ results from the solution of the inhomogeneous boundary value problem similar to (8.42) with homogeneous boundary conditions. A solution of this problem is tried in the form of the Fourier series in the system $\{V_n(\eta)\}$. We expand the function $\psi_0(\eta)$ in the Fourier series in the system $\{V_n\}$ and sum it with the obtained expression for $\tilde{\psi}$; then as a result we have the solution of the inhomogeneous boundary value problem (8.43) in the form

$$\psi(\eta) = \sum_{n=1}^{\infty} \psi_n V_n(\eta);$$

$$\psi_n = \frac{(\psi, V_n)}{\|V_n\|^2} = \frac{\int\limits_{-1}^{1} \left[\left(\chi \Omega_0^{-1} \right)' - \beta_{0,n} \hat{\rho}^{-1} \psi_0 \right] V_n \, d\eta}{\|V_n\|^2 (\beta - \beta_{0,n})}.$$

If we substitute $V_n(\eta) = \hat{\rho}(\eta) W_n'(\eta)$ instead of $V_n(\eta)$ into the integral in the numerator of the latter equation, then after integration by parts, taking into account that $W_n(\eta)$ is the solution for the boundary value problem (8.38) and the function ψ_0 is defined by Eq. (8.44), yields

$$\psi_n = \beta_{0,n} \frac{\int\limits_{-1}^{1} \hat{\rho} \chi W_n \, d\eta - (\hat{\rho}_+ W_n^+ \chi_+ - \hat{\rho}_- W_n^- \chi_-)}{\|V_n\|^2 (\beta - \beta_{0,n})}. \qquad (8.45)$$

Here $W_n^{\pm} = W_n(\pm 1)$ are obtained single-valued from the equation

$$V_n'(\eta) + \hat{\rho}(\eta) \Omega_0(\eta) W_n(\eta) = 0 \quad \text{for} \quad \eta \to \pm 1.$$

If the function $\psi(\eta)$ is already defined, then the unknown function f as a solution of the inhomogeneous boundary value problem (8.42) is obtained from the relation $f' = \psi \hat{\rho}^{-1}$ by only one quadrature.

In case when $\beta = \beta_{0,n}$ the solvability condition for problems (8.42), (8.48), as it follows from (8.45), has the form

$$\int\limits_{-1}^{1} \hat{\rho} \chi W_n \, d\eta - (\hat{\rho}_+ W_n^+ \chi_+ - \hat{\rho}_- W_n^- \chi_-) = 0. \qquad (8.46)$$

We now assume that the function $W(\eta)$ in (8.36) coincides with one of the orthogonal modes $W_n(\eta)$ of the problem (8.37) and β_0 with the according eigenvalue $\beta_{0,n}$. Consider $F_2^0(\eta, \xi) = U_2(\xi) W(\eta) + \tilde{F}_2^0(\eta, \xi)$, where $U_2(\xi)$ is the second approximation amplitude function obtained from the third approximation, and $\tilde{F}_2^0(\eta, \xi)$ is the solution of the inhomogeneous boundary value problem (8.35a) for $\beta = \beta_0$. For its solubility it is necessary and sufficient to meet the condition (8.46), where

$W_n \equiv W$, $\beta_{0,n} \equiv \beta_0$. Substituting $\chi(\eta, \xi)$, $\chi_\pm(\xi)$ defined by (8.35a) into (8.46) we obtain:

$$\tilde{\lambda}_+ \left.\frac{\partial F_1^+}{\partial \eta_+}\right|_{\eta_+=0} - \tilde{\lambda}_- \left.\frac{\partial F_1^-}{\partial \eta_-}\right|_{\eta_-=0} + \alpha\beta_1 U + \gamma U^2 = 0; \quad \lambda_\pm = \hat{\rho}_\pm W_\pm,$$

where α and γ are defined by

$$\alpha = \int\limits_{-1}^{1} \hat{\rho}\,\Omega_0 W^2\,d\eta > 0; \quad \gamma = 3\int\limits_{-1}^{1} \hat{\rho}\,\Omega_0 W'W\left(\beta_0 W - \frac{\delta}{2}W'\right)d\eta.$$

$$(8.47)$$

If we now substitute Eq. (8.36) into (8.34)–(8.34a) and introduce the following variables

$$\tilde{\eta} = \left\{\begin{array}{l} \eta_+\ (\tilde{\eta}>0) \\[2mm] \eta_-\ (\tilde{\eta}<0) \end{array}\right. ; \quad \Phi = \left\{\begin{array}{l} \Phi_+\ (\tilde{\eta}>0); \quad \lambda_\pm = \lambda_\pm W_\pm = \hat{\rho}_\pm W_\pm^2 > 0 \\[2mm] \Phi_-\ (\tilde{\eta}<0); \quad F_1^\pm = W_\pm\Phi_\pm, \end{array}\right.$$

we have the boundary value problem for the unknown amplitude function $U(\xi)$ in the first approximation :

$$\left.\begin{array}{l} \Delta\Phi = 0; \quad \Phi|_{\tilde{\eta}\to\pm\infty} = 0; \quad \Phi|_{\tilde{\eta}=0} = U; \\[3mm] \lambda_+ \left.\dfrac{\partial\Phi}{\partial\tilde{\eta}}\right|_{\tilde{\eta}=+0} - \lambda_- \left.\dfrac{\partial\Phi}{\partial\tilde{\eta}}\right|_{\tilde{\eta}=-0} + \alpha\beta_1 U + \gamma U^2 = 0; \\[3mm] \Delta = \dfrac{\partial^2}{\partial\xi^2} + \dfrac{\partial^2}{\partial\tilde{\eta}^2}; \quad \lambda_\pm = \hat{\rho}_\pm W_\pm^2 > 0. \end{array}\right\} \quad (8.48)$$

A solution of the Dirichlet problem for a harmonic function in the both half-planes $\tilde{\eta}>0$ and $\tilde{\eta}<0$ is the double-layer potential

$$\Phi_\pm(\tilde{\eta}, \xi) = \pm\frac{1}{\pi}\int\limits_{-\infty}^{\infty} \frac{\tilde{\eta}\,U(y)}{(\xi-y)^2 + \tilde{\eta}^2}\,dy. \quad (8.49)$$

Substitution of this solution into the last relation in (8.48) yields the unknown equation for the first approximation amplitude function

$$\frac{1}{\pi}(\lambda_+ - \lambda_-)\frac{d}{d\xi}\int\limits_{-\infty}^{\infty} \frac{U(y)\,dy}{(\xi-y)} = \alpha\beta_1 U(\xi) + \gamma U^2(\xi). \quad (8.50)$$

This equation in a slightly modified form was initially derived by Benjamin [21]. Here the integral in (8.49) is given with respect to its principal value.

Consider simple particular cases of the Eq. (8.50) solution. If the wave amplitude a is small, then we can try the solution in the form $\hat{U} = ae^{ik\xi}$. Then neglecting the quadratic term in Eq. (8.50) after some simple transforms yields

$$|k|\,(\lambda_+ - \lambda_-) = \alpha\hat{\beta}_1 \,. \tag{8.51}$$

From this formula follows that $\hat{\beta}_1 > 0$ because $\lambda_\pm > 0$, $\alpha > 0$. The latter shows that the propagation velocity of an infinitesimal disturbance relatively to the sound velocity (c_f^0) $\hat{c} \sim -\beta_1$ is subsonic, i.e., $\hat{c}_f < c_f^0$ similarly to the KdV equation. If the nonlinearity in (8.50) is sufficient, then the solution may be tried in the form of an 'algebraic' soliton [21] $U = a/(b^2 + \xi^2)$. Substitution of the latter expression into (8.50) yields:

$$U = \frac{a^2}{b^2 + \xi^2}\,; \quad a = -\frac{2}{\gamma}\alpha\beta_1\,; \quad |b| = -\frac{\lambda_+ + \lambda_-}{\alpha\beta_1} > 0\,. \tag{8.52}$$

These formulae show that the value $\beta_1 < 0$, i.e., for a soliton $c_f = c_f^0 + c > 0$ (because $c \sim -\beta_1 > 0$) and its motion, similarly to the KdV case, is supersonic. Thus for $\gamma > 0$ supersonic solitons propagate in the 'crest' form. For $\gamma < 0$ supersonic solitons have the 'trough' form.

The single sufficient difference between the considered problem and the KdV case (Section 8.2) is a bit different type of the dispersion relation following from Eq. (8.51)

$$\omega = kc_f = k\,(c_f^0 + \hat{c}) = kc_f^0 - r\,|k|\,k\,; \tag{8.53}$$

$$c_f^0 = \frac{N_0 d}{2\sqrt{\beta_0}}\,; \quad \hat{c} = -r\,|k| = \frac{N_0 dk_0 \hat{\beta}_1}{4\beta_0^{3/2}} < 0\,. \tag{8.54}$$

And this peculiarity leads to the property that, contrary to results of Section 8.2, in the present case 'negative' solitons (troughs) are supersonic.

Consider similarly to Section 8.1. the dispersion relation (8.54) as the operator relation. Then the inverse Fourier transform of (8.54) yields

$$\frac{\partial u}{\partial t} + c_f^0 \frac{\partial u}{\partial x} - \frac{r}{\pi} \frac{\partial^2}{\partial x^2} \int\limits_{-\infty}^{\infty} \frac{u(\alpha, t)}{(x - \alpha)} \, d\alpha = 0. \tag{8.55}$$

When adding to this equation the characteristic nonlinear term $\gamma u \, \partial u / \partial x$ we have

$$\frac{\partial u}{\partial t} + (c_f^0 + \gamma u) \frac{\partial u}{\partial x} - \frac{r}{\pi} \frac{\partial^2}{\partial x^2} \int\limits_{-\infty}^{\infty} \frac{u(\alpha, t)}{(x - \alpha)} \, d\alpha = 0. \tag{8.56}$$

This equation obtained in [87], [211] describes unsteady nonlinear internal waves in a narrow pycnocline. Stationary solutions of this equation were studied here. The unsteady problem can be solved by methods of the inverse problem of the scattering theory.

APPENDIX 8.A: Evolution of the Korteweg–de Vries equation

An exact solution of the KdV equation was obtained by Perelman et al. [296] and by Miles [244], [245]. Some more complicated forms of the 'classic' KdV equation are described in [286], where theoretical models of the propagation of nonlinear internal waves, including the model evolution equations and their soliton solutions, are discussed.

The influence of higher orders of nonlinearity on the KdV equation solutions was studied as well. The quadratic nonlinear term may vanish for certain combinations of parameters, for instance, when layer depths are equal in a two-layer fluid, and the role of the higher nonlinear terms in the asymptotic series is increased ('modified' KdV, the mKdV). Cubic correction α_1 to the KdV equation ('extended' KdV, the eKdV) were obtained for a two-layer fluid where the sign of the cubic nonlinearity coefficient is always negative ([84], [125], [151], [175], [244], [245]). The effect of mixed nonlinearity on solitary wave transformation was investigated in [111], [168].

The nonlinear theory of internal waves in an arbitrary ocean stratification was developed by Lamb and Yan [191] and Beardsley [202]. The eKdV equation was obtained by the perturbation method in the second order in the wave's amplitude and first order in wavelength. A detailed analysis of the cubic effects for steady state solitary waves in a fluid with linear variation of the Brunt–Väisälä frequency and for a flow with constant shear is given by Gear and Grimshaw [97]. Grimshaw et al. [110] and Talipova et al. [347] considered the solution of the eKdV equation for interfacial waves in a three-layer fluid with symmetric locations of the jump boundaries.

Modification of the eKdV by rotation ('rotated modified' KdV, the rmKdV) has been considered by Anderson and Gill [5], Grimshaw [108], [109], Leonov [204], Ostrovsky [284] and Stepanyants [287], Gerkema [98], Grimshaw et al. [109]. The combined equation ('rotated modified extended' KdV, the rmeKdV) looks like:

$$\frac{\partial}{\partial x}\left(\frac{\partial \eta}{\partial t} + (c + \alpha\eta + \alpha_1\eta^2)\frac{\partial \eta}{\partial x} + \beta\frac{\partial^3 \eta}{\partial x^3}\right) = \frac{F^2}{2c}\eta,$$

where β is a dispersion coefficient, α, α_1 are, respectively, quadratic and cubic nonlinearity coefficients, c is the phase velocity of the linear long wave determined by the boundary value problem (3.64), η is the wave profile, t is time, and $f = 2\Omega \sin \phi$ is the Coriolis parameter, where $\Omega = 7,29 \times 10^{-5}$ s^{-1} and ϕ is latitude.

Gerkema and Zimmerman [99], using a forced Boussinesq model for a two-layer ocean, showed that rotation influences a nonlinear internal wave's evolution by reducing the number of solitons generated.

The effect of slowly varying depth was taken into account by including the weak additional term as described by Zhou and Grimshaw [417]. Friction in the bottom layer was included in the rmeKdV by Holloway et al. [128].

The most resent investigations in the theory of nonlinear internal waves concern fully nonlinear solitary waves. Computations of wave profiles in a two-layer fluid of finite depth have been performed by Choi and Camassa [51], by Grue et al. [115], [116], by Tung, Chan and Kubota [366] and by Turner and Vanden-Broek [369]. A fully nonlinear theory of internal solitary waves of the first mode in continuously stratified shear-free both shallow and deep fluid was concerned by Brown and Christie [41].

The problem of the generation of nonlinear internal waves by topography in fluids with different stratification profiles ('forced' KdV, the fKdV) has been developed by Grimshaw and Smyth [113], Grimshaw and Yi [114], Hanazaki [118], Helfrich and Melville [125].

Notes

1 Obviously, the phase wave velocity as well as the 'mean flow' can not be single-valued without the assumption (8.23)

2 Other cases of ratios of stratification parameters and solitary wave length are considered in [3], [4], [182], [216], [279], [286], [306], [368], [369]

3 Numerical calculations were performed with the $Q - \mathcal{L}$ algorithm by R.E. Tamsalu.

4 Analytical expressions for the first and second approximations of a stationary solution of the KdV equation for internal waves are obtained in [372] for the case of a general law for continuous fluid stratification (e.g., a smooth seasonal or main pycnocline). In the first approximation the soliton is presented in terms of one mode; a correction of the basic solution in the second approximation is a sum of modes. The modes with numbers close to n $(n + 1, n - 1)$ have the largest amplitudes. The resulting analytical solution describing the internal wave soliton as a bound multi-modal formation is consistent with the results of numerical computations. The evolution of a finite amplitude unimodal soliton is accompanied by the partial transfer of the initial energy to a group of neighboring modes of a multi-modal isolated wave of solitary type.

Chapter 9

PROPAGATION OF WEAKLY NONLINEAR INTERNAL WAVE PACKETS IN THE OCEAN

The basic nonlinear effects caused by the propagation of narrow spectrum wave packets are considered in this chapter. The first of such effects interesting from the viewpoint of oceanology is that of a weakly nonlinear internal wave packet with the mode structure generating velocity, density, and pressure fields which do not vanish when averaged over the wave period. In Section 9.1 the main characteristics of such fields are obtained. In Section 9.2 the problem of the stability of an internal wave packet with respect to its longitudinal modulation is studied. The long wave packet is shown to be always stable, and wavelength reduction causes the appearance of instability zones, whose parameters depend upon the stratification profile. In Section 9.3 it is shown that if a weakly nonlinear packet is formed by internal waves of relatively short period, then induced mean fields oscillate strongly in the vertical direction. The theory of the formation of oceanic fine structure, based on this phenomenon, is proposed. A rather good agreement between the evidence and numerical results for certain of some of the stratification profiles was obtained. In Section 9.4 propagation of a weakly nonlinear wave packet in the fluid with a constant Brunt–Väisälä frequency is studied.

259

1. GENERATION OF BACKGROUND FIELDS BY WEAKLY NONLINEAR INTERNAL WAVE PACKET WITH MODE STRUCTURE

Consider propagation of a narrow spectrum weakly nonlinear internal wave packet. It was shown in Section 7.3 how to reduce the basic equation

$$\dot{a}_{\mathbf{k}}^{\nu} + i\,\frac{\delta H}{\delta \bar{a}_{\mathbf{k}}^{\nu}} = 0$$

to the linear form

$$\dot{b}_{\mathbf{k}}^{\nu} + i\omega_{k}^{\nu} b_{\mathbf{k}}^{\nu} = 0$$

going canonically to new variables $b_{\mathbf{k}}^{\nu}$ by a formal integral power series. This equation has the following solution

$$b_{\mathbf{k}}^{\nu} = b_{\mathbf{k}}^{\nu}(0)\,e^{-i\omega_{k}^{\nu} t}\,. \tag{9.1}$$

To the second-order accuracy the inverse transform to initial wave amplitudes $a_{\mathbf{k}}^{\nu}$ is according to (7.51):

$$
a_{\mathbf{k}}^{\nu} = b_{\mathbf{k}}^{\nu} + \int \left(\frac{-V_{\mathbf{k}\mathbf{k}_1\mathbf{k}_2}^{\nu\nu_1\nu_2}\,\delta\left(\mathbf{k}-\mathbf{k}_1-\mathbf{k}_2\right)}{\omega_{k}^{\nu}-\omega_{k_1}^{\nu_1}-\omega_{k_2}^{\nu_2}}\,b_{\mathbf{k}_1}^{\nu_1} b_{\mathbf{k}_2}^{\nu_2} \right.
$$

$$
+\,2\,\frac{\bar{V}_{\mathbf{k}_1\mathbf{k}\mathbf{k}_2}^{\nu_1\nu\nu_2}\,\delta\left(\mathbf{k}_1-\mathbf{k}-\mathbf{k}_2\right)}{\omega_{k_1}^{\nu_1}-\omega_{k}^{\nu}-\omega_{k_2}^{\nu_2}}\,b_{\mathbf{k}_1}^{\nu_1}\bar{b}_{\mathbf{k}_2}^{\nu_2}
$$

$$
\left. -\,\frac{\bar{U}_{\mathbf{k}\mathbf{k}_1\mathbf{k}_2}^{\nu\nu_1\nu_2}\,\delta\left(\mathbf{k}+\mathbf{k}_1+\mathbf{k}_2\right)}{\omega_{k}^{\nu}+\omega_{k_1}^{\nu_1}+\omega_{k_2}^{\nu_2}}\,\bar{b}_{\mathbf{k}_1}^{\nu_1}\bar{b}_{\mathbf{k}_2}^{\nu_2} \right)\,d\mathbf{k}_1\,d\mathbf{k}_2\,d\nu_1\,d\nu_2\,.
$$

$$\tag{9.2}$$

This formula will help us obtain a solution of the problem. Assume that internal waves form a mode and that a packet is infinitely wide. If we consider waves propagating along the x-axis it yields $b_{\mathbf{k}}^{\nu} = \delta\left(\nu - \nu_0\right)\delta\left(k_y\right)A_{\mathbf{k}}$, and $A_{\mathbf{k}}$ considerably differs from zero only in the ε-vicinity of the point $\mathbf{k}_0$, which is the wave number of the packet wave components. The spectrum spreading around this value means that there is some modulation of the internal waves.

Reveal whether the denominators in Eq. (9.2) may go to zero. As for internal waves on the surface quiet water $\omega_k^{\nu} > 0$, which is impossible for the third term of the equation. In the first term a singularity appears only when for any mode ν the relation $\omega^{\nu}_{2k_0} = 2\omega^{\nu_0}_{k_0}$ is valid. Such a situation corresponds to the resonant generation of the second harmonics and appears only for extraordinary k_0 values. Assume that $\omega^{\nu}_{2k_0} \neq 2\omega^{\nu_0}_{k_0}$ for all ν. If, e.g., $\omega^{\nu_0}_{k_0} > \frac{1}{2}N_{\max}$, where $N_{\max} = \max N_0(z)$, then this condition will certainly be met. Consider now the second term in the integrand of Eq. (9.2). Taking into account that the packet has a narrow spectrum, we may equate approximately

$$\omega^{\nu_1}_{k_1} \approx c_g(k_1 - k_0) + \omega^{\nu_0}_{k_0}, \quad \omega^{\nu_2}_{k_2} \approx c_g(k_2 - k_0) + \omega^{\nu_0}_{k_0},$$

where

$$c_g = \left. \frac{d\omega^{\nu_0}_k}{dk} \right|_{k=k_0}.$$

As the difference $k_1 - k_2 = k$ is of order of the packet width ε, then the second term is sufficient only for low $k \sim \varepsilon$. In this case $\omega^{\nu}_k \approx c_\nu|k|$, where

$$c_\nu = \lim_{k \to +0} \frac{\omega^{\nu}_k}{k}. \tag{9.3}$$

Finally,

$$\omega^{\nu_1}_{k_1} - \omega^{\nu}_k - \omega^{\nu_2}_{k_2} \approx (c_g \mp c_\nu)\,k.$$

So when considering $\omega^{\nu}_{2k_0} \neq 2\omega^{\nu_0}_{k_0}$ and $c_g \neq c_\nu$ for $\nu = 1,\ 2,\ \ldots$ peculiarities are absent in Eq. (9.2). As follows from the presence of corresponding δ-functions, the first and the third terms of this formula describe the generation of second harmonics, and the second term the generation of low frequency waves forming, as will be shown further, mean fields. Keeping only the second integrand term and substituting the a^{ν}_k obtained into (7.31a) for ρ_k and taking into account the packet sharpness, yields

$$-\frac{\rho_k}{\rho_0'} = \sqrt{2} \int \varphi^{\nu}_k \sqrt{\omega^{\nu}_k} \, \frac{c_g k\,(V^{\nu_0\nu_0\nu}_{k_2 k_1 -k} - V^{\nu_0\nu_0\nu}_{k_1 k_2 k}) - \omega^{\nu}_k\,(V^{\nu_0\nu_0\nu}_{k_1 k_2 k} + V^{\nu_0\nu_0\nu}_{k_2 k_1 -k})}{\omega^{\nu^2}_k - c_g^2 k^2}$$

$$\times A_{k_1}(0)\,\bar{A}_{k_2}(0)\,e^{-ic_g k t}\delta(k - k_1 + k_2)\,dk_1\,dk_2\,d\nu. \tag{9.4}$$

Proceed to the limit $k_1 \to k_0$, $k_2 \to k_0$, $k \to 0$ in the expression for $V_{k_1 k_2 k}$. As a result yields

$$V^{\nu_0\nu_0\nu}_{k_0 k_0 k} \approx \frac{1}{8\pi} \cdot \frac{1}{\sqrt{2}} \, \omega^{\nu_0}_{k_0} \sqrt{\omega^{\nu}_k} \left\{ \left(1 \pm \frac{c_f}{c_\nu}\right) \int N_0^2 \left(\varphi_0^2\right)' \varphi_0^{\nu}\,dz \right.$$

$$+ \left(\pm 2 \frac{c_\nu}{c_f} \pm \frac{c_g}{c_\nu} - 1 \right) \int N_0^2 \, \varphi_0^2 \, (\varphi_0^\nu)' \, dz \right\} ,$$

where the upper sign is valid for $k > 0$, and the lower sign for $k < 0$, and the following variables are introduced:

$$c_f = \frac{\omega_{k_0}^{\nu_0}}{k_0}, \quad \varphi_0 = \varphi_{k_0}^{\nu_0}, \quad \varphi_0^\nu = \lim_{k \to 0} \varphi_k^\nu (z).$$

Then substituting those limiting values of V into Eq. (9.4) and expanding the Fourier transform we have ρ versus spatial coordinates

$$-\frac{\rho}{\rho_0'} = \frac{1}{2} \, |A \, (x - c_g t)|^2 \, \omega_{k_0}^{\nu_0} \int \frac{\varphi_0^\nu \, (t)}{\frac{1}{c_g^2} - \frac{1}{c_\nu^2}} \int \left[\left(\frac{2}{c_f c_g} - \frac{1}{c_g^2} \right) \left(N_0^2 \varphi_0^2 \right)' \right.$$

$$\left. - \frac{1}{c_g^2} N_0^2 \, \left(\varphi_0^2 \right)' + \frac{1}{c_\nu^2} \left(\left(N_0^2 \varphi_0^2 \right)' - \frac{c_f}{c_g} N_0^2 \, \left(\varphi_0^2 \right)' \right) \right] \varphi_0^\nu \, dz \, d\nu , \quad (9.5)$$

where

$$A \, (x) = \frac{1}{2\pi} \int A_k \, (0) \, e^{ikx} \, dk$$

is the wave amplitude in a physical domain. The value $-\rho/\rho_0'$ has the meaning of the vertical displacement of fluid particles. Thus we may conclude that mean fields generated by a wave packet result from the 'almost resonant' interaction between spectral components forming this packet. We have used the term 'almost resonant' because the expression $\omega_{k_1}^{\nu_0} - \omega_k^\nu - \omega_{k_2}^{\nu_0}$ is of order of the packet's width but always nonzero.

The value c_g decreases rapidly for decreasing k_0, and in this case a minimum of the difference $\frac{1}{c_g^2} - \frac{1}{c_\nu^2}$ in the denominator of Eq. (9.5) occurs for large ν. As the respective eigenfunctions φ_0^ν oscillate frequently in the vertical plane, similar behaviour is quite possible for the addition to the density ρ. As mentioned in Section 7.2, the value $\omega_{k_0}^{\nu_0} |A|^2$ represents the energy density of internal waves per unit of the horizontal plane. So the value of the addition to ρ in a given point, as follows from Eq. (9.5), is a multiplication of the wave energy density by a fixed depth function, which will be defined as $c_f^2 \, R \, (z)$ and is independent of any characteristics of the packet's modulation.

To obtain certain numerical results it is convenient to reduce calculations of ρ to some boundary value problem solution. Assume that

$$\rho = 1/2\, E c_f^2 R\left(z\right), \tag{9.6}$$

where E is the energy density of the internal waves per unit horizontal plane. The expression for $R\left(z\right)$ will be written in the following form

$$gR\left(z\right) = -\frac{N_0^2}{c_f^2}\int \varphi_0^\nu \int \left\{\left(N_0^2\varphi_0^2\right)' - \frac{\bar{c}_f}{c_g}\, N_0^2\left(\varphi_0^2\right)'\right\}\varphi_0^\nu\, dz\, d\nu$$

$$-\frac{N_0^2}{c_g}\int \frac{\varphi_0^\nu}{\frac{1}{c_g^2}-\frac{1}{c_\nu^2}}\int \left\{-\frac{2}{c_f^3}\left(N_0^2\varphi_0^2\right)' + \frac{N_0^2}{c_g^2 c_f}\left(1+\frac{c_g}{c_f}\right)\left(\varphi_0^2\right)'\right\}\varphi_0^\nu\, dz\, d\nu.$$

$$\tag{9.7}$$

We integrate the first term over ν using Eq. (7.29):

$$gR\left(z\right) = \frac{1}{c_f c_g}\, N_0^2\left(\varphi_0^2\right)' - \frac{1}{c_f^2}\left(N_0^2\varphi_0^2\right)' - \frac{N_0^2}{c_g}\,\Phi\left(z\right), \tag{9.8}$$

where $\Phi\left(z\right)$ represents the second integral in Eq. (9.7). If we now use the operator

$$\left(\frac{d^2}{dz^2} + \frac{N_0^2}{c_g^2}\right)$$

and take into account the equality

$$\frac{d^2\varphi_0^\nu}{dz^2} + \frac{N_0^2}{c_\nu^2}\,\varphi_0^\nu = 0$$

for $\Phi\left(z\right)$ which follows from Eq. (7.26) when $k \to 0$, then the multiplier $\frac{1}{c_g^2} - \frac{1}{c_\nu^2}$ in this integral will be reduced and we may use again the orthogonality condition (7.29). Finally the boundary value problem for Φ yields:

$$\left(\frac{d^2}{dz^2} + \frac{N_0^2}{c_g^2}\right)\Phi = -\frac{2}{c_f^3}\left(N_0^2\varphi_0^2\right)' + \frac{1}{c_g^2 c_f}\left(1+\frac{c_g}{c_f}\right)N_0^2\left(\varphi_0^2\right)'\,; \tag{9.9}$$

$$\Phi\left(z\right) = 0 \quad \text{at} \quad z = 0,\ -H\,,$$

where the points $z = 0$ and $z = -H$ relate to the free surface and to the ocean bottom (in this section we consider the propagation of internal modes). Similarly we can obtain the second-order additions (caused by the propagation of the internal wave packet) for any hydrodynamic value. Those additions to velocity, density, and pressure fields (which will sometimes be called 'fluxes') were initially obtained in [30] by the direct asymptotical solution of the input hydrodynamics' equations (see also [381], [382]). As the source of those fluxes is clear we do not perform here the relevant calculations, but using the results of [30] we present the final formulae which omit the Boussinesq approximation, sometimes important for this problem (see [382]).

So the mean values of the second-order additions to the stream function, the horizontal velocity, the density and the pressure are formed by respective multiplications of $\Phi(z)$, $\Phi'(z)$, $R(z)$ and $P(z)$ by a multiplier $\frac{1}{2} E c_f^2$, where $E = \omega_{k_0}^{\nu_0} |A|^2$ is the energy density of internal waves per unit horizontal plane, where c_f is the wave phase velocity. The function Φ is the solution of the following boundary value problem

$$\frac{d}{dz}\left(\rho_0 \frac{d\Phi}{dz}\right) + \rho_0 \frac{\mu(z)}{c_g^2} \Phi$$

$$(9.10)$$

$$= -\frac{1}{c_f}\left(\frac{2}{c_f^2}\rho_0\mu W^2 + \rho_0'(W^2)'\right)'$$

$$+\frac{1}{c_f c_g^2}\rho_0\mu(W^2)' + \frac{1}{c_g}\left(\frac{1}{c_f^2}\rho_0\mu(W^2)' + \rho_0'(W^2)' + k_0^2\rho_0'W^2\right).$$

Here

$$\mu(z) = -\frac{g}{\rho_0}\frac{d\rho_0}{dz}$$

is the square of the Brunt–Väisälä frequency; $W(z)$ is the eigenfunction of the boundary value problem describing mode under study with the number ν_0

$$\frac{d}{dz}\left(\rho_0 \frac{dW}{dz}\right) + \rho_0\left(\frac{\mu}{c_f^2} - k_0^2\right)W = 0, \quad W|_{z=0,-H} = 0,$$

with the following normalization: $\int\limits_{-H}^{0} \rho_0 \mu W^2 \, dz = 1$.

The functions $R(z)$ and $P(z)$ can be obtained form the formulae:

$$gR(z) = -\rho_0 \frac{\mu}{c_g} \Phi(z) + \frac{1}{c_g c_f} \rho_0 \mu (W^2)' - \frac{1}{c_f^2} (\rho_0 \mu W^2)' ; \qquad (9.11)$$

$$P(z) = - c_g \rho_0 \Phi'(z) - \rho_0 \left[(W^2)' - \left(\frac{\mu}{c_f^2} - k_0^2 \right) W^2 \right]$$

$$+ \frac{c_g}{c_f} \left[\rho_0' (W^2)' + \frac{2}{c_f^3} \rho_0 \mu W^2 \right] . \qquad (9.12)$$

To pass to the Boussinesq approximation we should formally set $\rho_0 \mu = N_0^2$, $W = \varphi_0$, $\rho_0' = 0$ in Eqs. (9.10)–(9.12). Obviously, here the equations (9.8)–(9.9) go to (9.10), (9.11).

Remember that formulae (9.8)–(9.9) were obtained under the assumption that $c_g \neq c_\nu$, $\nu = 1, 2, \ldots$ [the same for (9.10)–(9.12)]. For $c_g \to c_\nu$ one of denominators in Eq. (9.5) goes to zero and $\Phi \to \infty$. In this case a special kind of resonance between the wave and the flow, representing a degenerated internal wave, occurs. And in spite of $k \to 0$ in the limit, it is a real flow with its velocity independent of x. It has a final value of the phase and the group velocity equal to c_ν and it has a vertical profile described by the function $\frac{d}{dz} \varphi_0^\nu$. In the vicinity of critical wave parameters satisfying the equation $c_g = c_\nu$, $\nu = 1, 2, \ldots$ the present consideration is invalid. But we may conclude that internal waves may have intensive energy exchange with such flows.

Thus a solution for any hydrodynamic variable Q consists of the main wave, of the second harmonics, and of fluxes, which may be expressed by the symbolic relation:

$$Q = A(x - c_g t) \, e^{i(k_o x - \omega_o t)} \, \varphi_0(z) + |A^2(x - c_g t)| \, e^{2i(k_o x - \omega_o t)} \, \chi(z) + \overline{Q} . \qquad (9.13)$$

Here $\overline{Q}$ is obtained from Eqs. (9.10)–(9.12), as mentioned above, and the expression for $\chi(z)$ is given by the first and the third terms in Eq. (8.2).

The form of the function A is defined by the initial conditions, i.e., by the shape of the packet given for $t = 0$. A meets the energy equation in the ray approximation derived in Chapter 3:

$$\frac{\partial}{\partial t} A^2 + \frac{\partial}{\partial x} c_g A^2 = 0. \tag{9.14}$$

In this equation the diffraction expansion of the wave packet caused by the finite width of the spectrum, and the influence of nonlinearity on its propagation have been omitted. The equation for the amplitude A taking into account all of these effects will be derived in the next section.

2.   NONLINEAR PARABOLIC EQUATION. ANALYSIS OF SELF-MODULATION INSTABILITY OF WEAKLY NONLINEAR INTERNAL WAVE PACKET

As shown in Section 7.2 and Section 8.1, the cubic terms may be omitted in the Hamiltonian for a narrow spectrum packet. But in this case H will have a number of fourth-order terms, which were not considered previously. But they relate to some important physical phenomena which will be studied in this section. Only one of the fourth-order terms appears to be important for the description of a narrow spectrum wave packet, and the corresponding Hamiltonian has the form [407]:

$$H = \int \omega_k^\nu a_\mathbf{k}^\nu \bar{a}_\mathbf{k}^\nu \, d\nu \, d\mathbf{k}$$

$$+ \frac{1}{2} \int T_{\mathbf{k}\mathbf{k}_1\mathbf{k}_2\mathbf{k}_3}^{\nu\nu_1\nu_2\nu_3} \bar{a}_\mathbf{k}^\nu \bar{a}_{\mathbf{k}_1}^{\nu_1} a_{\mathbf{k}_2}^{\nu_2} a_{\mathbf{k}_3}^{\nu_3} \, \delta\left(\mathbf{k} + \mathbf{k}_1 - \mathbf{k}_2 - \mathbf{k}_3\right) \Pi \, d\mathbf{k} \, d\nu \,,$$

where $T_{\mathbf{k}\mathbf{k}_1\mathbf{k}_2\mathbf{k}_3}^{\nu\nu_1\nu_2\nu_3}$ is some coefficient which will not be derived here. The equation of motion

$$\dot{a}_\mathbf{k}^\nu + i \, \delta H / \delta \bar{a}_\mathbf{k}^\nu = 0$$

yields

$$\dot{a}_k^\nu + i\omega_k^\nu a_k^\nu + i \int T_{\mathbf{k}\mathbf{k}_1\mathbf{k}_2\mathbf{k}_3}^{\nu\nu_1\nu_2\nu_3} \bar{a}_{k_1}^{\nu_1} a_{k_2}^{\nu_2} a_{k_3}^{\nu_3} \, \delta\left(k + k_1 - k_2 - k_3\right) \Pi \, dk \, d\nu = 0. \tag{9.15}$$

Assume that a packet is formed by waves of the mode ν_0 so that $a_k^\nu \sim \delta\left(\nu - \nu_0\right)$. Taking into consideration the narrowness of the packet in the vicinity of the point $k = k_0$ we may write:

$$\omega_k^\nu = \omega_0 + c_g\left(k - k_0\right) + 1/2 \, c_g'\left(k - k_0\right)^2 + \dots ,$$

where

$$\omega_0 = \omega_{k_0}^{\nu_0}, \ c_g = \frac{d\omega_k^{\nu_0}}{dk}\Big|_{k=k_0} \quad \text{and} \quad c_g' = \frac{d^2\omega_k^{\nu_0}}{dk^2}\Big|_{k=k_0}.$$

Now separate slow spatial amplitude variations:

$$A(x,t) = \frac{1}{2\pi} \int a_k^{\nu_0} \, e^{i(k-k_0)x} \, dk$$

and expand the inverse Fourier transform in Eq. (9.15). This yields [407]:

$$\frac{\partial A}{\partial t} + i\omega_0 A + c_g \frac{\partial A}{\partial x} - \frac{i}{2} c_g' \frac{\partial^2 A}{\partial x^2} + iT|A^2|A = 0, \tag{9.16}$$

where $T = (2\pi)^2 T_{k_0 k_0 k_0 k_0}^{\nu_0 \nu_0 \nu_0 \nu_0}$. When passing to the coordinates moving at the velocity c_g assuming $\xi = x - c_g t$ and replace $A \to A e^{-i\omega_o t}$, Eq. (9.16) will have the form:

$$\frac{\partial A}{\partial t} - \frac{i}{2} c_g' \frac{\partial^2 A}{\partial \xi^2} + iT|A^2|A = 0. \tag{9.16a}$$

This equation is called the nonlinear parabolic equation or the nonlinear Schrödinger equation.

Eq. (9.16) has an exact solution of the form:

$$A = A_0 e^{-i\omega t}; \quad A_0 = \text{const}, \tag{9.17}$$

where

$$\omega = \omega_0 + T|A_0|^2, \tag{9.18}$$

which represents the nonlinear dispersion relation. Eqs. (9.16), (9.18) were obtained for internal waves in [30] from the input hydrodynamics equations by the multi-scale expansions method.

Consider a stability of solution (9.17) representing a plane monochromatic wave within the linear approximation. To do this we substitute into Eq. (9.16) the expression $(A_0 + \delta A) e^{-i\omega t}$ assuming for simplification that $A = A_0$ and then linearize the resulting equation with respect to

δA. Introduction of the real functions $\alpha = \delta A + \overline{\delta A}$ and $\beta = i\,(\delta A - \overline{\delta A})$ instead of the complex value δA yields the real simultaneous equations

$$\alpha_t + c_g \alpha_x - 1/2\,c'_g\,\beta_{xx} = 0\,; \quad \beta_t + c_g \beta_x + 1/2\,c'_g\,\alpha_{xx} - 2TA_0^2 \alpha = 0\,.$$

Trying the solution of these equations in the form

$$\begin{pmatrix} \alpha \\ \beta \end{pmatrix} = \mathrm{Re} \begin{pmatrix} \alpha_0 \\ \beta_0 \end{pmatrix} e^{-i\Omega t + i \ae x}\,,$$

where Ω, $\ae$ are the frequency and the wave number of wave modulations, the following relation between Ω and $\ae$ yields:

$$(\Omega - c_g \ae)^2 = 1/2\,c'_g \ae^2 \left(1/2\,c'_g \ae^2 + 2TA_0^2 \right)\,. \tag{9.19}$$

If the input wave modulation, described by the addition δA, has quite slow spatial variations so that $c'_g\,\ae^2 \ll TA_0^2$, then Eq. (9.19) has the form

$$(\Omega - c_g \ae)^2 = c'_g TA_0^2 \ae^2\,.$$

Solutions of this equation may be real or complex depending on the sign of $c'_g T$. Hence the Lighthill condition follows: for $c'_g T > 0$ the wave (9.17) is stable with respect to the longitudinal modulation, and for $c'_g T < 0$ it is unstable, and the instability increment δ is given by

$$\delta = \mathrm{Im}\,\Omega = |\,A_0^2 \ae^2 c'_g T/, |^{1/2}\,. \tag{9.20}$$

In [30] it is shown that for $k_0 \to 0$ and for any stratification it will be always $c'_g < 0$ and $T < 0$ so that by the Lighthill condition [see also Eq. (9.19)] a packet formed by long waves will be stable with respect to longitudinal modulation. From this viewpoint internal waves appear to be similar to surface gravity waves [391]. In the case of a three-dimensional wave packet (anisotropic internal waves) the Lighthill condition is modified and the stability problem is more complicated [254].

For an arbitrary wave length $\lambda = 2\pi k_0^{-1}$ the values and the signs of T and c'_g depend only on the certain stratification profile $N_0\,(z)$. Consider, for example, the case:

$$N\,(z_0) = \begin{cases} 0 & -d \le z \le 0 \\ N_0 & -H \le z < -d\,. \end{cases} \tag{9.21}$$

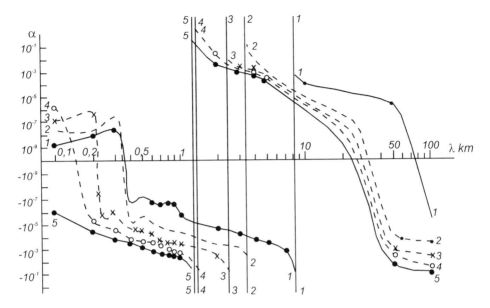

Figure 9.1. The dependence α on λ for five lower modes of a basic wave for the stratification model (9.21) (curve number equals to mode number).

Fig. 9.1 shows the dependences for T (or $\alpha = T/c_f^2\omega_0^2$) versus the wavelength λ for five lower modes. Long waves with $\lambda > \lambda_*$ ($\lambda_*^{(n)}$ has its own value for every n-th mode) are stable with respect to longitudinal disturbances ($\alpha < 0$). While the wavelength λ decreases, α changes its sign when $\lambda = \lambda_*$ and becomes unstable when $\lambda < \lambda_*$. Then for $\lambda = \lambda_0$, where λ_0 corresponds to the wave flow or wave's second harmonics resonance, discussed in Section 9.1, α changes its sign immediately, which again leads to internal waves stability for $\lambda = \lambda_0$ (but our asymptotic method becomes invalid in the vicinity of $\lambda = \lambda_0$). And, at least, very short waves with $\lambda < \lambda_*$ become unstable again. While increasing the basic wave mode number all of the zones shown in the figure are shifted into the region of shorter waves. Eq. (9.16a) belongs to the partial differential nonlinear equations, which may be integrated rigorously by the methods of the inverse scattering problem [412], [413]. Some particular solutions of Eq. (9.16a) are given by Karpman [158]. However, this equation describes modulating waves of the wave packet, and the term $-\frac{i}{2}c_g'\frac{\partial^2 A}{\partial\xi^2}$ describes high-order dispersion effects, limiting the appearance of shock waves in the modulation equations. It is typical that the basic nonlinear solution of this equation is again a solitary wave (or the N-soliton solution).

3.   DESCRIPTION OF MEAN HYDROPHYSICAL FIELDS INDUCED BY NONLINEAR INTERACTION OF INTERNAL WAVES. FORMATION OF FINE STRUCTURE OF OCEANIC HYDROPHYSICAL FIELDS

9.3.1. Consider Eqs. (9.10)–(9.12) describing fluxes, i.e., mean values of hydrodynamic parameters with respect to the wave phase θ, caused by propagation of a weakly nonlinear wave packet.

Analyze first the qualitative picture of the mean velocity field and of the mean addition to the undisturbed density. Stress once more that these parameters are proportional to the square amplitude A^2 which satisfies the evolution equations considered above.

While a wave packet running in an initially undisturbed medium (see the scheme in Fig. 9.2) when fluid particles meet the bow packet front they receive some mean horizontal velocity proportional to $\Phi'(z)$, and in addition, slightly deflect up or down by a distance $\xi \sim R(z)$. As the total consumption of fluid through the ocean cross-section is equal to zero, then the ocean's depth will be divided into horizontal layers where the mean velocity changes its direction alternately. While the back of the packet's front passes the fluid particles will return to the initial levels, so that the stratification will be restored. And, furthermore, their velocity will also become zero. But every particle will deflect in the horizontal plane to a distance l proportional to $L\Phi'/H$, where L is the packet's length. As a result the particle's trajectory will have the form shown in Fig. 9.2. Particles in the layers with mean velocities of opposite directions have open trajectories. Such a motion obviously causes an irreversible horizontal mixture of the ocean water. A similar process occurs with the density and the pressure, but their variations, being proportional to $|A|^2$, are reversible.

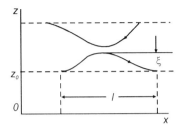

Figure 9.2.   Trajectories of fluid particles in a wave packet.

Now let us analyze the vertical structure of the mean fields by Eqs. (9.10)–(9.12). Let us consider qualitative peculiarities of the boundary value problem (9.10) for $\Phi(z)$. This solution contains, as its summands, the fundamental solution (9.10) which oscillates vertically at the frequency $\sim N/c_g^0$ (where $N = \sqrt{\mu}$ is the Brunt–Väisälä frequency) and another summand oscillating at the frequency proportional to the frequency of the r.h.s. of the equation. For $c_g^0 \to 0$ the vertical period will be negligibly small. This case corresponds to small wave lengths[1] filling the wave packet, because from the dispersion relation for internal waves follows $c_g^0 \to 0$ when $k \to \infty$. As for the component of the solution of the boundary value problem (9.10), corresponding to its r.h.s., it oscillates vertically with a period proportional to the mode number. The same properties are typical of the mean density and the pressure fields, as follows from Eqs. (9.11)–(9.12) for $R(z)$ and $P(z)$.

These results show that weakly nonlinear internal wave packets generate, together with large scale vertical variations of the background, the vertical fine structure of the mean fields caused by the interaction of nonlinear waves. In accordance with the analysis presented the vertical fine structure of hydrophysical fields manifests itself more evidently for smaller horizontal lengths of waves filling the wave packet.

9.3.2.   As the microstructure of oceanic fields is one of the most interesting discoveries, it is useful to give a brief discussion of the late evidence before the theoretical calculations.

It s known that the fine vertical structure of the hydrophysical fields is a universal phenomenon in the ocean which occupies almost its total depth. But, nevertheless, the theory explaining its nature has been lacking. There were only some qualitative conclusions and the hypothesis of its appearance [75] which did not explain its characteristic properties.

We list here the most typical properties of the oceanic fine structure observed in various regions of the World Ocean, ignoring the fine structure formed in some specific conditions: in the vicinity of shore and straits, in the Arctic zones covered with ice, in the vicinity of water fronts etc.

1. The fine structure is a relatively long lived formation. Its observations show that microstructural peculiarities, if the density field has a vertical scale $h \geq 1$ m, are maintained for a long time, have slow time evolution, and remain almost constant during and up to several days.

2. The characteristic horizontal dimensions of different components of the thermochalinic fine structure are considerably greater than the vertical ones. In lines with [75] $10^{-5} \leq h/L \leq 10^{-3}$, where L is the characteristic horizontal scale. The horizontal parameters of the fine structure vary slowly.

3. Thin structural peculiarities with scales $h \approx 1$ cm–1 m appear in the background of large scale formations with $h \approx 10 \ldots 50$ m. The vertical spatial fine structure spectrum, as a rule, has no maxima with respect to the vertical wave number, which evidences the absence of dominating vertical scales in the microstructural peculiarities [89], [88], [258].

4. The fine structure of the density, the salinity, and the temperature fields is always followed by the velocity field fine structure [199], [375], [264]. And, furthermore, the density and the velocity gradients correlate with each other [199].

5. Temperature and density deflections (ΔT, $\Delta \rho$) owed to the fine structure) from their mean values (T_0, ρ_0) depend on the mean vertical gradients (dT_0/dz, $d\rho_0/dz$). Large deflections resulting from the fine structure correspond to large values of mean gradients (see, e.g., [88], [258]). And, furthermore, the constancy of the value $\sigma_T/(dT_0/dz)$ with depth is observed [88], where σ_T is the mean square value of

ΔT, i.e., $\Delta T \sim dT_0/dz$. By analogy, ΔN grows as grows $N(z)$, and $\Delta N \sim N(z)$, where $N(z) = \sqrt{\mu(z)}$ is the Brunt–Väisälä frequency.

6. Both the vertical fine structure and internal waves were always observed during their simultaneous measurements in the ocean [53], [88], [199], [258], [272].

9.3.3. Following [382] we pass to examples of the fine structure calculations for particular oceanic conditions on the basis of Eqs. (9.10)–(9.12). Let us introduce the following dimensionless parameters

$$\tilde{z} = zH^{-1}; \quad \tilde{\omega} = \omega_0 N_0^{-1}; \quad k = k_0 H; \quad \Omega = \mu N_0^{-2};$$

$$\varphi = W(H^2 N_0)^{-1}; \quad f = \Phi(H^2 N_0)^{-1}; \quad r = R\rho_0^{-1}; \quad \delta = N_0^2 H g^{-1};$$

$$c_f = c_f^0 (HN_0)^{-1}; \quad \varepsilon = c_g^0 (HN_0)^{-1}, \tag{9.22}$$

where N_0 is the maximum value $N(z) = \sqrt{\mu(z)}$ of the Brunt–Väisälä frequency. Let us replace variables:

$$w = \varphi \exp\left(-\frac{\delta}{2}I\right); \quad y = f \exp\left(-\frac{\delta}{2}I\right); \quad I(\tilde{z}) = \int_0^{\tilde{z}} \Omega(z)\, dz.$$
$$\tag{9.23}$$

While taking into account (9.22), (9.23), Eqs. (9.10)–(9.12) for w, Φ, R will have the form:

$$w'' + \left[\frac{\Omega - \tilde{\omega}^2}{\tilde{\omega}^2} k^2 + \frac{\delta}{2}\Omega' + \frac{\delta^2}{4}\Omega^2\right] w = 0; \quad w(0) = w(-1) = 0; \tag{9.24}$$

$$\varepsilon^2 y'' + \tilde{\mu}(\tilde{z}) y = \exp\left(\frac{\delta}{2}I\right)[G_0 + \varepsilon G_1 + \varepsilon^2 G_2]; \quad y(0) = y(-1) = 0; \tag{9.25}$$

$$\tilde{r} = \frac{\delta}{\varepsilon}\Omega f + \left\{\frac{\delta}{\varepsilon}\Omega\frac{(\varphi^2)'}{c} - \frac{\delta}{c^2}[(\Omega\varphi^2)' + \delta\Omega^2\varphi^2]\right\}. \tag{9.26}$$

Here

$$
\begin{aligned}
\tilde{\mu}(\tilde{z}) &= \Omega + \varepsilon^2 \frac{\delta\Omega'}{2} + \varepsilon^2 \frac{\delta^2\Omega^2}{4}; \quad G_0 = -\frac{\Omega}{c}\left[(w^2)' + \delta\Omega w^2\right]; \\
G_1 &= \left(\frac{\Omega^2\delta^2}{c^2} + \delta\Omega k^2 + \frac{\delta^3\Omega^3}{4}\right) w^2 + \left(\frac{\Omega}{c^2} + \frac{\delta\Omega}{2}\right)(w^2)' + \delta\Omega(w')^2; \\
G_2 &= -\left(\frac{2}{c^3}\Omega' + \frac{2\delta}{c^3}\Omega^2 + \frac{\delta^4\Omega^3}{c} + \delta^3\frac{\Omega\Omega'}{c}\right) w^2 \\
&\quad - \left(\frac{2\Omega}{c^3} + \frac{2\delta^3\Omega^2}{c}\right)(w^2)' - \frac{\delta^2\Omega}{c}(w^2)''.
\end{aligned}
\tag{9.27}
$$

Example 1. To solve the boundary value problems (9.24), (9.25) we may use the WKB method for small ε, i.e., for $c_g \ll HN_0$ (which is always valid for high k) and for a smooth function $\Omega(z)$.

For natural ocean conditions $\delta \ll 1$ (e.g., for $H = 3 \times 10^5$ cm, $g = 10^3$ cm/s^2, $N_0 = 3 \times 10^{-3}$ s^{-1} yields $\delta \approx 3 \times 10^{-3} \ll 1$) because $\varepsilon \ll 1$, and we can equate in (9.27) $\tilde{\mu}(\tilde{z}) \approx \Omega(\tilde{z})$. Then omitting the 'standard' WKB solution (9.24) (see Section 3.3) we may write the general WKB solution (9.25) with an accuracy to $\mathcal{O}(\varepsilon^2)$ in the form

$$
y = \Omega^{-1/4}\left[C_1 \sin(\varepsilon^{-1} J) + C_2 \cos(\varepsilon^{-1} J)\right] + F; \quad J = \int_0^{\tilde{z}} \Omega^{1/2}\, dz, \tag{9.28}
$$

where C_1 and C_2 are the constants obtained from the boundary conditions for (9.25), and F is the particular solution of the inhomogeneous equation (9.25), which has the following form

$$
F = \Omega^{-1} \exp\left(\frac{\delta}{2} I\right)(G_0 + \varepsilon G_1) + \mathcal{O}(\varepsilon^2). \tag{9.29}
$$

Using the boundary conditions for (9.25) and $w(0) = w(-1) = 0$ and $w'(0) \neq 0$, $w'(-1) \neq 0$ we derive C_1 and C_2:

$$
C_1 = \varepsilon\delta D; \quad C_2 = -\varepsilon\delta B; \quad B = \Omega^{-1/4}(0)[w'(0)]^2;
$$

$$
D = [\sin(\varepsilon^{-1} J_0)]^{-1}\{\Omega^{1/4}(0)[w'(0)]^2 \cos(\varepsilon^{-1} J_0) - \Omega^{-1/4}(-1) \times
$$

$$\times e^{\delta I_o/2} \left[w'\left(-1\right)\right]^2 \};$$

$$(9.30)$$

$$J_0 = \int\limits_0^{-1} \Omega^{1/2} \, dz \, ; \quad I_0 = \int\limits_0^{-1} \Omega \, dz \, .$$

From Eqs. (9.28)–(9.30) it follows that the general solution of (9.25) consists of the summand F slowly varying with $\tilde{z}$ (the number of oscillations of $F(\tilde{z})$ with depth is proportional to the mode number of the basic wave) and of the summands varying rapidly with depth with a vertical oscillation period proportional to ε. These terms describe the fine structure. The coefficients C_1 and C_2 in Eq. (9.30) have the terms $\Omega^{1/4}(0)$ and $\Omega^{-1/4}(-1)$ as multipliers. If they are zero then the WKB approximation is invalid and a more thorough solution should be used.

Write now the formulae for the rapidly varying summands of the dimensionless horizontal velocity $\tilde{U} = U / \left(H N_0 \, |a/\lambda|^2 \right) = f'$ and of the dimensionless additions to the density r and to the density gradient r':

$$f' = \delta\Omega^{1/4} \exp\left(\frac{\delta}{2} I\right) \left\{ A^{-1} + \varepsilon A^+ \left(\frac{\delta}{2}\Omega^{1/2} - \frac{1}{4}\frac{\Omega'}{\Omega^{2/3}}\right)\right\};$$

$$r = \delta^2 \Omega^{3/4} A^+ e^{\frac{\delta}{2} I};$$

$$r' = \frac{\delta^2}{\varepsilon} \Omega^{5/4} e^{\frac{\delta}{2} I} A^- + \delta^2 e^{\frac{\delta}{2} I} A^+ \left(\frac{3}{4}\Omega^{-1/4}\Omega' + \frac{\delta}{2}\Omega^{7/4}\right); \qquad (9.31)$$

$$A^- = D \cos\left(\varepsilon^{-1} J\right) - B \sin\left(\varepsilon^{-1} J\right); \quad A^+ = D \sin\left(\varepsilon^{-1} J\right) + B \cos\left(\varepsilon^{-1} J\right).$$

In Eqs. (9.31) the terms slowly varying over depth and relating to large-scale variations are omitted.

From (9.31) the induced flow and the addition to the density r are small in the WKB approximation (are proportional to δ and δ^2 respectively). But additions to the density gradient and, which is the same, to the Brunt–Väisälä frequency are considerable because of their proportionality to δ^2/ε. Now using Eqs. (9.31) we obtain these estimates of the dimensional parameters:

$$U \sim H N_0 \left|\frac{a}{\lambda}\right|^2 \delta\Omega^{1/4} e^{\delta I/2} A^-; \quad U' \sim \frac{N_0 \, |a/\lambda|^2 \delta}{\varepsilon} \Omega^{3/4} e^{\delta I/2} A^+$$

$$\tilde{\rho} \sim \rho_0 \left|\frac{a}{\lambda}\right|^2 \delta^2 \Omega^{3/4} e^{\delta I/2} A^+ ; \quad \rho' \sim \frac{|a/\lambda|^2 \delta^2}{H\varepsilon} \Omega^{5/4} e^{\delta I/2} A^- . \qquad (9.32)$$

From (9.32) follows

$$\frac{\tilde{\rho}}{\rho_0'} \sim \left|\frac{a}{\lambda}\right|^2 H\delta A^+ \Omega^{-1/4} e^{\delta I/2} ; \quad \frac{\Delta\mu}{\mu} \sim \frac{|a/\lambda|^2 \delta}{\varepsilon} \Omega^{1/4} A^- e^{\delta I/2} . \qquad (9.33)$$

Eq. (9.32) shows that the density addition $\tilde{\rho} \sim \Omega^{3/4}$ and $\rho' \sim \Omega^{5/4}$, i.e., while the Brunt–Väisälä frequency grows the microstructural peculiarities manifest themselves strongly, which was observed experimentally (see Subsection 9.3.2, property 5). As $\Omega(\tilde{z})$ is a smooth function of depth, the value $\tilde{\rho}/\rho_0'$ (the analog of the Cox number) is almost the same at any depth, which is also typical for the fine structure of the oceanic density field.

Example 2. Now we calculate the fine structure parameters for the simplest stratification $\mu = \mu_0 = $ const, i.e., in this case in Eqs. (9.27)–(9.29) we have $\Omega \equiv 1$. Then the solution of the boundary value problem (9.24) yields:

$$\left.\begin{array}{l} w = \sin lz; \quad l = n\pi; \quad \tilde{\omega} = k\left(k^2 + l^2 + \delta^2/4\right)^{-1/2}; \\[2mm] c = \left(k^2 + l^2 + \delta^2/4\right)^{-1/2}; \quad \varepsilon = \left(k^2 + \delta^2/4\right)\left(k^2 + l^2 + \delta^2/4\right)^{-3/2}; \end{array}\right\} \tag{9.34}$$

For $k \gg 1$ inequalities $\varepsilon \ll c \ll 1$ and $\delta \ll 1$ are valid. Taking this into consideration and using (9.34) Eqs. (9.28)–(9.30) for $\Omega \equiv 1$ yield

$$f \approx \delta l^2 \varepsilon e^{\delta \tilde{z}/2} \left[\cos\left(\varepsilon^{-1}\tilde{z}\right) - \frac{1 - \delta/2 - \cos\left(\varepsilon^{-1}\tilde{z}\right)}{\sin\left(\varepsilon^{-1}\right)}\right]$$

$$+ \frac{\delta}{c} e^{\delta \tilde{z}} \left[\frac{1}{2} + \frac{l}{\delta} \sin\left(2l\tilde{z}\right) - \left(1/2 + 2l^2 \varepsilon\right)\cos\left(2l\tilde{z}\right)\right] . \qquad (9.35)$$

Numerical calculations of Eqs. (9.34) have been performed. For the typical oceanic parameters we have assumed $N_0 = 3 \times 10^{-3}$ 1/s; $H = 3 \times 10^5$ cm;

Table 9.1.   Results of calculation for the fine structure parameters from Eqs. (9.34)–(9.35)

k	λ (m)	c	c_f (cm/s)	ε	c_g (cm/s)	h (m)	$\frac{\Delta N}{N}$	U' (1/s)
100	200	10^{-2}	10	10^{-5}	10^{-2}	0.1	0.3	10^{-3}
70	300	$1.43 \cdot 10^{-2}$	14.3	$3 \cdot 10^{-5}$	$3 \cdot 10^{-2}$	0.3	0.1	$3.3 \cdot 10^{-4}$
50	400	$2 \cdot 10^{-2}$	20	$8 \cdot 10^{-5}$	$8 \cdot 10^{-2}$	0.8	0.038	$1.25 \cdot 10^{-4}$
20	1000	$5 \cdot 10^{-2}$	50	$1.25 \cdot 10^{-3}$	$1.25 \cdot 10^{-1}$	12.5	0.0024	$0.8 \cdot 10^{-4}$

$g = 10^3$ cm $\times$ s^{-2}, i.e, $\delta = 3 \times 10^{-3}$, $|a|/\lambda = 0.01$. In addition it was assumed that the basic wave belongs to the first mode $n = 1$, i.e., $l = \pi$. The calculation results are given in Table 9.1, where $\lambda = 2\pi/k_0$ is the wave length and $h = \pi H \varepsilon$ is the characteristic vertical scale of the fine structure.

As follows from Table 9.1, the pronounced fine structure of the density and the velocity fields appears for quite small wave lengths filling the wave packet ($\lambda \leq 1000$ m) even when $N^2 = \mu = $ const. Table 9.1 contains values for the vertical gradient of the horizontal mean velocity $U'(z)$ because the mean velocity itself is almost independent of the wave length λ.

From Eqs. (9.32), (9.35) it follows that the amplitude of the horizontal mean velocity is proportional to $HN_0 |a/\lambda|^2 \delta l^2 \approx 3 \times 10^{-3}$ cm/s and is rather small. But for $\varepsilon^{-1} \to n\pi$ (though the exact equality is not possible) from Eq. (9.35) it follows that some resonant conditions between the wave and induced mean flow (with the formal wave length $\lambda = \infty$) appears. In the vicinity of this point the asymptotic expression in the right of Eqs. (9.30), (9.35) are invalid (formally they go to infinity), but it shows that in this vicinity of ε^{-1} the flow may be quite intense.

Example 3. Examples 1 and 2 relate to cases when the Brunt–Väisälä frequency varies slowly with depth or is constant, and the Boussinesq approximation should be omitted. If $N(z)$ has sufficient depth variations, then the Boussinesq approximation does not change anything. In such a case we should equate $\rho_0 \equiv 1$ in Eqs. (9.10)–(9.12). In the Boussinesq approximation for $N^2 = \mu = $ const the fine structure does not appear, Eq. (9.35) passes into $f = (l/c) \sin(2l\tilde{z})$, and for slow variations of $N(z)$ (the WKB approximation) the fine structure is present, but is less pronounced. In this case the constants in Eq. (9.30) $C_1 \sim C_2 \sim \varepsilon^3$. But when $N(z)$ varies rapidly enough strongly pronounced microstructural peculiarities appear under the Boussinesq approximation.

In this connection it is interesting to consider the stratification (9.21) under the Boussinesq approximation, when analytical solutions of corresponding boundary value problems may be obtained. Along the lines of the formulae obtained the calculations of $U(z)$ and $\Delta N^2/N_0^2$ were

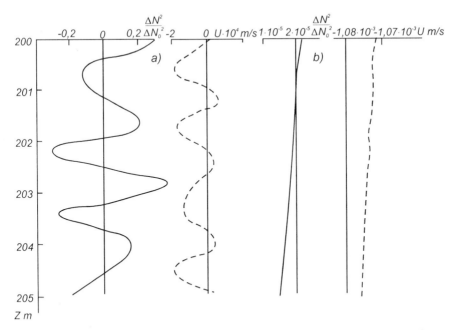

Figure 9.3. An example of the vertical fine structure distribution for the velocity U and the normalized mean addition to N^2 for the first mode ($n = 1$) and the stratification model (9.21): a — $\lambda = 200$ m; b — $\lambda = 3000$ m.

performed for two modes ($n = 1, 2$) and for two basic wavelengths $\lambda = 200$ m and $\lambda = 3000$ m. Here we assume $H = 3 \times 10^5$ cm; $N_0 = 3 \times 10^{-3}$ 1/s, the vertical displacement amplitude $|a| = 10$ m and the parameter d in the distribution (9.21) was equal to $d = 10^4$ cm. From Figs. 9.3,

9.4 it is obvious that the fine structure appears only for a long enough wavelength forming the wave packet.

Example 4. Analyze now the natural stratification relating to [164]. In this work the evidence of the density field (the Brunt–Väisälä frequency) vertical structure in one point of the Arkona Basin were presented. The measurements were performed with an instrument of a high vertical resolution (of about 0.5 m) every half an hour. In Fig. 9.5 the curve 1 represents the 10 h average of a dimensionless profile of the square Brunt–Väisälä frequency $\Omega(z)$. In such a way we extract the fluctuating part of the density field (proportional to $e^{i\theta}$, $e^{2i\theta}$) and

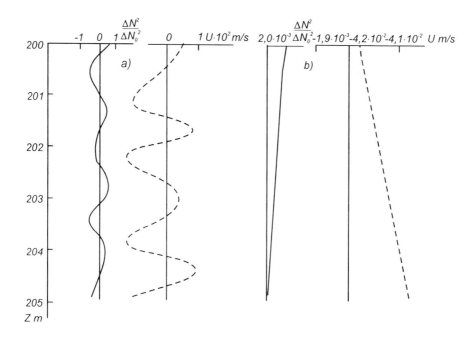

Figure 9.4. An example of the vertical fine structure distribution for the velocity U and the normalized mean addition to N^2 for the second mode ($n = 2$) and the stratification model (9.21): a — $\lambda = 200$ m; b — $\lambda = 3000$ m.

the microstructural peculiarities may be considered to be caused by the phase θ averaged fields induced by the passage of the wave packet.

It is interesting to calculate the fine structure of the density and the velocity fields by Eqs. (9.10)–(9.12) while giving the 'smoothed' undisturbed profile $\mu(z)$ contained in those equations. In Fig. 9.5 the dimensionless profile $2\Omega_0(z) = N_0^2(z)/\max N_0^2(z)$ obtained from the smoothed profile $\Omega(z) = N_0^2/\max N_0^2(z)$ is presented as well. The smoothing was performed such that the profile $N_0^2(z)$ will be free of the fine structure peculiarities with a vertical scale $h \leq 10$ m. This smoothed profile was given for the boundary value problems calculations.

All the calculations were carried out by numerical methods under the Boussinesq approximation ($\rho \equiv 1$). In Fig. 9.5 (curve 3) the first eigenfunction $W_1(z, \omega)$ for $\omega = 4.15 \times 10^{-2}$ 1/s normalized so that $\int_{-H}^{0} W_1^2(z, \omega)\, dz = 1$ is given as an example. The chosen frequency allowed us to obtain the wave solution only for a narrow waveguide of

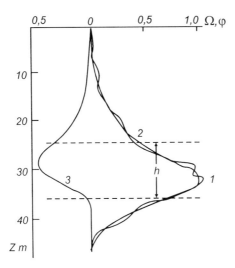

Figure 9.5. The vertical distribution of the dimensionless Brunt–Väisälä frequency, according to [209], and the first eigenfunction of the vertical operator for internal waves: $1 — \Omega(z)$ distribution; 2 — the smoothed distribution $\Omega_0(z)$; 3 — the profile of $\varphi_0(z)$ cm/s.

Table 9.2. Wave parameters

ω (s^{-1})	$4.84 \cdot 10^{-2}$	$4.15 \cdot 10^{-2}$	$3.28 \cdot 10^{-2}$	$2.15 \cdot 10^{-2}$
$T = 2\pi\omega^{-1}$ (min)	2.2	2.5	2.2	4.9
$\lambda = 2\pi k^{-1}$ (m)	11	24	60	125
c_f (cm/s)	8.7	15.9	24.3	43.0
c_g (cm/s)	1.2	3.8	15.0	26.0

a width ~ 10 m.; $W_1(z,\omega)$ decays exponentially below and above the waveguide. In Table 9.2 the parameters of the basic internal wave for the first mode and different ω values are given. They were obtained as the solution of the boundary value problem for W under the Boussinesq approximation.

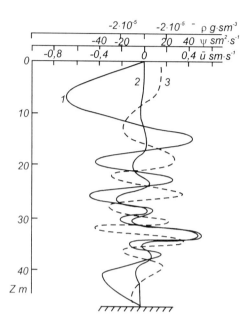

Figure 9.6. The vertical distribution of the fine structure: $1 - \psi$ cm^2/s; $2 - \bar{\rho}$ g/cm^3; $3 - \bar{u}$ cm/s.

The calculated values of $W_1(z, \omega)$, c_f and c_g were substituted into Eqs. (9.10)–(9.12) which then were solved numerically. In Fig. 9.6 the functions $\psi_2^{(0)}$, $\rho_2^{(0)}$ and the horizontal mean-velocity component $U = \psi_2^{(0)}$ for $\omega = 4.15 \times 10^{-2}$ s^{-1} ($\lambda = 24$ m) and for the wave amplitude $|a| = 40$ cm are shown. For these λ and $|a|$ the wave slope $|a/\lambda| \approx 0.02 \ll 1$ and the weakly nonlinear theory is valid. The values of ω and $|a|$ mentioned were chosen such that the best agreement with the evidence was obtained. As the internal wave parameters in the work [164] are omitted then ω and $|a|$ are in this case in fact approximate ones.

It is obvious from Fig. 9.6 that all of the geophysical fields presented have fine vertical structural peculiarities. In Fig. 9.7 the dimensionless square Brunt–Väisälä frequency profile $\Omega(z)$ obtained from

$$\Omega(z) = \frac{N^2(z)}{\max N_0^2(z)} = \frac{1}{\max N_0^2} \left[N_0^2(z) + \frac{g}{\rho} \frac{d\bar{\rho}}{dz} \right];$$

$$\bar{\rho} = |a|^2 R(z),$$

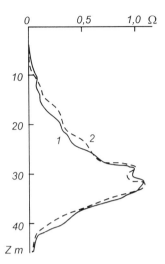

Figure 9.7. The vertical distribution of the Brunt–Väisälä frequency: 1 — experimental [164]; 2 — theoretical.

is shown as well as the experimental one. The calculated curve contains the most typical peculiarities of the experimental distribution. We could reach better coincidence between the numerics and the experiment while varying the parameters ω and $|a|$ and adding, e.g., the solution of Eqs. (9.10)–(9.12) for the second eigenfunction $W_2(z,\omega)$, because the superposition of modes is still valid in the approximation considered.

9.3.4. Now compare results of the theory, presented in this section, with the characteristic peculiarities of the oceanic fine structure mentioned in 9.3.2.

1. From the theory it follows that the 'life time' of the fine structure at a fixed point in the ocean is approximately equal to the the wave packet's 'life time' $t_w \sim L/c_g$. This time may vary from several days up to a month because L may be several kilometers. Time evolution of the fine structure is defined for the wave packet while solving an equation similar to Eq. (9.16). This equation may have unsteady solutions for low wavelengths (see Section 9.2) corresponding to the wave packet breaking up into wave trains with the solitary type of envelope wave.

2. The value of $h/L \sim 10^{-4} \ldots 10^{-5}$, where $L = m\lambda$ (m is the number of waves in the packet) should be considered to be the wave packet's length. The evolution of the horizontal fine structure is again governed by the solution of Eq. (9.16).

3. As follows from Eqs. (9.28), (9.35) the fine structure appears on a background of large scale vertical fluctuations of the ocean fields with a period proportional to the mode number of the basic wave. The spatial vertical spectrum of the fine structure is discrete in the examples presented. But if there is a superposition of wave packets of different wave lengths or modes numbers, then the fine structure rapidly becomes stochastic and forms a continuous vertical spectrum.

The fine structure peculiarities 4–6, described in 9.3.2, directly follow from the theory presented [see Eqs. (9.31)–(9.35), Table 9.1 and Figs. 9.3, 9.4].

This mechanism of the formation of oceanic fine structure is already effective for small amplitudes of internal waves (slopes $\left|\frac{a}{\lambda}\right| \ll 1$). It differs favourably from models based on the assumption of the breaking of internal waves or of the appearance of fine structure as a result of a shear instability [391], [281]. The latter processes require large wave slopes ($|a/\lambda| \sim 1$). But as the microstructure is spread within the ocean's volume those models suppose some 'internal storm' always present inside the ocean, which is far from being real. The hypothesis of Garrett and Munk [94] suggested that the fine structure is a result of the passage of a *linear* internal wave through the point of observation is, from our viewpoint, also unreal because it contradicts the evidence showing that the fine structure is present in the long time average profiles. More than that, to produce the fine structure within the whole volume of the ocean the internal wave field should be formed from a great number of modes (of about 100–1000) or internal waves should have a vertical scale of about 1 m. The latter seems to be doubtful and is refuted by the evidence.

In the theory presented the fine structure formation is reversible, i.e., after the wave packet's (or packets') passage the initial stratification profile is restored. But the appearance of *secondary instability* is possible, which can produce irreversible deformations of the mean fields.

The Richardson number for the mean density and the velocity fields is expressed by the formula

$$Ri = [\mu_0(z) + \Delta\mu](U')^{-2} = \left[\Omega + \left|\frac{a}{\lambda}\right|^2 \Omega r + \frac{r'}{\delta}\right] \left|\frac{a}{\lambda}\right|^{-4} (f'')^{-2}. \quad (9.36)$$

Taking into consideration Eqs. (9.31), (9.32) for Ri yield

$$Ri \sim \left[\varepsilon^2 \left|\frac{a}{\lambda}\right|^{-4} \delta^{-2} + \varepsilon^2 \left|\frac{a}{\lambda}\right|^{-2} + \varepsilon\,\delta^{-1} \left|\frac{a}{\lambda}\right|^{-2}\right] \Omega^{-1/2} D^{-1}, \quad (9.37)$$

where D is derived from (9.30). Under the corresponding combination of parameters, e.g., for $\varepsilon < 0.2\,\delta\,|a/\lambda|^2$ (quite typical for the ocean) Ri may be less than $Ri_* = \frac{1}{4}$, which will cause an instability. Ri decreases rapidly in the vicinity of the resonant points (for $N = \text{const}$ and $\varepsilon^{-1} \to n\pi$), where $D \gg 1$. The other mechanism resulting in irreversible processes (in a water mixture) may be the breaking of very short waves propagating through the fine structure. But this assumption needs to be proved by an appropriate theory.

In concluding all the above we may assume that the mechanism presented of the formation of fine structure is 'basic' for more complicated secondary phenomena of hydrodynamic instability.

For more detailed comparison of this theory with the evidence simultaneous horizontal measurements should be provided of the fine structure's characteristics and of internal waves (e.g., the frequency spectrum $S(\omega)$ allowing to estimate $|a|$ and ω). Rather useful from this viewpoint may be joint measurements of not only the velocity and the density profiles, but also of the pressure profile, which should also have microstructure, according to Eq. (9.27) for $P(z)$. The corresponding formulae for $P(z)$ can be easily obtained from the results of this section.

4. PROPAGATION OF A THREE-DIMENSIONAL WAVE PACKET IN A MEDIUM WITH CONSTANT BRUNT–VÄISÄLÄ FREQUENCY

Under the Boussinesq approximation consider the second-order effects appearing while the propagation of a three-dimensional wave packet in an unrestricted medium with constant Brunt–Väisälä frequency $N_0 =$ const (see [381]). The problem solution is again given by Eqs. (9.1) and (9.2) where the superscript ν, referring to the mode number, should be omitted, and the vectors $\mathbf{k}$ are considered to be three-dimensional.

So we study the expression for the second-order addition caused by the fluxes' presence:

$$a_{\text{æ}} = 2 \int \frac{V_{\text{æ}_1\text{ææ}_2}}{\omega_{\text{æ}_1} - \omega_{\text{æ}} - \omega_{\text{æ}_2}} \, b_{\text{æ}_1} \bar{b}_{\text{æ}_2} \, \delta \left(\text{æ}_1 - \text{æ} - \text{æ}_2 \right) d\text{æ}_1 \, d\text{æ}_2;$$

$$b_{\text{æ}} = b_{\text{æ}}(0) \, e^{-i\omega_{\text{æ}} t} ; \quad \omega_{\text{æ}} = N_0 \frac{|\mathbf{k}|}{|\text{æ}|} , \tag{9.38}$$

where $\mathbf{k}$ again means the horizontal component of æ.

Extracting the dominant terms with respect to the packet's narrowness parameter in the expression for the coefficient V (7.42), from Eqs. (9.38), (7.45) yields the expression for the vertical particle displacement $\xi = -\rho/\rho_0'$:

$$\xi = \frac{1}{(2\pi)^3} \frac{i}{2N_0^2} \int \frac{\omega_{\text{æ}}^2 \left(\nu_0 - \frac{\mathbf{k}\mathbf{k}_0}{k^2} \right)}{(\mathbf{c}_g \text{æ})^2 - \omega_{\text{æ}}^2} \, \mathbf{c}_g \text{æ} B_{\text{æ}}(0) \, e^{-i\text{æ}(\mathbf{r} - \mathbf{c}_g t)} \, d\text{æ} . \tag{9.39}$$

Here ν_0 is the vertical projection of the vector æ; $\mathbf{c}_g = \left. \frac{\partial \omega}{\partial \text{æ}} \right|_{\text{æ}=\text{æ}_o}$;

$$B_{\text{æ}}(0) = \int |\psi(\mathbf{r}, 0)|^2 \, e^{-i\text{æ}\mathbf{r}} \, d\mathbf{r} \quad \text{and} \quad \psi(\mathbf{r}, t) = \frac{1}{(2\pi)^{3/2}} \int b_{\text{æ}} \, e^{-i\text{æ}\mathbf{r}} \, d\text{æ} .$$

In this section $\mathbf{r}$ means the three-dimensional radius vector of a point in a physical domain. The denominator in Eq. (9.39) goes to zero for some æ, which corresponds to the resonant interaction of spectral packet components. The integration contour may arbitrarily pass around the integrand singularities. The multiplicity appearing is usual for the Green's function, which is in fact studied here. It is caused by the opportunity of addition of any solution of the linearized equations to the solution obtained from (9.39). This addition should be, naturally, of the same order of smallness. The multiplicity mentioned disappears in the expression for ξ under definite initial conditions.

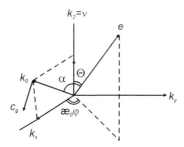

Figure 9.8.   Coordinates and variables used in the problem of a three-dimensional wave packet's propagation for $N^2 = $ const.

Introduce coordinates in the **k**-domain as shown in Fig. 9.8 and assume that the envelope wave is infinite in the direction normal to some unit vector **e**. In the direction **e** for $t = 0$ the packet is finite: $\psi(\mathbf{r}, 0) = 0$ for $|\mathbf{re}| > L$, where $2L$ is the packet's thickness. In this case $B_{\mathbf{k}}(0)$ on the plane normal to **e** has the form of δ-functions multiplied by $(2\pi^2)$. As a result we have in Eq. (9.39) only one integration over **k** directed along **e**. In this integral we pass over its denominator's poles through the upper half-plane. Then integrating it by means of residues yield:

$$\xi(\mathbf{r}, t) = 0; \quad \mathbf{e}(\mathbf{r} - \mathbf{c}_g t) > L$$

and

$$\xi(\mathbf{r}, t) = \frac{k_0}{N_0} \frac{\omega_e}{c_e} (\cos \alpha \sin \theta - \sin \alpha \cos \theta \cos \varphi)$$

$$\times \operatorname{Re} \left\{ \int_{-L}^{L} |\psi(\xi, 0)|^2 \, e^{-i \frac{\omega_e}{c_e} \xi} \, d\xi \, e^{i \left(\frac{\omega_e}{c_e} \mathbf{er} - \omega_e t \right)} \right\} \qquad (9.40)$$

for $\mathbf{e}(\mathbf{r} - \mathbf{c}_g t) < -L$. Here $c_e = \mathbf{c}_g \mathbf{e}$ and $\omega_e = N_0 \sin \theta$ is the frequency of a wave with the wave vector directed along **e**. The meaning of the angles α, θ and φ is clear from Fig. 9.8. So in the region $\mathbf{e}(\mathbf{r} - \mathbf{c}_g t) < -L$ the solution represents a free internal wave with wave vector $\frac{\omega_e}{c_e}$ **e** and frequency ω_e.

Let us now set the initial problem, requiring that at $t = 0$ the non-linear addition to the density field also goes to zero for $\mathbf{er} < -L$. In this case [381] the considered internal wave packet has in the region $-L < \mathbf{er} < -L + c_e t$ a 'wake' of internal waves described by Eq. (9.40).

Contrary to the propagation of a mode, studied in the previous section, when the generation of fluxes was absolutely reversible and the state of rest was restored in the medium, the present situation causes irreversible loss of energy from the packet. From the physical viewpoint it may be explained by the exact synchronism between radiated waves and spectral components of the three-dimensional packet, and only 'quasi-resonant' interaction in case of a mode's propagation takes place.

Taking into consideration the packet's narrowness, $æ_0 L \gg 1$, from Eq. (9.40) it follows that only low frequency waves are intensely radiated (which may be supposed in advance), and the packet should be modulated in the direction inclined to the vertical at the angle $\theta \leq (\sin 2\alpha)/k_0 L \ll 1$. When setting $\theta \sim (\sin 2\alpha)/k_0 L$ from Eq. (9.40) it follows that ξ is of order

$$\xi \sim \frac{k_0}{N_0^2} E \cos \varphi,$$

where E is the energy density of the packet. From this formula it follows that if the envelope wave has a cylindrical symmetry, then the directional characteristic of radiated waves is a figure of eight curve with the axis directed along the horizontal component k_0 of the wave number of the packet's filling. As has been already mentioned, the radiation of vertical waves is accumulated in a narrow cone with a half angle of about $(k_0 L)^{-1}$, where L is the vertical dimension of the wave packet.

Some qualitative conclusions may be made for $N_0 \neq$ const proceeding from the fact that the generation of 'fluxes' is a result of the interaction of the packet's spectral components. The most favourable conditions in this case will have waves with

$$\omega_{æ_1} - \omega_æ - \omega_{æ_2} \approx \mathbf{c}_g^{\nu_o} \mathbf{k} + \frac{\partial \omega_æ}{\partial \nu} (\nu_1 - \nu_2) - \omega_æ \approx 0.$$

Here $\mathbf{k}$, as always, is of the order of the packet's width. If $\nu_1 - \nu_2 \approx 0$, then this condition may be rewritten in the form $c_g \cos \beta \approx c_\nu$, where β is the angle between the vectors $\mathbf{c}_g^{\nu_o}$ and $\mathbf{k}$, and c_ν is the phase velocity of the ν-th wave for $\mathbf{k} = 0$. The eigenfunctions of those long waves satisfy the equation

$$\frac{d^2}{dz^2} \varphi_0^\nu - \frac{N_0^2}{c_0^\nu} \varphi_0^\nu = 0 \,. \tag{9.41}$$

Generally speaking, within a spectrum there are always components $\mathbf{k}_1$ and $\mathbf{k}_2$ such that $\mathbf{c}_g$ will be normal to $\mathbf{k}_1 - \mathbf{k}_2$, i.e., $\cos \beta \approx 0$. So as a result of interaction there always may be waves with ν such that $c_\nu \to 0$. The vertical structure of such waves, as follows from Eq. (9.41), oscillates frequently along z. Thus while the three-dimensional packets propagate in a medium with $N_0(z) \neq$ const, which is typical for the ocean, the appearance of fluxes with small z scales is possible. And more than that, these motions may be maintained after the packet's passage because of the exact synchronism between the packet's spectral components and induced fluxes.

Those qualitative estimations are valid for the propagation of modes as well, when the packet is not infinitely wide, as was assumed in the previous section. And therefore it also has some spectrum spreading along the k_y-axis.

Notes

1 It will follow from the further analysis that such 'small' wavelengths have values of some hundreds (!) meters in natural conditions.

Chapter 10

RESONANT INTERACTIONS OF WAVE TRIADS AND KINETIC EQUATION FOR OCEANIC INTERNAL WAVES' SPECTRUM

Nonlinear effects while internal waves propagate in the ocean are caused not only by the self-modulation of almost monochromatic internal wave packets, but also by the interaction of internal waves with different wave numbers, frequencies and mode numbers.

The latter are also manifested in the form of so called resonant interactions. In Section 10.1 resonant interactions of wave triads in a fluid at rest will be considered. It is shown that such an interaction causes an exchange of energy between waves forming a resonant triad. But the situation changes for the propagation of internal waves in the presence of a plane current. In Section 10.2 it is shown that in this case amplitudes of each of triple waves may grow in time taking energy from the mean flow. Under the weak nonlinearity approximation their amplitudes go to infinity in a finite time. Such a phenomenon is called 'explosive instability'.

The presence of many disturbances among propagating internal waves, most of which are not taken into account in the internal waves' theory, makes their statistical description necessary. One of the most representative statistical characteristics for such a description is a wave's energy spectrum. Its equation, called the kinetic equation, together with its solution's analysis is given in Section 10.3.

1.   RESONANT INTERACTION OF INTERNAL WAVE TRIADS IN FLUID AT REST

Equations describing the interaction of wave triads are frequently yielded from the Hamiltonian formalism developed in previous sections. But it will be useful to discuss a qualitative side of the problem before its mathematical description.

A superposition of triple waves with amplitudes A_1, A_2, A_3, of frequencies ω_1, ω_2, ω_3 and wave vectors æ_1, æ_2, æ_3, is given in the linear approximation. And the synchronism conditions

$$\left.\begin{array}{r} \omega_1 + \omega_2 = \omega_3\,, \\[2mm] \text{æ}_1 + \text{æ}_2 = \text{æ}_3\,. \end{array}\right\} \tag{10.1}$$

are satisfied. We can try to improve the solution by the iteration method if the condition of infinitely small amplitudes, when all of the waves propagate independently, is not met. Then by the synchronism conditions and because of the nonlinearity of the equations an interaction of, e.g., the first and the second waves results in a 'source', which is in resonance with the third wave, the amplitude of which grows linearly in time while A_1 and A_2 remain constant. The velocity of the amplitude's growth should be proportional to the product of amplitudes $A_1 A_2$. When introducing the proportionality coefficient $-iV_3$ the amplitude of the third wave for the time t will be expressed by

$$A_3 = -iV_3 A_1 A_2 t\,. \tag{10.2}$$

Here we have a resonance usual in linear systems, similar to the one considered in Chapter 7 for the generation of internal waves by fluctuations of the atmospheric pressure field.

As the interaction of the first and the second, and the second and the third waves affects in turn the second and the first wave accordingly, their amplitudes will also vary. This variation is described by two other equations similar to Eq. (10.2). All of these equations are valid only in the vicinity of an arbitrary t. As a result, for slowly varying complex amplitudes the following combined equations yield

$$\left.\begin{array}{l} \dot{A}_1 = -iV_1 \bar{A}_2 A_3 \,, \\[6pt] \dot{A}_2 = -iV_2 \bar{A}_1 A_3 \,, \\[6pt] \dot{A}_3 = -iV_3 A_1 A_2 \,. \end{array}\right\} \qquad (10.3)$$

These equations are well known in nonlinear optics [25], [148], [149]. For Hamiltonian systems, as shown later, the equality should be valid

$$\overline{V}_1 = \overline{V}_2 = V_3 = V \qquad (10.4)$$

(while considering $\omega_n \left| A_n \right|^2$ to be the energy of the n-th wave). As our system is Hamiltonian the equality (10.4) is valid for internal waves as well. The numerical value of the coefficient V allows us to estimate characteristic times of interactive processes.

Interaction of internal wave triads has been studied for a long time. After [352], where this problem was stated for the first time, a number of works was written (see, e.g., [55], [59], [120], [166], [167], [230], [239], [336], [406]), where the internal waves' interaction was studied theoretically, numerically and experimentally. Laboratory tests show that the energy exchange may occur as a result of effective interaction of wave triads. In particular, in [59], [235], [236] initial (both running and standing) internal wave transformation owed to the wave triads' interaction, but not owed to the shear instability, was proved experimentally. In [59], [230] experiments on the resonant generation of the internal wave by two other waves were performed. It was shown that one wave of moderate amplitude can produce not only two other waves forming a resonant triad, but also several (4–5) resonant triads. Almost all of the numerous experimental results are well explained by the theory of resonant wave triads interaction. A detailed enough theoretical study of this process by ordinary asymptotic methods (omitting the Hamiltonian formalism) is presented in [379].

Now derive Eqs. (10.3). We proceed from Eq. (7.56) in Section 7.3. Assume that $a_{\mathbf{k}}^{\nu}$ has three spectral bands described by the functions $b_1(k)$, $b_2(k)$, $b_3(k)$ respectively in the vicinity of the points $\mathbf{æ}_1$, $\mathbf{æ}_2$, $\mathbf{æ}_3$ for the modes μ_1, μ_2, μ_3:

$$a_{\mathbf{k}}^{\nu} = b_1 \, \delta \left(\nu - \mu_1 \right) + b_2 \, \delta \left(\nu - \mu_2 \right) + b_3 \, \delta \left(\nu - \mu_3 \right) .$$

Consider that the wave vectors $æ_n$ and the frequencies $\omega_n = \omega_{æ_n}^{\mu_n}$; $n = 1, 2, 3$ meet the synchronism conditions (10.1). Then in Eq. (7.56) for $\mathbf{k} \simeq æ_3$ the second integral vanishes and the first integral is supported by two narrow regions $\mathbf{k}_1 \simeq æ_1$, $\mathbf{k}_2 \simeq æ_2$ and $\mathbf{k}_1 \simeq æ_2$, $\mathbf{k}_2 \simeq æ_1$. In these regions the coefficient $V_{\mathbf{k}\mathbf{k}_1\mathbf{k}_2}^{\nu\nu_1\nu_2}$ may be considered to be constant and equal to

$$V = V_{æ_3 æ_1 æ_2}^{\mu_3\mu_1\mu_2} = V_{æ_3 æ_2 æ_1}^{\mu_3\mu_2\mu_1}$$

owing to the symmetry conditions (7.46). So Eq. (7.56) has the form:

$$\dot{b}_3 + i\omega_{\mathbf{k}}^{\mu_3} b_3 + 2iV \int b_1(k_1) b_2(k_2) \delta(k - k_1 - k_2) \, dk_1 \, dk_2 = 0. \quad (10.5)$$

Extract the slow time amplitude variations by setting $b_n = a_n e^{-i\omega_n t}$; $n = 1, 2, 3$. Substitution of these equations into (10.5) under the first synchronism condition (10.1) yields

$$\dot{a}_3 + i(\omega_{\mathbf{k}}^{\mu_3} - \omega_3) a_3 + 2iV \int a_1(k_1) a_2(k_2) \delta(k - k_1 - k_2) \, dk_1 \, dk_2 = 0.$$

Owing to the packet's narrowness we may set $\omega_{\mathbf{k}}^{\mu_3} - \omega_3 \simeq c_g^{(3)} (\mathbf{k} - æ_3)$, where

$$c_g^{(3)} = \left. \frac{d\omega_{\mathbf{k}}^{\mu_3}}{dk} \right|_{k=æ_3}.$$

It is convenient to perform the inverse Fourier transform on the resulting equation

$$\psi_n(\mathbf{r}, t) = \frac{1}{(2\pi)} \int e^{i(\mathbf{k}-æ_n)\mathbf{r}} a_n(\mathbf{k}) \, d\mathbf{k}.$$

The multipliers $e^{-iæ_n \mathbf{r}}$ select the slow spatial amplitude variations for the corresponding wave, so that the functions $\psi_n(\mathbf{r}, t)$ vary slightly within the period and the wavelength. In terms of ψ_n Eq. (10.5) may be rewritten:

$$\dot{\psi}_3 + (c_g^3 \nabla) \psi = -i4\pi V \psi_1 \psi_2, \quad (10.6)$$

where $\nabla = \left(\frac{\partial}{\partial x} \frac{\partial}{\partial y} \right)$ is the horizontal gradient. The function $\omega_n |\psi_n|^2$, as shown in Section 7.2 represents the energy density of the n-th wave per unit horizontal plane. While studying by analogy the vicinity of the points $\mathbf{k} = æ_1$, $\mathbf{k} = æ_2$ we find that now only the second integral is sufficient in Eq. (7.56). After the extraction of low components and the inverse Fourier transform yield two other equations forming together with Eq. (10.6) the following system

$$\left.\begin{array}{l} \dot{\psi}_1 + (c_g^{(1)}\nabla)\,\psi_1 = -i4\pi\bar{V}\bar{\psi}_2\psi_3\,; \\[2mm] \dot{\psi}_2 + (c_g^{(2)}\nabla)\,\psi_2 = -i4\pi\bar{V}\bar{\psi}_1\psi_3\,; \\[2mm] \dot{\psi}_3 + (c_g^{(3)}\nabla)\,\psi_3 = -i4\pi V\psi_1\psi_2\,, \end{array}\right\} \qquad (10.7)$$

where

$$\mathbf{c}_g^{(n)} = \left.\frac{\partial\omega_k^{\mu n}}{\partial\mathbf{k}}\right|_{k=\mathbf{æ}_n}.$$

It is possible to make V real multiplying ψ_n by the same phase term. Eq. (7.38) for the coefficient V is slightly simplified as a result of the first synchronism condition (10.1). When vectors are collinear these equations may be integrated by the method of the inverse scattering problem. But such investigations will be omitted here, and some simple physical conclusions will be made.

Here and later, ψ_n will be considered to be independent of (x,y). Then the combined Eqs. (10.7) transform into the six combined real ordinary equations. These equations, as may be easily shown, have two integrals:

$$\sum_{n=1}^{3}\omega_n\,|\psi_n|^2 = \text{const}\,; \qquad \sum_{n=1}^{3}\mathbf{æ}_n\,|\psi_n|^2 = \text{const}\,. \qquad (10.8)$$

The first is the energy conservation law within the triple waves system, and the second may be considered to be the momentum conservation law.

By Eqs. (10.7) we may study the stability of finite amplitude waves with frequency ω_3 and with wave vector $\mathbf{æ}_3$ relative to a disturbance formed by the superposition of two satellite waves with frequencies ω_1 and ω_2 and wave vectors $\mathbf{æ}_1$ and $\mathbf{æ}_2$ satisfying the synchronism conditions (10.1). While setting $|\psi_3|\,(0)| \gg \psi_{1,2}\,(0)|$ yields from Eqs. (10.7) for low t $\psi_3 \approx \text{const}$. Another two equations may be reduced to the form

$$\ddot{\psi}_{1,2} - |4\pi V|^2\,|\psi_3|^2\psi_{1,2} = 0\,.$$

Thus at least for the initial time period disturbances grow exponentially $\psi_{1,2} = \psi_{1,2}(0) e^{t/\tau}$, where

$$\tau^{-1} = |4\pi V| |\psi_3| . \tag{10.9}$$

The wave instability with respect to breaking into two other waves with lower frequencies is proved. Let us estimate the characteristic time τ of this process. From the formula for the vertical velocity component (7.36a) we find that for a wave with amplitude A it will be $\psi \sim a_0 \sim A\omega^{-1/2}\varphi^{-1}$, where φ is some mean value of the eigenfunction φ_k^ν. Assuming that all of the frequencies in (10.1) have the same orders yields from Eq. (7.38) for V: $V \sim \omega^{3/2}\bar{N}_0^2\varphi^3$. The normalizing condition (7.27) yields: $\bar{N}_0^2\varphi^2 l_z \sim 1$, where l_z is the characteristic vertical scale of the eigenfunctions variation, and $\bar{N}_0^2$ is the mean Brunt–Väisälä frequency.

In the case $N_0 = \text{const}$ l_z^{-1} is equal to the vertical component of the wave number. As a result

$$\tau^{-1} \sim V\psi_3 \sim A\omega\bar{N}_0^2\varphi^2 \sim \omega \frac{A}{l_z} .$$

Thus the ratio of the wave amplitude to the characteristic vertical scale appears to be a small parameter of the problem. To satisfy Eqs. (10.7) it should be $A_n \ll l_z$. The estimation of the time τ performed is now rather rough; some numerical results are presented below. Further details may be found in [379].

Consider now a disturbance of the basic wave with the parameters ω_1 and $\mathbf{æ}_1$ by two resonant waves, one of them has a greater frequency than the basic wave. Assume the synchronism conditions (10.1) to be still satisfied and $|\psi_1| \gg |\psi_{2,3}|$ at $t = 0$. The above procedure for the disturbance amplitudes yields

$$\psi_{2,3} + |4\pi V|^2 |\psi_1|^2 \psi_{2,3} = 0 ,$$

from which it follows that $\psi_{2,3} = \psi_{2,3}(0) e^{\pm i|4\pi V||\psi_1|t}$. As the disturbance amplitudes do not grow versus time, then the wave is stable with respect to such disturbances. We confine ourselves here to the most important

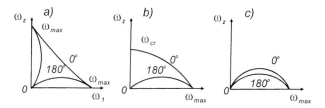

Figure 10.1. The resonant region of the interaction of wave triads on the plane ω_i. Angles between the vectors $\mathbf{k}_1$ and $\mathbf{k}_2$ corresponding to the points at the region borders are marked. Here $\omega_{\max} = (\max N_0(z))/z$.

phenomena from the physical viewpoint following from Eqs. (10.7). A more thorough study is performed in, e.g., [25], [148], [149].

The most important point of all above studies is an opportunity to meet conditions (10.1). The dispersion relations for which it is or is not the solution $\omega_{\mathbf{k}}$, are called, respectively, dissipative or nondissipative. Internal waves have a dissipative spectrum.

We represent here results obtained in [379] for modes propagation. By a simple geometrical construction, using the dispersion curves' properties valid for any stable stratification, it is possible to prove the following statement: to make a wave with frequency ω_3 break into any waves relating to the modes μ_1 and μ_2 it is necessary and sufficient to meet the condition $\mu_3 < \mu_{\max} = \max(\mu_1, \mu_2)$. So any internal wave is unstable relative to disturbance by a pair of resonant waves, and at least one of their modes' numbers should be greater than the mode number of the basic wave.

Let us now present some forms of characteristic frequency regions, where interaction of waves occur, versus modes' numbers of interacting waves (see Fig. 10.1). The case a corresponds to $\mu_3 < \mu_{\min} = \min(\mu_1, \mu_2)$, the case b to $\mu_3 = \mu_{\min}$. The value of ω_* on the ω-axis satisfies the equation

$$\frac{d\omega^{\mu_3}}{dk}\bigg|_{\omega=\omega_3} = c_{\min}(0) \quad \text{where} \quad c_{\min}(0) = \lim_{k\to 0}\frac{\omega_{\min}^{\mu}}{k}.$$

Fig. 10.1 c relates to the case $\mu_{\min} < \mu_3 < \mu_{\max}$. If the Brunt–Väisälä frequency has only one maximum, then dispersion curves have, in general, a humped shape: $d^2\omega_k^{\mu}/dk^2 < 0$. In this case the interaction of triple internal waves related to this mode is impossible.

Now consider the instability increment $\delta = \tau^{-1}$. Numerical results are performed for the stratification

$$N_0^2\,(z) = N_{\max}^2\,\mathrm{ch}^{-2}\left(\frac{2z}{d}\right)\,, \qquad (10.10)$$

where $N_{\max}$ is the maximum value of the Brunt–Väisälä frequency and d is the characteristic thermocline thickness. When fixing the frequency of the breaking wave ω_3 Eqs. (10.1) may be satisfied by fixing, e.g., the angle θ between the vectors $\mathbf{æ}_1$ and $\mathbf{æ}_2$. Numerical calculations for the stratification (10.10) shows that the maximum increment $\delta_{\max}$ is reached, as a rule, when $\theta = 0°$ or $\theta = 180°$. The dependence of $\delta_{\max}$ on the frequency of the breaking wave ω_3 for different mode numbers of interacting waves is presented in Fig. 10.2. Time is measured here in periods of the breaking wave and its amplitude is equal to the thermocline thickness d (according to Eq. (10.9) the increment is proportional to the wave's amplitude). The value of the increment is high in the case of breaking into higher modes. And an instability is also rather high; e.g., amplitudes of two meeting waves of the fourth and the fifth modes resonantly generated by the first-mode (the lowest) wave of frequency $\omega_3 = 0.7\,N_{\max}$ and of amplitude $d/10$ grow e times in about seven periods of the basic wave.

In this section we confine ourselves to the investigation of wave stability. Long time solutions of Eq. (10.7) show that a periodic energy exchange between internal waves occurs, which also follows from the conservation laws (10.8).

2.   INTERACTION OF INTERNAL WAVE TRIADS IN A SHEAR FLOW. EXPLOSIVE INSTABILITY

Consider the interaction of triple internal waves in the presence of shear flow (see [18], [44], [55], [285], [312], [365], [383]). A complete description of the problem's statement is given in Section 7.1. The corresponding Hamiltonian has the form

$$H = \int \left\{1/2\,\mathbf{v}^2 - 1/2\,\mathbf{u}_0^2\,(\rho_0 + \rho) + \pi\,(\rho, z)\right\}\,dV\,, \qquad (10.11)$$

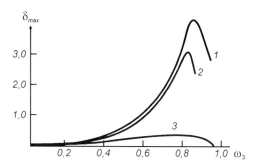

Figure 10.2.   The dependence of $\delta_{\max}$ on the frequency of a breaking wave in the case of breaking into different modes.   Different curves correspond to the following numbers of interacting waves: 1 — $\nu_1 = 5$, $\nu_2 = 4$, $\nu_3 = 1$, $1 \to 5 + 4$; 2 — $\nu_1 = 2$, $\nu_2 = 3$, $\nu_1 = 1$, $1 \to 2 + 3$; 3 — $\nu_1 = 1$, $\nu_2 = 2$, $\nu_3 = 1$, $1 \to 1 + 2$.

where $\mathbf{u}_0\,(\rho)$ is the given undisturbed flow profile versus ρ_0, $\pi\,(\rho, z) = \Pi\,(\rho_0 + \rho) - \Pi\,(\rho_0) + \rho g z$ is the potential energy density (see Eq. (7.12)) and $\mathbf{v}$ is the fluid velocity

$$\mathbf{v} = \mathbf{u}_0\,(\rho_0 + \rho) + \nabla\,\Phi + \lambda\,\nabla\,(\rho_0 + \rho)\,.$$

Here Φ is obtained meeting the incompressibility condition div $\mathbf{v} = 0$:

$$\Phi = \Delta^{-1}\{-\nabla\,(\lambda\,\nabla\,(\rho_0 + \rho)) - \nabla u_0\,(\rho_0 + \rho)\}\,.$$

The canonical equations $\dot{\lambda} = \delta H/\delta\rho$; $\dot{\rho} = -\delta H/\delta\lambda$ are omitted, but for a harmonic wave with a frequency ω and a wave vector $\mathbf{k}$ they are reduced to the boundary value problem for the vertical velocity component $\omega_{\mathbf{k}}^\nu$ (see Eq. (5.93)):

$$\frac{d^2\omega_{\mathbf{k}}^\nu}{dz^2} + \left(\frac{N_0^2 k^2}{\omega_d^2} - \frac{\omega_d''}{\omega_d} - k^2\right)\omega_{\mathbf{k}}^\nu = 0\,, \tag{10.12}$$

where $\omega_d = \omega - \mathbf{k}\mathbf{u}_0$ is a local Doppler frequency depending on z. The canonical variables may be expressed through $\omega_{\mathbf{k}}^\nu$:

$$\rho = \rho_0'\,\frac{\omega_{\mathbf{k}}^\nu}{i\omega_d} + c.c.\,;\quad \lambda = \frac{1}{\rho_0' k^2}\left(\frac{N_0^2 k^2}{\omega_d^2}\,\omega_{\mathbf{k}}^\nu - \left(\frac{\omega_d'}{\omega_d}\right)^2\omega_{\mathbf{k}}^\nu + \frac{\omega_d'}{\omega_d}\,\frac{\partial\omega_{\mathbf{k}}^\nu}{\partial z}\right) + c.c.\,. \tag{10.13}$$

Define the energy density for a harmonic wave. As H is the homogeneous quadratic functional of ρ and λ for infinitely small internal waves amplitudes, when using the typical equality for quadratic operators, yields:

$$H = \frac{1}{2} \int \left(\frac{\delta H}{\delta \rho} \rho + \frac{\delta H}{\delta \lambda} \lambda \right) dV .$$

As H coincides with the system's energy, taking into consideration the canonical equation yields:

$$E = \frac{1}{2} \int (\rho \dot{\lambda} - \dot{\rho} \lambda) \, dV = \int \rho \dot{\lambda} \, dV - \frac{1}{2} \frac{d}{dt} \int \rho \lambda \, dV .$$

Integrating over the wave period we obtain the relation between the energy and the adiabatic invariant of the harmonic oscillator

$$\frac{dE}{dV} = \omega \frac{dI}{dV}$$

where V is a volume;

$$\frac{dI}{dV} = \frac{1}{2\pi} \int \rho \lambda_\theta \, d\theta$$

is the adiabatic invariant density; θ is the phase and ω is the real wave frequency. When the flow u_0 varies slowly versus z we may assume

$$\lambda \approx \frac{1}{\rho_0'} \frac{N_0^2}{\omega_d^2} w_\mathbf{k}^\nu + \text{c.c.} ,$$

this yields the known equation for the adiabatic invariant density [37], [38].

$$\frac{dI}{dV} = \overline{\rho i \lambda} = \frac{1}{2} \frac{N_0^2 W^2}{\omega_d^3} = \frac{1}{\omega_d} \frac{dE}{dV} ,$$

where W is the real amplitude of the vertical velocity and

$$\frac{dE}{dV} = \frac{N_0^2}{2} \left(\frac{W}{\omega_d} \right)^2$$

is the energy density (for $\rho_0 = 1$; to obtain dimensional values in the latter equation the multiplier $\bar{\rho}_0$ should be introduced). Using the exact equality (10.13) and additionally integrating over z we find the expression for the adiabatic invariant density per a unit horizontal plane:

$$\frac{dI_{\mathbf{k}}^{\nu}}{dS} = \int \frac{dI}{dV} dz = \int \frac{1}{2\pi} \int_0^{2\pi} \rho i \lambda \, d\theta \, dz = \frac{1}{4k^2} \int \left(\frac{2N_0^2 k^3}{\omega_{\alpha}^3} - \frac{\omega_d''}{\omega_d^2} \right) \omega_{\mathbf{k}}^{\nu^2} \, dz \,.$$

$$(10.13a)$$

This equation for $dI_{\mathbf{k}}^{\nu}/dS$, which will be further replaced by $dI_{\mathbf{k}}^{\nu}$ or I, was already used in Section 5.3 (Eq. (5.100)); the difference is only in the multiplier $\frac{1}{2}$. And this particular variable (see [380]) is maintained during the wave packet propagation in the ocean, when the packet's parameters vary slowly. The conservation law obtained in [380] is the direct result of the corresponding theorem of classical mechanics.

Let us return to Eq. (10.12). Each value of $\mathbf{k}$ has the respective frequency ω for the fixed mode number ν. The substitution $\mathbf{k} \to -\mathbf{k}$, $\omega \to -\omega$ does not change the equation, but the adiabatic invariant changes its sign $I \to -I$, as follows from (10.13). A simultaneous sign change for $\mathbf{k}$ and ω does not lead to a new physical state and describes the same wave motion. Which pair $(\mathbf{k}, \omega)$ or $(-\mathbf{k}, -\omega)$ should be chosen for the wave description? Remember (Section 7.1) that while introducing the normal variables the energy density of the wave of unit amplitude should be equal to its frequency. This defines a proper choice: we should choose the pair $(\mathbf{k}, \omega)$ which yields $I > 0$. If it results in $I < 0$ the values $(-\mathbf{k}, -\omega)$ should be chosen for the wave description. The wave energy's sign does not depend on the choice. But it is sufficient for resonant triads. In case of the propagation of internal waves in a quiet fluid

$$I = \frac{1}{2\omega^3} \int N_0^2 \, \omega_{\mathbf{k}}^{\nu^2} \, dz \,,$$

and according to the rule defined above for the wave's description, we should use the positive frequencies branch.

We have chosen three waves for which the synchronism conditions are satisfied

$$\left. \begin{array}{l} \mathbf{k}_1 + \mathbf{k}_2 = \mathbf{k}_3 \,; \\ \omega_1 + \omega_2 = \omega_3 \,. \end{array} \right\} \qquad (10.14)$$

In addition, the pair of the first and the second waves and the third wave have different adiabatic invariant signs. In this case, to choose the 'proper' parameters for the description of waves we should change the frequency and the signs of the wave vectors for waves with $I < 0$. After that those terms are replaced on the opposite side of Eq. (10.14). As a result the synchronism conditions (7.54) are satisfied for the 'proper' parameters.

$$\left.\begin{aligned} \mathbf{k}_1 + \mathbf{k}_2 + \mathbf{k}_3 = 0\,; \\ \omega_1 + \omega_2 + \omega_3 = 0\,. \end{aligned}\right\} \tag{10.15}$$

To reveal the result of such a triplet wave interaction we consider the Hamiltonian (7.37). The terms containing the coefficient U will be considered now, and the terms with V may be neglected by the transform (7.49). The description of interaction of these three waves coincides the derivations of Section 10.1. As a result the following simultaneous equations yield

$$\left.\begin{aligned} \dot{\psi}_1 + (c_g^{(1)}\nabla)\,\psi_1 = -i2\pi\dot{U}\bar{\psi}_2\psi_3\,; \\ \dot{\psi}_2 + (c_g^{(2)}\nabla)\,\psi_2 = -i2\pi\dot{U}\bar{\psi}_1\psi_3\,; \\ \dot{\psi}_3 + (c_g^{(3)}\nabla)\,\psi_3 = -i2\pi\dot{U}\psi_1\psi_2\,. \end{aligned}\right\} \tag{10.16}$$

These equations also have two conservation laws (10.8). But the presence of negative frequencies does not restrict the waves' amplitudes ψ_n. Indeed, Eqs. (10.16) may have a solution:

$$\psi_1 = \psi_2 = \psi_3 = \psi\,(t) = \frac{1}{2\pi U\,(t - t_0)}\,e^{i\theta},$$

where $\theta = \frac{1}{3}\left(\frac{2}{\pi} - \arg U\right)$ and the constant t_0 is governed by the initial amplitude $\psi\,(0)$ and by the coefficient $U = V_{\mathbf{k}_1\mathbf{k}_2\mathbf{k}_3}^{\mu_1\mu_2\mu_3}$. Obviously for $t \to t_0$ $|\psi| \to \infty$. Owing to such a solution of Eqs. (10.16) (infinite waves amplitude during a finite time) this interaction is frequently called 'explosive'. Eqs. (10.16) are exactly integrable and their general solution also has singularities. Certainly, when the amplitudes ψ_n grow considerably the higher-order nonlinear effects will be considerable. So the real ψ_n values remain finite for $t \to t_0$. But the solution's behaviour over relatively short time periods allows to estimate efficiency of the process.

Thus we have shown that the triple-wave system satisfying Eqs. (10.14) has an explosive instability when the sign of the adiabatic invariant of the third wave is opposite to the other two. Eqs. (10.14) are similar to Eqs. (10.1), but in this case all of the waves should have $I > 0$. So for resonant triads the sign of the adiabatic invariant is now decisive as well as the dispersion dependence versus the wave vector $\omega^\nu(\mathbf{k})$. It appears that information about the sign of I is contained in the dispersion relation $\omega^\nu(\mathbf{k})$ and we should not solve Eq. (10.13). As follows from Eqs. (5.100), (5.101) the term I is in the denominator of the expression for the group velocity of internal waves at a shear flow. Therefore I changes its sign at the points where the dispersion curve has a vertical tangential line:

$$\mathbf{c}_g = \frac{d\omega^\nu(k)}{d\mathbf{k}} = \infty.$$

If $\omega^\nu(\mathbf{k}) \neq 0$ small variations of $\mathbf{k}$ with $\omega l < 0$ cause the negative energy waves to appear. Such a critical point is a bifurcation point when small variations of $\mathbf{k}$ lead to the appearance of complex roots instead of a pair of real roots. Therefore this point is singular with respect to the system's linear stability. Thus a strong connection between the negative energy waves and the linear instability of the system occurs.

The conclusions above [383] are valid not only for an internal wave at a shear flow, but for any Hamiltonian systems which are invariant with respect to simultaneous time and space inversion. From such invariance it follows that both $(\mathbf{k},\omega)$ and $(-\mathbf{k},-\omega)$ satisfy the dispersion equation, and I changes its sign as a result of that substitution (since the wave energy ωl should not change its sign). Another useful fact: if $c_g = \infty$, then $I = 0$. It may be shown if we use the mean Lagrangian $\mathcal{L}(\omega, \mathbf{k})$. Calculation of its mean value with respect to a period $L = \int \lambda \dot{\rho} dv - H$ yields $\overline{L} = \mathcal{L}(\omega, \mathbf{k}) = 0$ (the virial theorem):

$$
\begin{aligned}
\mathcal{L}(\omega, \mathbf{k}) &= \frac{1}{T} \int_0^T \left\{ \int (\lambda \dot{\rho} dv - H) \right\} dT \\
&= \frac{1}{T} \int_0^T \int \left(\lambda \dot{\rho} - \frac{1}{2} \rho \frac{\delta H}{\delta \rho} - \frac{1}{2} \lambda \frac{\delta H}{\delta \lambda} \right) dt dV \\
&= \int_0^T \frac{1}{2T} \frac{d}{dt} \int \lambda \rho dV \, dt = 0 \,.
\end{aligned}
$$

Hence it directly follows that

$$\frac{d\omega}{d\mathbf{k}} = -\frac{\partial \mathcal{L}/\partial \mathbf{k}}{\partial \mathcal{L}/\partial \omega}.$$

But $I = \partial \mathcal{L}/\partial \omega$ (this equality becomes trivial for normal variables; it is always valid, owing to the invariant character of contained terms). All the above conclusions on the dispersion curves and the explosive instability are valid for different systems.

Let us consider the flow velocity and the Brunt–Väisälä frequency profiles:

$$U_0(z) = \begin{cases} 0, & |z| < h, \\ u_0, & |z| > h, \end{cases} \qquad \rho_0 = \begin{cases} \rho_1, & z > 0, \\ \rho_2, & z < 0, \end{cases} \quad ;$$

$$N_0^2 = 2g \frac{\rho_2 - \rho_1}{\rho_2 + \rho_1} \delta(z).$$

There are two tangential discontinuities (the points $z = \pm h$) stabilized by a bisecting density jump. Eq. (10.12) should be soluble for continuous

$$\frac{w}{\omega_d} \quad \text{and} \quad \frac{\omega_d^2}{k^2} \left(\frac{w}{\omega_d}\right)' - g\rho_0 \frac{w}{\omega_d}$$

denoting the vertical displacement and the continuity of pressure. This example has the characteristic dimensionless parameter

$$R = \frac{\rho_2 - \rho_1}{\rho_2 + \rho_1} \frac{gh}{u_0^2}$$

(the Richardson number's analog). Direct derivations give the dispersion equation:

$$\frac{k}{R} x^2 (x^2 \tanh kh + 1) = (x - 1)^2 (x + \tanh kh); \quad k > 0,$$

where $x = \omega/(\omega - k u_0)$, and the expression for I:

$$I = \frac{2}{k^2 u_0} \left(1 - \frac{1}{x}\right) \left(\frac{1}{2}\left(1 + \frac{1}{x}\right)^2 \sinh 2kh + e^{-2kg} + x - 1\right).$$

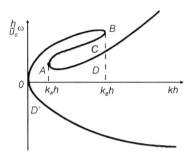

Figure 10.3. The general dispersion curve for the considered model for $Ri \leq 0.1$. Regions with $I < 0$ are dashed.

For the values $R < R_* \approx 0.1$ the dispersion curve is schematically shown in Fig. 10.3. The curve's parts relating to the negative adiabatic invariant values are dashed. The region AB corresponds to the negative energy waves. The dispersion equation has the fourth order with respect to the frequency. The region $k_A < k < k_B$ corresponds to absolute linear stability when four roots of the dispersion equation are real. The points A and B are the bifurcation points; for $k > k_B$ and $0 < k < k_A$ there is a pair of complex roots corresponding to the instability of the tangential discontinuity. From Fig. 10.3 follows an opportunity to meet the synchronism conditions (10.15): the lower branch of the dispersion curve should be shifted so that the point O goes to the C point. Assuming $(\omega_1, k_1) = \mathbf{OC}$, $(\omega_2, k_2) = \mathbf{CD}$ and $(\omega_3, k_3) = \mathbf{OD}$ we shall satisfy conditions (10.15) and (10.16).

The time of the instability evolution depends on the value of the coefficient U. But the expression for U appears to be rather cumbersome and the corresponding calculations are omitted. But from dimensional analysis we may conclude that the order of the evolution time of the instability

$$t_{expl} \sim \frac{1}{\omega} \frac{A}{l},$$

where A is the wave amplitude, ω is the frequency and l is the characteristic scale which may be replaced by h or by the wavelength λ. Possibly the explosive instability mechanism considered causes the appearance of oceanic turbulence in the presence of mean stratified currents (observed, e.g., by Woods [392], [393]).

3.   STATISTICAL DESCRIPTION OF INTERACTING WEAKLY NONLINEAR INTERNAL WAVES

All wave fields in the real ocean have random time and space transformations caused by numerous sources which are excluded from consideration in the study of internal waves.

When internal waves are linear their statistical characteristics will be totally defined by properties of disturbing sources. But owing to waves' interactions their nonlinearity may lead to the formation of a wide band spectrum even when the disturbing source has a narrow spectrum. In this section we derive the kinetic equation describing the dynamics of the spectrum of internal waves caused by nonlinear processes. Its successive derivation needs diagram methods and is rather complicated [23], [411]. So for convenience we present the corresponding procedure from [407].

Assume that the ensemble of internal waves has the horizontal statistical inhomogeneity:

$$\langle a_{\mathbf{k}}^{\nu} \bar{a}_{\mathbf{k}_1}^{\nu_1} \rangle = n_{\mathbf{k}}^{\nu, \nu_1} \delta \left(\mathbf{k} - \mathbf{k}_1 \right), \quad \langle \bar{a}_{\mathbf{k}}^{\nu} a_{\mathbf{k}_1}^{\nu_1} a_{\mathbf{k}_2}^{\nu_2} \rangle = I_{\mathbf{k}\mathbf{k}_1\mathbf{k}_2}^{\nu\nu_1\nu_2} \delta \left(\mathbf{k} - \mathbf{k}_1 - \mathbf{k}_2 \right), \quad \text{etc.}$$

Additionally assume that different wave modes have weak cross-correlation, so that $n_{\mathbf{k}}^{\nu, \nu_1} = n_{\mathbf{k}}^{\nu} \delta \left(\nu - \nu_1 \right)$. In the case $N_0 = \text{const}$ this hypothesis follows from the statistical inhomogeneity with depth of the internal waves' field and is arbitrary in general. Consider further that $n_{\mathbf{k}}^{\nu}$, $I_{\mathbf{k}\mathbf{k}_1\mathbf{k}_2}^{\nu\nu_1\nu_2}$ have slow time variations relative to the phases of corresponding waves. And these phases are 'chaotic' so that the correlators $\langle a_{\mathbf{k}}^{\nu} a_{\mathbf{k}_1}^{\nu_1} \rangle$ may be omitted (the 'chaotic phases' hypothesis). The last assumption is that the random field $a_{\mathbf{k}}^{\nu}$ is close to the normal field and the fourth-order correlators may be expressed through the double ones. We proceed from Eq. (7.56).

$$\dot{a}_{\mathbf{k}}^{\nu} + i\omega_{\mathbf{k}}^{\nu} a_{\mathbf{k}}^{\nu} + i \int V_{\mathbf{k}\mathbf{k}_1\mathbf{k}_2}^{\nu\nu_1\nu_2} a_{\mathbf{k}_1}^{\nu_1} a_{\mathbf{k}_2}^{\nu_2} \delta \left(\mathbf{k} - \mathbf{k}_1 - \mathbf{k}_2 \right) d\Gamma$$

$$+ 2i \int \bar{V}_{\mathbf{k}_1\mathbf{k}\mathbf{k}_2}^{\nu_1\nu\nu_2} a_{\mathbf{k}_1}^{\nu_1} \bar{a}_{\mathbf{k}_2}^{\nu_2} \delta \left(\mathbf{k}_1 - \mathbf{k} - \mathbf{k}_2 \right) d\Gamma = 0, \qquad (10.17)$$

where $d\Gamma = d\mathbf{k}_1 \, d\mathbf{k}_2 \, d\nu_1 \, d\nu_2$.

Multiplying this equation by $\bar{a}_{\mathbf{k}}^{\nu}$, then summing with the conjugate complex and averaging with respect to **k'** yields:

$$
\dot{n}_{\mathbf{k}}^{\nu} = 2\,\mathrm{Im}\left\{ \int V_{\mathbf{k}\mathbf{k}_1\mathbf{k}_2}^{\nu\nu_1\nu_2} I_{\mathbf{k}\mathbf{k}_1\mathbf{k}_2}^{\nu\nu_1\nu_2} \, \delta\left(\mathbf{k} - \mathbf{k}_1 - \mathbf{k}_2\right) d\Gamma \right.
$$

$$
\left. - 2 \int V_{\mathbf{k}_1\mathbf{k}\mathbf{k}_2}^{\nu_1\nu\nu_2} I_{\mathbf{k}_1\mathbf{k}\mathbf{k}_2}^{\nu_1\nu\nu_2} \, \delta\left(\mathbf{k}_1 - \mathbf{k} - \mathbf{k}_2\right) d\Gamma \right\} . \tag{10.18}
$$

Consider now the equation for I. Calculation of $\frac{\partial}{\partial t}\left(\bar{a}_{\mathbf{q}}^{\mu} a_{\mathbf{q}_1}^{\mu_1} a_{\mathbf{q}_2}^{\mu_2}\right)$ in accordance with the equation of motion (10.17) and taking its averaging result in

$$
\frac{\partial}{\partial t} I_{\mathbf{q}\mathbf{q}_1\mathbf{q}_2}^{\mu\mu_1\mu_2} + i\left(\omega_{\mathbf{q}_1}^{\mu_1} + \omega_{\mathbf{q}_2}^{\mu_2} - \omega_{\mathbf{q}}^{\mu}\right) I_{\mathbf{q}\mathbf{q}_1\mathbf{q}_2}^{\mu\mu_1\mu_2} = \sum, \tag{10.19}
$$

where $\sum$ contains different fourth-order correlators. Using the sets above we conclude that only mean terms in the form $\langle \bar{a}\bar{a}aa\rangle$ should be retained in $\sum$:

$$
\langle \bar{a}_{\mathbf{k}}^{\nu} \bar{a}_{\mathbf{q}}^{\mu} a_{\mathbf{k}_1}^{\nu_1} a_{\mathbf{q}_1}^{\mu_1} \rangle = n_{\mathbf{k}}^{\nu} n_{\mathbf{q}}^{\mu} \left(\delta_{\mathbf{k}-\mathbf{k}_1}^{\nu-\nu_1} \delta_{\mathbf{q}-\mathbf{q}_1}^{\mu-\mu_1} + \delta_{\mathbf{k}-\mathbf{q}_1}^{\nu-\mu_1} \delta_{\mathbf{q}-\mathbf{k}_1}^{\mu-\nu_1} \right),
$$

where $\delta_{\mathbf{a}-\mathbf{b}}^{\alpha-\beta} = \delta\left(\alpha - \beta\right)\delta\left(\mathbf{a} - \mathbf{b}\right)$. Hence it follows that

$$
\sum = 2i\bar{V}_{\mathbf{q}\mathbf{q}_1\mathbf{q}_2}^{\mu\mu_1\mu_2} \left(n_{\mathbf{q}_1}^{\mu_1} n_{\mathbf{q}_2}^{\mu_2} - n_{\mathbf{q}}^{\mu} n_{\mathbf{q}_1}^{\mu_1} - n_{\mathbf{q}}^{\mu} n_{\mathbf{q}_2}^{\mu_2} \right).
$$

Neglecting slow time variations of I from Eq. (10.19) it follows that:

$$
I_{\mathbf{q}\mathbf{q}_1\mathbf{q}_2}^{\mu\mu_1\mu_2} = 2 \frac{\bar{V}_{\mathbf{q}\mathbf{q}_1\mathbf{q}_2}^{\mu\mu_1\mu_2}}{\left(\omega_{\mathbf{q}_1}^{\mu_1} + \omega_{\mathbf{k}_2}^{\mu_2} - \omega_{\mathbf{q}}^{\mu}\right)} \left(n_{\mathbf{q}_1}^{\mu_1} n_{\mathbf{q}_2}^{\mu_2} - n_{\mathbf{q}}^{\mu} n_{\mathbf{q}_1}^{\mu_1} - n_{\mathbf{k}}^{\nu} n_{\mathbf{q}_2}^{\mu_2} \right) \tag{10.20}
$$

Now substitute the expression obtained into Eq. (10.18). Its integrand contains poles caused by the denominator's zeros in Eq. (10.20). Introduce the weak system's decay by small negative frequency additions. Taking into account that when $\varepsilon \to 0$ $\lim \frac{1}{x-i\varepsilon} = \pi\delta\left(x\right)$, finally we have:

$$
\frac{1}{4\pi} \frac{\partial}{\partial t} n_{\mathbf{k}}^{\nu} = \int |V_{\mathbf{k}\mathbf{k}_1\mathbf{k}_2}^{\nu\nu_1\nu_2}|^2 \left(n_{\mathbf{k}_1}^{\nu_1} n_{\mathbf{k}_2}^{\nu_2} - n_{\mathbf{k}_1}^{\nu_1} n_{\mathbf{k}}^{\nu} - n_{\mathbf{k}_2}^{\nu_2} n_{\mathbf{k}}^{\nu} \right)
$$

$$\times \, \delta \left(\mathbf{k} - \mathbf{k}_1 - \mathbf{k}_2 \right) \delta \left(\omega_{\mathbf{k}}^{\nu} - \omega_{\mathbf{k}_1}^{\nu_1} - \omega_{\mathbf{k}_2}^{\mu_2} \right) d\Gamma$$

$$+ \, 2 \int |V_{\mathbf{k}_1 \mathbf{k} \mathbf{k}_2}^{\nu_1 \nu \nu_2}|^2 \, \left(n_{\mathbf{k}_1}^{\nu_1} \, n_{\mathbf{k}_2}^{\nu_2} + n_{\mathbf{k}}^{\nu} \, n_{\mathbf{k}_1}^{\nu_1} - n_{\mathbf{k}}^{\nu} \, n_{\mathbf{k}_2}^{\nu_2} \right) \delta \left(\mathbf{k}_1 - \mathbf{k} - \mathbf{k}_2 \right)$$

$$\times \, \delta \left(\omega_{\mathbf{k}_1}^{\nu_1} - \omega_{\mathbf{k}}^{\nu} - \omega_{\mathbf{k}_2}^{\nu_2} \right) d\Gamma \, .$$

$$(10.21)$$

In the case $N_0 = \text{const}$ the indices ν and $\mathbf{k}$ are replaced by the three-dimensional vector æ, and Eq. (7.42) is used for $V_{æ æ_1 æ_2}$. The kinetic equation (10.21) for internal waves was obtained in [290] from the hydrodynamics equations by omitting the Hamiltonian formalism. A discussion of the statistical basis of the derivation of the kinetic equation for random nonlinear wave fields in application to the geophysical hydrodynamics is presented in [8], [23]. Plasma application of the kinetic equation, together with detailed references, are given in [148]. Strict grounds of the derivation of the kinetic equation for a weakly nonlinear ensemble of random wave fields may be found in [411].

4.   LOW TURBULENCE OF INTERNAL WAVES IN THE OCEAN

In this section we pass to the analysis of the kinetic equation (10.21). A manifold of phenomena caused by interaction of weakly-nonlinear random wave fields, when the nonlinearity level is too small to affect a wave's dispersive properties, is called 'weak turbulence' and is described by the kinetic equation.

A quite representative statistical characteristics for the weak (and sometimes for rather strong) turbulence description is the energy spectrum

$$\varepsilon_{\mathbf{k}}^{\nu} = \omega_{\mathbf{k}}^{\nu} n_{\mathbf{k}}^{\nu} \quad \varepsilon_{æ} = \omega_{æ} n_{æ} \, . \qquad (10.22)$$

So under the weak turbulence approximation the problem is reduced to the solution of the kinetic equation (10.21). This integro-differential equation is similar to the Boltzmann equation and the integral term in its r.h.s. is frequently called the collision integral $I \, (n_{\mathbf{k}})$. In spite of the kinetic equation's complexity it does not describe completely the weak turbulence produced by internal waves because it does not consider dissipative phenomena and numerous factors causing the generation of internal waves (the atmospheric pressure fluctuations, the wind's influence, the unevenness of the bottom, currents, other waves, etc.). That is why this equation should be added with the sources (the summands characterizing the weak turbulence induction) and with the sinks as the characterisations of the dissipation.

Such an incompleteness as well as an extreme complexity of the kinetic equation obtained, the solution of which is no more simple than the solution of the initial hydrodynamics equations, makes the situation almost unsolvable. Nevertheless, from some additional physical conclusions the stationary solutions of the kinetic equation, defining the spectra of weak turbulence, may be obtained [290], [405], [410].

Consider the stationary solutions of Eq. (10.21) when $I(n_æ) = 0$; $I(n_k^\nu) = 0$. Eq. (10.21) has the following exact solution corresponding to the Rayleigh–Jeans equilibrium distribution

$$\left.\begin{array}{l} n_æ = C_1 \omega_æ^{-1}\,; \quad \varepsilon_æ = \text{const}\,; \\[2mm] n_k^\nu = C_2 \left(\omega_k^\nu\right)^{-1}\,; \quad \varepsilon_k^\nu = \text{const}\,, \end{array}\right\} \tag{10.23}$$

where C_1, C_2 are arbitrary positive constants.

But in the presence of sources (the disturbance) and sinks (the dissipation) such a distribution is impossible for the whole of the wave number domain $\mathbf{k}$. Assuming that sources and sinks are far apart with respect to their wave numbers (which is the meaning of the physical hypothesis) the weak turbulence may be similar to the strong turbulence. In particular, the cascade mechanism of transfer of the wave energy from lower to higher wave numbers (where the dissipation is considerable) may be valid. This mechanism was suggested for the first time by Kolmogorov and Obukhov for the description of strong turbulence. If this hypothesis is applicable to weak turbulence, then we may try to find the universal inertial spectrum, separating the sources and the sinks, by some considerations (concerning mainly dimensions theory) valid for the usual turbulence. The required (but not sufficient!) condition initially set by Zakharov [405], [410] is the locality of $I(n_æ)$, $I(n_k^\nu)$. According to this condition the interaction coefficients $V_{\mathbf{k}\mathbf{k}_1\mathbf{k}_2}^{\nu\nu_1\nu_2}$ are considerably reduced when k_1, $k_2 \ll k$, and k_1, $k_2 \gg k$. Then the term of the spectrum of the energy flux may be introduced, and the nonequilibrium stationary spectrum distributions may be tried. In experimental conditions one-dimensional energy spectra $E_{\text{æ}}$ and E_{k}^{ν} proportional to the 3D spectrum $\varepsilon_{\text{æ}}$ and the 2D spectrum $\varepsilon_{\text{k}}^{\nu}$ are usually measured:

$$E_{\text{æ}} \sim \text{æ}^2 \varepsilon_{\text{æ}} \,; \quad E_{\text{k}}^{\nu} \sim k \varepsilon_{\text{k}}^{\nu} \,. \tag{10.24}$$

In particular, from Eqs. (10.24) it follows that for the Rayleigh–Jeans distribution (10.23) $E_{\text{æ}} \sim \text{æ}^2$, $E_{\text{k}}^{\nu} \sim k$, which is out of agreement with the evidence.

We now try to obtain by the Kolmogorov hypothesis the nonequilibrium stationary spectra for small scale internal waves. To do this we introduce the spectral energy flux

$$\varepsilon_{\text{æ}} \sim \varepsilon_{\text{æ}} \text{æ}^3 \tau_{\text{æ}}^{-1} = \text{const} \,, \tag{10.25}$$

which is constant in the inertial interval studied. Here $\tau_{\text{æ}}$ is the characteristic time of nonlinear wave interaction in the vicinity of wave numbers æ. From Eqs. (10.17), (10.21) this value is estimated as

$$\tau_{\text{æ}}^{-1} \sim \frac{I\left(n_{\text{æ}}\right)}{n_{\text{æ}}} \sim |V_{\text{æææ}}|^2 \frac{n_{\text{æ}} \text{æ}^3}{\omega_{\text{æ}}} \sim n_{\text{æ}} \text{æ}^5 \,. \tag{10.26}$$

From (10.25), (10.26), taking into consideration Eqs. (10.24) and the conditions $\varepsilon_{\text{æ}} \sim n_{\text{æ}} \omega_{\text{æ}}$, $\omega_{\text{æ}} = N \cos\theta$, where θ is the vertical inclination of the wave vector æ, we have

$$\varepsilon_{\text{æ}} \sim \sqrt{\dot{\varepsilon}_{\text{æ}} N} \, \text{æ}^{-4} \,; \quad E_{\text{æ}} \sim \sqrt{\dot{\varepsilon}_{\text{æ}} N} \, \text{æ}^{-2} \,. \tag{10.27}$$

Here the question of the distribution of the spectrum $\varepsilon_{\text{æ}}$ versus the angle θ is still open. So in the general case

$$\varepsilon_{\text{æ}} = \sqrt{\dot{\varepsilon}_{\text{æ}} N} \, \text{æ}^{-4} F\left(\cos\theta\right) \,; \quad \cos\theta = \frac{\omega_{\text{æ}}}{N} \,. \tag{10.27a}$$

By analogy with the plasma physics in [290] the energy spectrum of weak turbulence was tried under the assumption that the flux Q of the z-component of momentum $\mathcal{P}_z = \int \text{æ}_z n_{\text{æ}} d^3 \text{æ}$ is constant over the wave number spectrum instead of the energy flux $\dot{\varepsilon}_{\text{æ}}$. And by analogy with Eqs. (10.27)

$$\varepsilon_{\text{æ}} = N \sqrt{\theta} \text{æ}^{-3/2} \, \Phi\left(\cos\theta\right) \,; \quad E_{\text{æ}} = N \sqrt{\theta} \text{æ}^{-5/2} \, \Phi_1\left(\cos\theta\right) \,. \tag{10.28}$$

The angular distribution, i.e., an anisotropy of the small scale internal waves spectrum $\varepsilon_{\ae}$, remains unknown. But the spectra (10.27), (10.28) meet the locality condition: for $\omega_{\ae} \to 0$ Φ grows slowly than ω^{-2} and Φ_1 grows slowly than $\omega^{-3/2}$.

The above estimations of dimensions may be hardly used for the spectrum of large scale turbulence, generated by internal waves with a mode structure, obtained from the stationary solution of the kinetic equation (10.21) $I\left(n_{\mathbf{k}}^{\nu}\right) = 0$. The interaction coefficients $V_{\mathbf{k}\mathbf{k}_1\mathbf{k}_2}^{\nu\nu_1\nu_2}$ and $\omega_{\mathbf{k}}^{\nu}$ have complicated dependences on the wave vectors determined by the stratification profile. So we consider the long wave case (in shallow water or in a thin pycnocline) when $\omega_{\mathbf{k}}^{\nu} \approx c_{\nu}|\mathbf{k}|$ and the integrals in Eq. (7.38) are independent of $\mathbf{k}_i$.

In this case, when assuming the energy flux $\dot{\varepsilon}_{\mathbf{k}}$ over the spectrum to be constant for each wave mode

$$\dot{\varepsilon}_{\mathbf{k}} \sim \varepsilon_{\mathbf{k}} k^2 \tau_{\mathbf{k}}^{-1} = \text{const}, \tag{10.29}$$

for $\tau_{\mathbf{k}}$ from Eqs. (10.17), (7.38) the following estimation yields

$$\tau_{\mathbf{k}}^{-1} \sim \frac{k^4 n_{\mathbf{k}}}{\rho_0 L} \sim \frac{\varepsilon_{\mathbf{k}} k^3}{N_0 L}. \tag{10.30}$$

Here L is the characteristic thickness of the stratified layer (e.g., the pycnocline thickness), N_0 is the characteristic value (e.g., the maximum) of the Brunt–Väisälä frequency. Substitution of Eq. (10.30) into Eq. (10.29) yields

$$\varepsilon_{\mathbf{k}} \sim \sqrt{\dot{\varepsilon}_{\mathbf{k}} N_0 L} k^{-5/2}; \quad E_{\mathbf{k}} \sim \sqrt{\dot{\varepsilon}_{\mathbf{k}} N_0 L} k^{-3/2}. \tag{10.31}$$

Let us discuss the question of the validity of the presented approach with respect to oceanic turbulence, taking into consideration that the weak turbulence concept is valid only when the ratio of the interaction time $\tau_{\mathbf{k}}$ to the wave period $T \sim \omega_{\mathbf{k}}^{-1}$ is much greater than unity ($\mu = (\omega_{\mathbf{k}}\tau_{\mathbf{k}})^{-1} \ll 1$).

For the small scale weak turbulence from Eqs. (10.25)–(10.27) yields

$$\mu \sim \ae \dot{\varepsilon}_{\ae}^{1/2} N^{-3/2}; \quad E_{\ae} \sim \mu N^2 \ae^{-3}. \tag{10.32}$$

From the expressions (10.32) it follows that the condition $\mu \ll 1$ is invalid for quite small scale waves when $æ \gg 1$. From the estimations of Section 2.1 it also follows that the dissipation is sufficient for the small scale waves with a wave length $\approx (1 \ldots 10)$ cm. Let us now estimate the scale value of $E_{æ_0}$ when $\mu \approx 10^{-1}$, $æ_0 \approx 0.1$ cm^{-1}, $N_0 \approx 10^{-2}$ s^{-1}. From Eqs. (10.32) results $E_{æ_0} \approx 1$ cm$^3 \times$ s^{-2}, which is in a good agreement with the evidence [135].

For the large scale turbulence when $\tau_\mathbf{k} \sim \left(\dot{\varepsilon}_\mathbf{k} = \frac{k}{N_0 L} \right)^{1/2}$, $\omega_\mathbf{k} \sim NkL$ yields

$$\mu \sim (\dot{\varepsilon}_\mathbf{k} N L)^{-1/2} k^{-3/2} \, ; \quad E_\mathbf{k} \sim \mu \, (NL)^2 k^{-1} \, . \tag{10.33}$$

From Eqs. (10.33) it follows that μ grows versus decreasing wavelength. If, e.g., $k_0 \approx 10^{-4}$ cm^{-1}, $\mu \approx 10^{-2}$, $(NL)^2 \approx 10^3$ cm$^2 \times$ s^{-2}, then $E_{k_0} \approx 10^5$ cm$^3 \times$ s^{-2}, which was also observed in the ocean [135].

As for the application of the concept of weak turbulence to long waves (when $\omega_\mathbf{k} \approx c_\nu k$) their self-modulation effects, omitted in this consideration (see Chapter 9), may be considerable. The question on the prevalence of either the self-modulating instability, or the resonant interactions is still far from being clear and needs more adequate experimental investigations.

The problem of the derivation of the universal energy spectra for the internal waves scales is under discussion as well. And applicability of the basic hypothesis on the inertial interval (i.e., on the difference of sources' and sinks' scales) for the real ocean is still vague, being unproved by the evidence (see Chapter 12).

IV

SOME INFORMATION ON INTERNAL WAVE OBSERVATIONS IN THE OCEAN

Chapter 11

ON STATISTICAL DESCRIPTION OF NATURAL INTERNAL WAVES DATA

In this chapter the methods of measurements of oceanic internal waves both in a fixed point (by an anchored buoy) and in moving coordinates (by a towed chain of thermistors) are described. In the final section the problem of the discernment of internal waves and turbulence is considered. Existing statistical methods of the description of internal waves' field as well as the methods of interpretation of evidence are valid only for small amplitude (linear and weakly nonlinear) waves. When natural internal waves have strong slopes and nonlinear effects take place, traditional spectral and correlation statistical methods are hardly applicable.

1. STATISTICAL DESCRIPTION OF LOW AMPLITUDE ENSEMBLE OF INTERNAL WAVES

Consider a statistical small amplitude ensemble well described by the linear theory. Assume that the characteristics of internal waves (e.g., the vertical displacement field $\zeta(\mathbf{x}, z, t)$) are statistically uniform with respect to the horizontal coordinates $\mathbf{x}$ and are stationary versus time t. Let us statistically describe such a wave ensemble [263].

A stationary and uniform field may be described by the Fourier–Stieltjes integral

$$\zeta(\mathbf{x}, z, t) = \int e^{i(\mathbf{kx} - \omega t)} \hat{\zeta}(k, z, \omega) \, dZ(\mathbf{k}, \omega), \qquad (11.1)$$

315

where dZ are components of the Fourier–Stieltjes field ζ, and $\hat{\zeta}$ is the amplitude function satisfying the basic boundary value problem of the theory of internal waves along the z-coordinate [see Eqs. (3.7), (3.8)]. This problem has the dispersion relations $\omega = \omega_n(k)$ $(n = 1, 2, \ldots)$ for each normal internal wave mode. Let us define inverse dispersion functions $k = k_n(\omega)$ and corresponding eigenfunctions $\zeta_n(z, \omega) = \zeta(k_n(\omega), z, \omega)$ of the boundary value problem (3.7), (3.8) orthonormalized by the condition $(\zeta_m, \zeta_n) = \delta_{nm}$ (the scalar product defined by Eq. (3.16)). For the dispersion relations $\omega = \omega_n(k)$ corresponds

$$dZ(\mathbf{k}, \omega) = \sum_n \delta(\omega - \omega_n(k)) \, d\omega \, dZ_n(\mathbf{k}), \qquad (11.2)$$

For the stationary and horizontally uniform random field (11.1) a discorrelation of the random spectral measures is required and sufficient:

$$\langle dZ_m(\mathbf{k}_1) \, dZ_n^*(\mathbf{k}_2) \rangle = \delta_{nm} \delta(\mathbf{k}_1 - \mathbf{k}_2) E_n(\mathbf{k}_1) \, d\mathbf{k}_1 \, d\mathbf{k}_2, \qquad (11.3)$$

where $E_n(\mathbf{k})$ is the spatial spectral density. Obtain the correlation function for two points of the field (11.1)

$$B(\mathbf{r}, z_1, z_2, \tau) = \langle \zeta(\mathbf{x} + \mathbf{r}, z_1, t + \tau) \, \zeta(\mathbf{x}, z_2, t) \rangle.$$

Taking into account (11.1)–(11.3) yields

$$B(\mathbf{r}, z_1, z_2, \tau) = \int e^{i(\mathbf{kr} - \omega\tau)} \hat{\zeta}(k, z_1, \omega) \, \hat{\zeta}(k, z_2, \omega) \, E(\mathbf{k}, \omega) \, d\mathbf{k} \, d\omega, \qquad (11.4)$$

$$E(\mathbf{k}, \omega) = \sum_n \delta(\omega - \omega_n(k)) E_n(\mathbf{k}). \qquad (11.5)$$

Here $E(\mathbf{k}, \omega)$ is the space time spectrum of the displacement of levels. The correlation function $\langle \varphi(M_1) \psi^*(M_2) \rangle$ of two linear ζ-field functionals is expressed by Eqs. (11.4), (11.5) when replacing $\zeta(k, z_1, \omega) \zeta(k, z_2, \omega)$ by $\varphi(k, z_1, \omega) \psi^*(k, z_2, \omega)$. Using this property and Eqs. (3.2) we may express spectral densities of the fields u, v, w, ρ at a depth z and of the internal waves' energy per unit horizontal plane through $E(\mathbf{k}, \omega)$ introducing the corresponding multipliers

$$\frac{k_1^2\omega^2 + k_2^2 f^2}{k^4} \left(\frac{\partial\hat{\zeta}}{\partial z}\right)^2 ; \quad \frac{k_2^2\omega^2 + k_1^2 f^2}{k^4} \left(\frac{\partial\hat{\zeta}}{\partial z}\right)^2 ; \quad \omega^2\hat{\zeta}^2 ;$$

$$\left(\frac{\omega^2 - f^2}{k^2}\rho_0\frac{\partial\hat{\zeta}}{\partial z}\right)^2 ; \quad \frac{\omega^2}{\omega^2 - f^2}\left[g\rho_0\hat{\zeta}^2\Big|_{z=0} + \int\limits_0^{-H}(N^2 - f^2)\hat{\zeta}^2\rho_0\,dz\right],$$

$$(11.6)$$

and for the spectral density of the fields u and v^* the multiplier

$$[k_1 k_2 (\omega^2 - f^2) - ik^2\omega f]\,k^{-4}\left(\frac{\partial\hat{\zeta}}{\partial z}\right)^2 .$$

Consider a particular case of the isotropic internal wave field when all E_n depend only on k and the correlation function (11.4) has the form

$$B(r, z_1, z_2, \tau) = \int\limits_f^{N_m} e^{-i\omega\tau}\sum_n J_0[rk_n(\omega)]\,\hat{\zeta}_n(z_1, \omega)\,\hat{\zeta}_n(z_2, \omega)\,G_n(\omega)\,d\omega ;$$

$$G_n(\omega) = 2\pi k_n(\omega)\,k_n'(\omega)\,E_n[k_n(\omega)],$$

$$(11.7)$$

where $N_m = \max N(z)$. The inversion of this formula

$$G_n(\omega) = \zeta_n(z_1, \omega)\,\zeta_n(z_2, \omega)\,\Phi(0, z_1, z_2, \omega),$$

$$\Phi(r, z_1, z_2, \omega) = \frac{1}{2\pi}\int e^{i\omega\tau}B(r, z_1, z_2, \tau)\,d\tau .$$

$$(11.8)$$

Thus the statistical structure of the isotropic internal waves field, characterized by spectra $E_n(k)$, is completely defined by vertical field parameters in one point [for a known solution of the boundary value problem (3.7), (3.8)].

Consider for example a case $N^2 = \text{const}$, when

$$\hat{\zeta}_n(z) = c_n \sin\left(\frac{\pi n z}{H}\right) ; \quad k_n(\omega) = \frac{\pi n}{\beta(\omega)H} .$$

Using this solution and the integral Bessel function $J_0(x)$, Eq. (11.8) yields [see [263]]

$$\Phi(r, z_1, z_2, \omega)$$

$$= \frac{1}{8\pi} \int_0^{2\pi} \left[\varphi\left(\alpha H + \frac{z_1 + z_2}{2}, \ \omega\right) + \varphi\left(\alpha H - \frac{z_1 + z_2}{2}, \ \omega\right) \right.$$

$$\left. - \varphi\left(\alpha H + \tfrac{z_1 - z_2}{2}, \ \omega\right) - \varphi\left(\alpha H - \tfrac{z_1 - z_2}{2}, \ \omega\right) \right] d\theta, \quad (11.9)$$

where

$$\alpha = \frac{r \cos\theta}{2H\beta(\omega)}; \quad \beta^2 = \frac{N^2 - \omega^2}{\omega^2 - f^2}$$

and $\varphi(z, \omega) = \Phi(0, z_1, z_2, \omega)$ is the frequency spectrum at the certain depth. From Eq. (11.9) it follows that asynchronous depth measurements will be enough to define completely the ζ-field statistical structure.

A number of useful relations between different internal waves' field components for more complicated $N(z)$ profile are contained in [92]. Here one can also find formulae for other statistical characteristics, such as cross-spectra, vertical and horizontal coherence, and the semi-empirical model of the mean 'climate' spectrum of natural internal waves. Details of this model are discussed in Section 11.3.

If the vertical scale of internal waves is small and the Brunt–Väisälä frequency varies slowly versus z, then the WKB method may be used for the analysis of internal waves. Such an analysis and corresponding relations are given in the work by Fofonoff [80].

The spatial spectrum $E_n(\mathbf{k})$ defined by Eq. (11.5) is symmetric, i.e., $E_n(\mathbf{k}) = E_n(-\mathbf{k})$. It is convenient to introduce a so called asymmetric spectrum, which allows us to define the dominant direction of the propagation of waves (see, e.g., [119]). Indeed, the internal waves' dispersion relations are $\omega = \pm\omega_n(k)$ and $\omega_n(k) = \omega_n(-k)$. Then the spatial spectrum $\Phi_n(\mathbf{k}, \omega_n(k))$ may be expressed through the general space time spectrum $E(\mathbf{k}, \omega)$ [by analogy with Eq. (11.5)]

$$E(\mathbf{k}, \omega) = \sum_{n=-\infty}^{\infty} \Phi_n(\mathbf{k}, \omega_n(k)) \left[\delta(\omega + \omega_n(k)) + \delta(\omega - \omega_n(k))\right]. \quad (11.10)$$

Then $\Phi_n(\mathbf{k}, \omega_n(k))$ and $\Phi_n(\mathbf{k}, -\omega_n(k))$ denote the spectral density of waves propagating in the positive and negative $\mathbf{k}$ directions, respectively, because owing to the statistical stationarity and uniformity of the random field $\zeta(\mathbf{x}, z, t)$ the following equalities are valid:

$$\Phi_n(\mathbf{k}, \omega_n) = \Phi_n(-\mathbf{k}, -\omega_n), \quad \Phi_n(\mathbf{k}, -\omega_n) = \Phi_n(-\mathbf{k}, \omega_n). \quad (11.11)$$

Let us introduce new variables

$$\left. \begin{array}{l} E_n(\mathbf{k}, \omega_n) = \frac{1}{2}\left[\Phi_n(\mathbf{k}, \omega_n) + \Phi_n(-\mathbf{k}, \omega_n)\right], \\[2mm] c_n(\mathbf{k}, \omega_n) = \frac{1}{2}\left[\Phi_n(\mathbf{k}, \omega_n) - \Phi_n(-\mathbf{k}, \omega_n)\right], \end{array} \right\} \quad (11.12)$$

where $E_n(\mathbf{k}, \omega_n)$ is the symmetric part of $\Phi_n(\mathbf{k}, \omega_n)$ equal to the spatial spectral density $E_n(\mathbf{k})$ defined by Eq. (11.5), and $c_n(\mathbf{k}, \omega_n)$ is the asymmetric part determining the direction of propagation of waves. Spectra (11.5) are used, as a rule, for processing evidence because of limited data on the spatial structure of the internal waves' field. The spectra $\Phi_n(k, \omega_n)$ are more informative, but require more empirical information.

2. ON INTERPRETATION OF INTERNAL WAVE MEASUREMENTS MADE BY MOVING SENSORS

A study of the spatial structure of internal waves is quite difficult for natural conditions as it needs a dense grid of buoys, which can be hardly supported in the open ocean. The most practical way for a spatial study is to use a chain of towed sensors. As a rule, chains of thermistors towed by a research vessel are widely used for such a purpose (see, e.g., [159], [256]). In this connection the development of methods of measurement and data processing are necessary. The main difficulty is that the sensor's velocity is of the order of the measured phase velocity of the internal waves. And the statistical characteristics obtained are distorted by the 'Doppler effect'. The problem of the derivation of real statistical parameters of a field from the natural data distorted by a sensor's motion was solved in [197].

As a result of the measurements of the towed sensors we have random field characteristics $f(\mathbf{x}, t, z_0)$ at a fixed horizon z_0 in one direction.

Consider the methods of improvement for the Doppler distortion of one-dimensional statistical characteristics of an initial scalar field (e.g., the temperature T or the level displacement ζ).

If the research vessel moves at a velocity V in a given direction, then the equation for the resulting function:

$$f(x, t, z_0) = f(x, t(x), z_0) = f(x_0 + Vt, t, z_0);$$ (11.13)

$$x = x_0 + Vt; \quad t = \frac{x - x_0}{V},$$

where the x-direction is collinear with the vessel's direction. The one-dimensional correlation function for a uniform and stationary random field:

$$B(r, z_0) = \lim_{L \to \infty} \frac{1}{2L} \int_{-L}^{L} f(x, t_0, z_0) f(x + r, t_0, z_0) \, dx,$$ (11.14)

where L is the spatial scale, r is the shift of the correlation function collinear with the ship's direction. Then the correlation function obtained for the moving sensor's (distorted) data:

$$\tilde{B}(r, V, z_0) = \lim_{L \to \infty} \frac{1}{2L} \int_{-L}^{L} \left\{ f\left(x, \frac{x - x_0}{V} + t_0, z_0\right) \right.$$

$$\left. \times f\left(x + r, \frac{x + r - x_0}{V} + t_0, z_0\right) \right\} dx.$$
(11.15)

This one-dimensional function using Eqs. (11.4), (11.5) for $z_1 = z_2 = z_0$ and Eq. (11.13) may be rewritten:

$$\tilde{B}(r, V, z_0) = \sum_{n=1}^{\infty} \int_{-\infty}^{\infty} \exp\left\{ i\left(kr - \frac{\omega_n r}{V}\right) \hat{f}_n^2(k, z_0, \omega_n) \right\} E_n(k) \, dk.$$

(11.16)

where k is the wave number in the x-direction, $\hat{f}_n$ is the vertical eigenfunction of the linear internal waves' operator corresponding to the field

f. From Eqs. (11.15), (11.16) it follows that $\lim\limits_{V\to\infty} \tilde{B}\,(r,V,z_0) = \tilde{B}\,(r,z_0)$. Correlation functions of physical fields are smooth continuous functions of V, so $\tilde{B}$ may be expanded in a convergent power series of $\nu = 1/V$:

$$\tilde{B}\,(r,V,z_0) = B\,(r,z_0) + \nu \left.\frac{\partial\tilde{B}}{\partial\nu}\right|_{\nu=0} + \frac{\nu^2}{2}\left.\frac{\partial^2\tilde{B}}{\partial\nu^2}\right|_{\nu=0} + \ldots . \qquad (11.17)$$

If the direction studied is tested several times at different velocities V_i, then the series' coefficients may be expressed through the corresponding $\tilde{B}_i\,(r,V_i,z_0)$ and $B\,(r,z_0)$ may be obtained from Eq. (11.17). The formulae for the linear $(i = 2)$ and the quadratic $(i = 3)$ approximations

$$B\,(r,z_0) = \frac{1}{2}\left(\tilde{B}_1 + \tilde{B}_2\right) + \frac{1}{2}\frac{V_1+V_2}{V_1-V_2}\left(\tilde{B}_1 - \tilde{B}_2\right) ; \qquad (i=2); \quad (11.18)$$

$$B\,(r,z_0) = D^{-1}\left[\tilde{B}_1 V_1^2\,(V_2-V_3) - \tilde{B}_2 V_2^2\,(V_1-V_3)\right.$$
$$\left. + \tilde{B}_3 V_3^2\,(V_1-V_2)\right] ; \qquad (i=3), \qquad (11.19)$$

where $D = V_1^2\,(V_2-V_3) + V_2^2\,(V_1-V_3) + V_3^2\,(V_1-V_2)$. Higher approximations may be provided by repetition of the measurements.

A convenient parameter of the accuracy for the above method may be chosen from Eq. (11.16). Evaluating the frequency $\omega_0 = \max \omega_n < N_m$ (N_m is the maximum Brunt–Väisälä frequency) we choose the spatial scale X for several V_j according to the condition $\omega_0 X/V_j = 1$. Then setting $r \ll X$ and expanding Eq. (11.16) in the Taylor series of $\beta = \omega_0 r/V \ll 1$ yield

$$\tilde{B}\,(r,V,z_0) = \sum_{j=0} \tilde{B}_j\,\frac{\beta^j}{j!} ; \quad \tilde{B}j = (-i)^j \sum_n \int\limits_{-\infty}^{\infty} e^{ikr}\left(\frac{\omega_n}{\omega_0}\right)^j \hat{f}^2\,E_n\,(k)\,dk .$$

$$(11.20)$$

For $\beta \ll 1$ (as $\omega_n/\omega_0 < 1$) the series (11.20) converges quickly and the estimation

$$|Bj| \leq \sum_n \int\limits_{-\infty}^{\infty} E_n\,(k)\,\hat{f}^2\,dk = \sum_n \sigma_n^2 = \sigma^2$$

is valid. Here σ^2 is the complete dispersion of a random process which the Doppler effect does not distort. It may be calculated from data obtained by a sensor moving at any velocity V_j. Then for the residual of the approximation $B_* = \tilde{B} - \sum\limits_{j=0}^{m} \tilde{B}_j \frac{\beta^j}{j!}$ yields

$$|B_*| \le \sigma^2 \sum\limits_{j=m+1}^{\infty} \frac{\beta^j}{j!} = \sigma^2 \left[e^\beta - \sum\limits_{j=0}^{m} \frac{\beta^j}{j!} \right]. \qquad (11.21)$$

Basing on the inequality (11.21) we may find the number of expansion terms in (11.20) or, what is the same, the number of velocities V_j, satisfying the given accuracy of approximation, ω_0, V, and r.

Consider illustrative examples.

Example 1. Set $f(x,t) = A \cos[k(x-ct)]$, where c is the wave velocity, A is the random amplitude. The correlation function of such random field is $B(r) = \sigma^2 \cos kr$, $\sigma^2 = \langle A^2 \rangle / 2$. In the case of measurements of the characteristics of this field with a moving sensor we have $t = (x - x_0)/V$, and

$$f(x, V) = A \cos\{k[x - (x - x_0)c/V]\},$$

with the correlation function

$$\tilde{B}(r, V) = \sigma^2 \cos\{kr(1 - c/V)\}. \qquad (11.22)$$

for $V \to \infty$ or $c/V \ll 1$ distortions may be neglected and $B(r)$ yields. As the velocity c is unknown *a priori*, then to obtain $B(r)$ we should use the above expansion in powers of $\nu = 1/V$.

In Fig. 11.1 a the correlation functions, normalized by dispersions, calculated in accordance with Eq. (11.22) for different V_j are shown (the function $B(r)$ for $V \to \infty$ is also given for comparison). The value of c for all calculations was equal to the characteristic phase velocity of natural internal waves $c \approx 20$ cm/s. The correlation functions for different V have considerable discrepancies. The correlation functions 'improved' by Eq. (11.18) (the linear approximation) and by Eq. (11.19) (the quadratic approximation) are presented in Fig. 11.1 b. The linear approximation provides a quite good agreement with the correlation functions (curve 1). And the quadratic approximation (curve 2) yields an almost complete coincidence with $V \to \infty$.

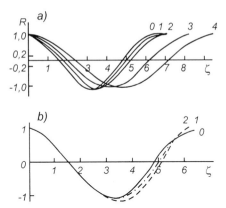

Figure 11.1. The calculation example for the initial normalized correlation function (a) and 'improved' normalized correlation functions (b) for different V and $c = 0.2$ m/s: a: calculation by Eq. (11.22): $0 - V = \infty$; $1 - V = 4$ m/s; $2 - V = 2$ m/s; $3 - V = 0.67$ m/s; $4 - V = 0.4$ m/s. b: $0 - V = \infty$; $1 -$ the linear approximation [Eq. (11.18)] ($V_1 = 2$ m/s; $V_2 = 0.4$ m/s); $2 -$ the quadratic approximation [Eq. (11.19)] ($V_1 = 4$ m/s; $V_2 = 2$ m/s; $V_3 = 0.67$ m/s).

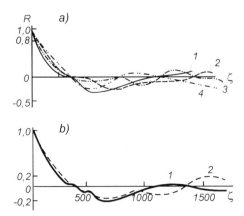

Figure 11.2. Normalized experimental correlation functions (a) and comparison of calculated functions [Eq. (11.18)] with experimental ones (b): a: $1 - V = 5$ kn; $2 - V = -6.1$ kn; $3 - V = 9.5$ kn; $4 - V = 10.6$ kn; b: $1 - V_1 = 5$ kn; $V_2 = -10.6$ kn; $2 - V_1 = -6.1$ kn; $V_2 = 9$ kn.

Example 2. The temperature field in the upper 200 meter layer of the equatorial Atlantic Ocean (5-th cruise of 'Academician Kurchatov') was measured at different backwards and forwards ship's velocities. In Fig. 11.2 a the calculated correlation functions of temperature variations at a depth of 90 m for four ship traverses are presented. All the functions have a serious discrepancy, especially when the sensors' velocities were considerably different. The improved correlation functions were then calculated on the basis of the linear approximation (11.18). The new functions are very close Fig. 11.2 b. Some discrepancy remains for large V, which is in line with the theory. Using Eq. (11.21) it is possible to estimate the value of $\tilde{B}$ satisfying the given accuracy of approximation. When requiring $|B_*/\tilde{B}| < 15\%$ for $\omega_0 = 0.2 \times 10^2$ s^{-1} and $V = 8$ knots (mean experimental values) yields from Eq. (11.21) $r \leq 1000$ m. This relates to r values corresponding to considerable discrepancy in the correlation functions (see Fig. 11.2 b).

Eqs. (11.18), (11.19) may be obtained for other statistical field characteristics as well (spectral, structural functions, etc.). And while expanding distorted statistical field characteristics in series of ν^{-1} (i.e., extrapolating with respect to $V \to \infty$) pure time statistical characteristics may be obtained (e.g., frequency spectra).

3.   ON DISCERNMENT OF INTERNAL WAVES AND TURBULENCE

Fluctuations of hydrological fields determining the spectral densities $S(\omega)$ and $E(k)$ may be both a result of the passage of an ensemble of internal waves or the transfer of turbulent vortices at high Reynolds numbers. So for the interpretation of evidence of space and time fluctuations in the ocean thermocline it is useful to know if we deal with internal gravity waves or with turbulence (horizontal two- or three-dimensional). Though this question is far from being new for stratified media investigation, it still has no simple solution.

When internal waves are weakly nonlinear they are governed by a definite dispersion relation which is almost independent of the waves' amplitude and close to the linear theory. So comparing the linear to the empirical dispersion relation we can discern internal waves and turbulence. An example of such a procedure is contained in [256]. Using the relation between the frequency and the spatial spectra [similar to the

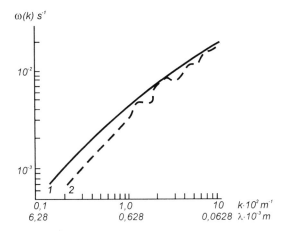

Figure 11.3. The dispersion relation for the lowest internal waves mode calculated for the three-layer model (1) and for experimental data (2) [256].

second formula of Eqs. (11.5)] the empirical dispersion relation $\omega_1(k)$ for spectra of natural temperature fluctuations $S(\omega)$ and $E(k)$ was obtained. Fig. 11.3 shows that it agrees well with the three-layer model, which shows the predominance of internal waves.

If we have temperature and velocity fluctuations caused by low amplitude internal waves, then some linear relations between internal waves characteristics may be obtained. For example, the potential energy's spectral density $S_\xi = S_T (d\overline{T}/dz)^{-2}$ versus the kinetic energy of horizontal wave velocity components $S_E = 1/2 (\bar{u}^2 + \bar{v}^2)$ under the vertical WKB approximation [80] is:

$$\frac{2N^2 S_\xi}{S_E} = \frac{2N^2 S_T}{\left(\frac{d\overline{T}}{dz}\right)^2 S_E} = \varphi(\omega), \qquad (11.23)$$

where

$$\varphi(\omega) = N^2 \left[(N^2 - \omega^2) \frac{\omega^2 - f^2}{\omega^2 + f^2} \right]^{-1}. \qquad (11.24)$$

This relation may be verified by the evidence. In Fig. 11.4 [258] the values of $\varphi(\omega)$ at different horizons of the Timor Sea for corresponding $N^2(z)$ are presented. The experimental points obtained for $r(\omega) =$

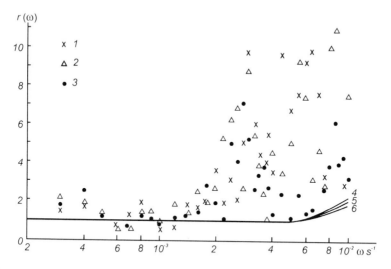

Figure 11.4. The dependences $r(\omega)$ on ω [258] $(1-3)$ and $\varphi(\omega)$ $(4-6)$: $1 - 225$ m; $2 - 125$ m; $3 - 175$ m; $4 - 225$ m; $5 - 125$ m; $6 - 175$ m.

$2N^2 S_T / \left(d\overline{T}/dz\right)^2 S_E$ at the same horizons are shown as well. Eq. (11.23) yields good agreement for a wide frequency range up to $\omega_s \approx 2 \times 10^{-3}$ s^{-1}. As this equation is valid for linear internal waves we may suggest that the fluctuation field is formed by internal waves in the low frequency range $(\omega \leq \omega_s)$. For $\omega > \omega_s$ the experimental points have a considerable discrepancy with Eq. (11.23). This may be caused both by high frequency nonlinear waves' interactions at the background of small scale turbulence as well as by distorting influence of the fine temperature structure on the temperature measurements at a fixed point. Indeed, according to [91] fine structural distortions play a decisive role at the frequency $\tau_\lambda^{-1} \sim 2\pi^{-1/2} N \left[\mathit{æ}(\omega)\right]^{-1}$, where $\mathit{æ}(\omega)$ is determined by the ratio of the spectral density of the temperature variations, caused by the fine structure, to the spectral density caused by the internal waves. According to the estimations [91], [272] $\mathit{æ} > 1$ and is of the order 1–10. Then estimations for τ_λ^{-1} show that the fine structure's influence will be sufficient for frequencies at least several times greater the local Brunt–Väisälä frequency, i.e., for $\omega \geq \omega_s$ where a considerable discrepancy was observed. An indirect proof of the fine structural distorting influence on relatively high frequency temperature fluctuations is the position of the experimental points relative to the dependence (11.23) (above the curve) at $\omega \geq \omega_s$, because the fine structure increases natural values of S_T. The examples also show the predominant role of internal waves in temperature and velocity fluctuations.

The most promising method of discerning of the turbulence and internal waves is the phase relation (the phase shift spectrum) between fluctuations of different velocity components and scalar fields, which are fixed for small amplitude internal waves and arbitrary for the turbulence.

One more useful statistical characteristic of the random wave field is the cross-spectrum $E_{\xi\eta}(\omega)$ (see [262]) of the variations $\xi(t)$ and $\eta(t)$ of two hydrodynamic parameters (the velocity, the temperature, the salinity, etc.), i.e., the Fourier transform of their cross-correlation function $B_{\xi\eta}(\tau) = \langle \xi(t)\,\eta(t+\tau) \rangle$. Here ξ and η may denote the same parameter at different points in the ocean or different parameters at the same or different points. The real part $C_{\xi\eta}(\omega)$ of the function $E_{\xi\eta}(\omega)$ (i.e., the Fourier transform of the even part $\frac{1}{2}[B_{\xi\eta}(\tau)+B_{\xi\eta}(-\tau)]$ of the cross-correlation function) is called the co-spectrum. And the imaginary part $Q_{\xi\eta}(\omega)$ (i.e., the Fourier transform of the uneven part $-\frac{1}{2}[B_{\xi\eta}(\tau)-B_{\xi\eta}(-\tau)]$) is called the quadrature spectrum of variations $\xi(t)$ and $\eta(t)$. When setting

$$E_{\xi\eta}(\omega)\,[E_{\xi\xi}(\omega)\,E_{\eta\eta}(\omega)]^{-1/2} = C_{0\xi\eta}(\omega)\,e^{i\Phi_{\xi\eta}(\omega)}, \qquad (11.25)$$

the modulus $C_{0\xi\eta}(\omega)$ of the r.h.s. is called the coherence and $\Phi_{\xi\eta}(\omega)$ is the spectrum of the phase shift between $\xi(t)$ and $\eta(t)$. For example, according to the theory of linear internal waves, oscillations of the field parameters u, v, ξ, p, ρ, T in the entire frequency range have a phase shift $\pi/2$ to w oscillations at the same fluid point, which is typical for internal waves only.

Even when internal waves are nonlinear their cross-spectra should be considerably different from those for turbulence. Turbulent co-spectra should be rather wide and more intensive than quadrature spectra, see, e.g., [288]. For internal waves both co- and quadrature spectra are narrow and have the same orders. In [288], this spectral property was used for the interpretation of laboratory data on the generation of turbulence by internal waves.

Chapter 12

BASIC EXPERIMENTAL FACTS OF INTERNAL WAVES' BEHAVIOUR IN THE OCEAN

In this chapter characteristic properties of natural internal waves and some problems of the interpretation of evidence are considered[1]. A brief list of basic properties of frequency and spatial spectra of internal waves measured in the ocean is given at the beginning. Then local properties of measured short period internal waves (in particular, wave packets) are analyzed. One of the important characteristics of an internal wave field is the probability partition function of wave components. Peculiarities of empirical partition functions are considered in Section 12.3. In Section 12.4 brief data from laboratory experiments on the study of internal waves are given. Hypothetical mechanisms of internal waves' breaking and dissipation, based on the evidence, are discussed in the final Section 12.5.

1. SPATIAL AND FREQUENCY SPECTRA OF INTERNAL WAVES

In this section a brief survey of spatial and frequency spectra of internal waves measured in different World Ocean regions is presented. We will consider long time (about a day) and long length (dozens of kilometers) averaged spectra.

Measurements of natural internal waves are rather difficult, but numerous data on hydrologic parameters of the scale of internal waves has been collected in recent years. Most of them concern temperature

329

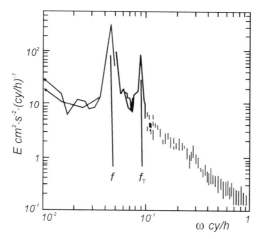

Figure 12.1. The frequency spectrum of the horizontal velocity variations measured in the Western part of the Sargasso Sea [106]: f — the local variational frequency, f_T — the half-day tidal frequency.

fluctuations in time at a fixed ocean point by anchored buoys. Such measurements allow us to obtain statistical characteristics of the frequency of internal waves. Temperature frequency spectra were measured, e.g., in the tropical Atlantics [165], [256], near Bermuda [398], between the Island of Madeira and the Azores [123], in the central part of the Bay of Biscay [122], in the Pacific Ocean near California [415], in the Norway [317], the Timor [258], the Japan [324], the Black [139], [143], [257] and the Caspian [170] Seas. As a rule, the measurements covered an upper ocean layer (up to 600 m depth) including the pycnocline layer. For internal waves induces by different sources the frequency range observed was $f \leq \omega \leq N$; the periods from minutes up to months were studied in [398].

Especially should we consider the long time experiment on investigating internal waves (IWEX) performed in 1975 in the Bermudas region [39], [40]. These measurements showed that although being different in detail, the frequency spectra of internal waves are quite common. First, they are almost continuous within the whole frequency range without pronounced maxima, excluding the inertial and the tidal frequencies. And the energy relating to the latter periods exceeds other periods' energy (see Fig. 12.1). Second, the spectral density level $S_T (\omega)$ decreases versus a growing frequency according to the power law $S_T \sim \omega^{-\alpha}$, where $1 \leq \alpha \leq 4$. The absence of a maximum frequency means that a statistical ensemble of internal waves with random phases and amplitudes is typical for the ocean. Some information about the behaviour of frequency spectra is contained in Fig. 12.2, where temperature oscillations in the thermocline core are shown.

Besides temperature measurements there is a number of velocity observations on the frequency range of internal waves [80], [258], [302], [377], [389]. The behaviour of the velocity spectra $S_u(\omega)$ is generally similar to that of temperature spectra: they also have no frequency maxima (excluding inertial and tidal maxima) and decrease steadily versus increasing ω as $S_u(\omega) \sim \omega^{-\alpha}$ ($1 \leq \alpha \leq 3$). But inertial and tidal bands are more pronounced compared with temperature bands. In Fig. 12.3 an example of the oscillations of temperature $S_T(\omega)$ and the horizontal kinetic energy $S_E(\omega) = \frac{1}{2}(S_u + S_v)$ (where S_u and S_v are the meridianal and the zonal velocity components) at different horizons at a fixed point of the Timor Sea [258] are given. As follows from the figure, the spectral functions' behaviour is quite similar.

Contrary to the case of time fluctuations few information about the spatial internal waves structure was obtained because of their complexity. Results of spatial investigations performed by towing thermistors are presented in [138], [157], [232], [256], [313], [315], [316], [378].

One-dimensional spatial spectra $E(k)$ as well as frequency spectra usually have no wave number maxima and decrease monotonically versus k as $E(k) \sim k^{-\beta}$ ($1 \leq \beta \leq 4$). In Fig. 12.4 the typical one-dimensional spatial spectra of thermocline perturbations measured by a towed chain of thermistors over several ship passages are shown [165].

The two-dimensional horizontal internal waves structure is quite different. The dependence of one-dimensional spectra on direction was not revealed during towing measurements by Charnock in the Straits of Gibraltar, which showed the horizontal field isotropy. Two-dimensional temperature spectra obtained by Voorhis and Perkins [378] and by Kitaigorodsky, Miropol'sky and Filushkin [165] are also quasi-isotropic. Later results, however, were quite contradictory. The data of Miropol'sky and Filushkin [256] in one of the Atlantic ocean test regions confirmed

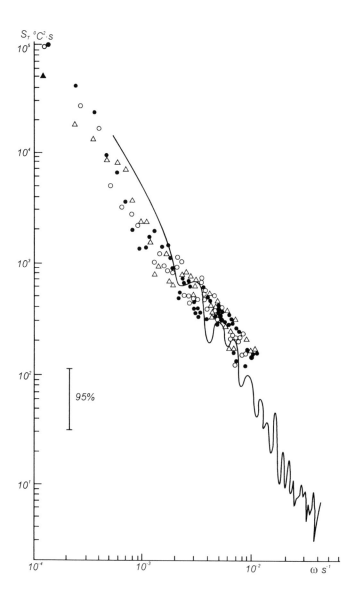

Figure 12.2. The time spectral function of temperature fluctuations measured by photothermographs (dots) and by a radio buoy (curve) at 70 m depth [165].

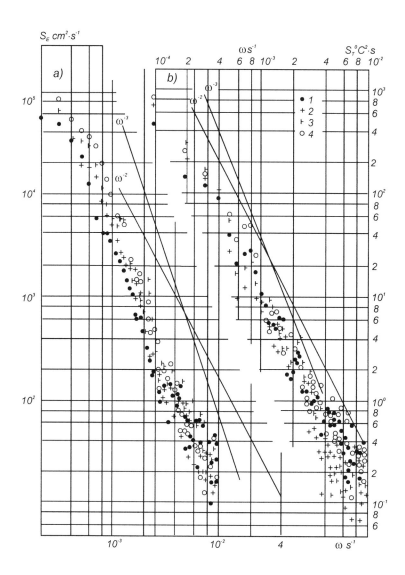

Figure 12.3. The spectral functions of the horizontal kinetic energy S_E (a) and the temperature S_T (b) pulsations [258]: 1 — 75 m; 2 — 125 m; 3 — 175 m; 4 — 225 m.

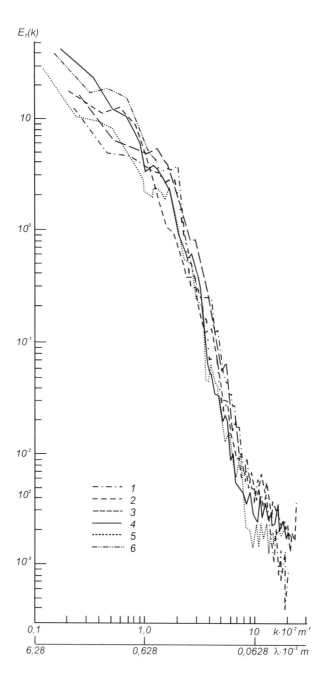

Figure 12.4. The spatial one-dimensional spectral functions of temperature variations in different directions [256].

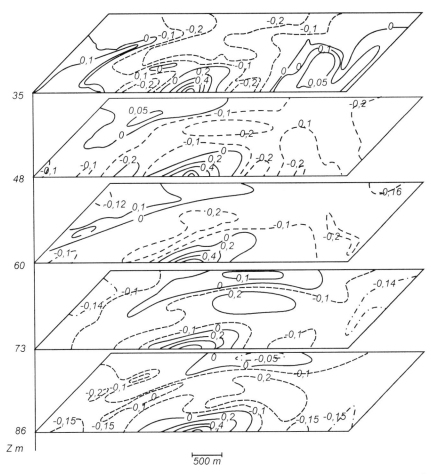

Figure 12.5. The two-dimensional spatial correlation functions of temperature fluctuations at different depths.

the internal waves' isotropy, but other test data were anisotropic (e.g., [141]).

Measurements by Sabinin [313] showed a pronounced directivity of high frequency internal waves, but in [157] short waves were isotropic and long waves had directivity. Garrett and Munk [92], while considering numerous data both for spatial cross-sections and for fixed points, concluded that the assumption on horizontal internal waves' isotropy is, in general, valid. But, evidently, both situations may occur in the ocean under certain factors.

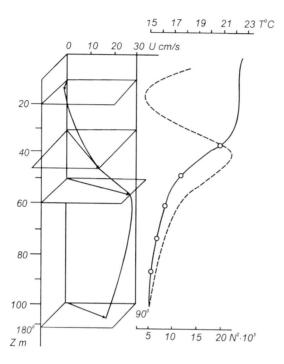

Figure 12.6. The depth distribution of the temperature T, the velocity U and the flow direction (sensors position is shown by dots).

Two-dimensional correlation functions and spectra are convenient for the study of spatial fields. A method of derivation of two-dimensional statistical characteristics from a finite number of one-dimensional random scalar field cross-sections is developed in [197], [198]. It consists in interpolation of one-dimensional correlation functions obtained for one-dimensional cuts along the measurement level with a further two-dimensional Fourier transform. Such correlation functions at five ocean horizons derived by four one-dimensional temperature measurements and reconstructed in Cartesian coordinates are shown in Fig. 12.5 [178]. The mean hydrological conditions for the region studied (the central part of the Atlantic Ocean) as well as the depth variations of the mean velocity are presented in Fig. 12.6. From Fig. 12.5 it follows that at small shifts (about 200–300 m) the internal waves' field is isotropic at all horizons excepting 35 m. But while decreasing the correlation function shift (for long lengths of waves) it becomes anisotropic. An interesting peculiarity of two-dimensional correlation functions is the variation of depth in the direction of the maximum correlation radius, following the mean current vector turn (Fig. 12.6). Such a behaviour may be explained by the linear the-

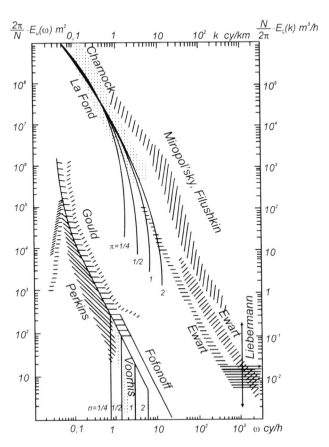

Figure 12.7. The combined spectra [92]: $E_u(\omega)$ is the frequency spectra of the longitudinal velocity component (anchored buoys); $E_\zeta(\omega)$ is the one-dimensional spatial spectra of vertical displacement of isotherms (towed thermistors). The curves correspond to Eq. (12.1).

ory of internal waves propagating at a shear flow.

The most complete list of natural data on spectra of internal waves as well as their vertical and horizontal coherence is presented by Garrett and Munk ([92], Fig. 12.7). On the basis of these data they suggested the simplified model for the spectral energy density $E(\mathbf{k}, \omega)$ which considers internal waves to be multi-mode (because a one-mode model can not describe the observed decay of coherence at vertical distances of about 10–100 m) and replacing a discrete mode manifold $\omega_n(k)$ by a continuous set $f \leq \omega \leq \omega_1(k)$, where $\omega_1(k)$ relates to the highest mode and f is the inertial frequency (the Coriolis parameter).

Remembering that resonant triple-wave interactions provide waves with isotropy, we consider the spectrum to be isotropic:

$$E(\mathbf{k}, \omega) = (2\pi)^{-1} E(k, \omega).$$

The vertical coherence calculation shows that the band width for internal wave numbers k with a fixed frequency ω is a function of the frequency $\mu(\omega)$. Garrett and Munk allowed the spectrum self-similarity $E(k, \omega) \sim \mu^{-1} A(k/\mu) E(\omega)$ and set $A(\chi)$ as an elementary function equal to unity for $0 \leq \chi \leq 1$ and to zero for $\chi > 1$ (though it should not be constant for $0 \leq \chi \leq 1$ and must go to zero only for $k > k_1(\omega) > \mu(\omega)$, where $k_1(\omega)$ relates to the lowest mode). For the frequency multiplier they chose the power law $E(\omega) \sim \omega^{-p+2s}(\omega^2 - f^2)^{-s}$ where it should be $0 < s < 1$ for the energy integral's convergence; the value $s = \frac{1}{2}$ was chosen, for example. A similar power formula was introduced for $\mu(\omega) \sim j\pi (\omega/f)^{r-1} (\omega^2 - f^2)^{1/2}$ (j denotes the equivalent modes number at the inertial frequency); the one-dimensional energy spectrum appears to be approximately proportional $E(k_1) \sim k_1^{-(p+r-1)/2}$. The exponents $p = 2$, $r = 1$ were chosen by Garrett and Munk to be in accordance with the data of natural spectra. Thus their dimensionless energy spectrum model has the form

$$\left. \begin{array}{l} E(k, \omega) = 2Ef\, j^{-1}\pi^{-2}\omega^{-1}(\omega^2 - f^2)^{-1}; \\[2mm] f \leq \omega \leq N; \quad 0 \leq k \leq j\pi \; (\omega^2 - f^2)^{1/2}, \end{array} \right\} \qquad (12.1)$$

where the frequencies are measured in N_m (c/s) and the wave numbers in M (c/cm) so that the total wave energy per unit ocean surface is equal to $\rho E/2\pi M^3 N_m$.

Garrett and Munk recommended $E = 2\pi \times 10^{-5}$ and $j = 20$. Then for $M = 1.22 \times 10^{-6}$ c/cm and $N_m = 0.83 \times 10^{-3}$ c/s the total energy is equal to 0.4 j/cm^2.

In their next work [94] Garrett and Munk modified their spectrum model by assuming that the main contribution to the vertical temperature spectra, measured by submerging sensors, is made by internal waves instead of the fine structure. They considered a three-dimensional spatial spectrum $E(k_1, k_2, \beta)$, where β is the vertical internal wave number. Proceeding, by analogy, from the self-similarity hypothesis they suggested the following spectrum formula:

$$E(k, \beta) = \frac{2\pi^{-1} f EN(z)(\beta/\beta^*) A(\beta/\beta^*)}{N^2(z)\alpha^2 + f^2\beta^2}; \quad 0 \leq \alpha \leq \beta \left[1 - \frac{f^2}{N^2}\right]^{1/2},$$

(12.2)

where β^* is the characteristic scale of the vertical wave number. Under a descent choice of $A(\beta/\beta^*)$ the spectrum (12.2) describes well the horizontal and vertical spectra of natural internal waves.

The spectra models (12.1) and (12.2) may be, seemingly, considered to be the proper approximations for the mean internal waves' spectra resulting from a great amount of evidence. But the hypothesis chosen for their derivation does not relate to any real physical mechanism of the formation of universal internal waves spectrum in the ocean. Some particular contradictions of this model were mentioned by Wunsch [397].

The above general description of the behaviour of space time hydrologic spectra in the scale of internal waves range does not reveal a number of fine structural peculiarities of the propagation of internal waves, on their interaction with the turbulence and the fine structure, on their mode content, etc.. These questions are considered below.

2.   LOCAL PROPERTIES OF SHORT PERIOD INTERNAL WAVES PROPAGATING IN THE UPPER THERMOCLINE

Recent field measurements have shown (see, e.g., [36]) that short period internal waves propagate mainly in the form of quasi-sine trains, and outer high frequency oscillations are, as a rule, less intense and probably have a turbulent nature. Such a picture of short period oscillations is contrary to the assumption of the stationarity of internal wave fields and homogeneity. Characteristic features of the field are omitted in the mean energy spectrum while internal waves propagate in the train form.

Koniaev and Sabinin [171], [172] suggested calculating the spectrum of oscillations in the framework of separate trains. Measurements by distributed sensors in different ocean regions and in the Black and Caspian

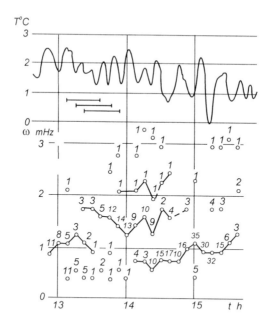

Figure 12.8. The data for internal wave train and wave periods recorded in the tropical Atlantics [92]. Periods were calculated for 30 minute records having 15 minute relative shift.

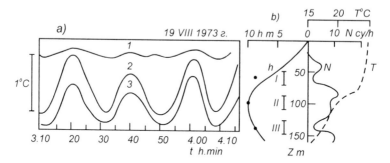

Figure 12.9. The example of the distributed sensors record (*a*); sensors positions (Roman figures), the temperature $T(z)$ and the Brunt–Väisälä frequency $N(z)$ profiles, wave amplitudes (dots) in comparison with the theory (solid line) for the first eigenfunction (*b*) [36]. 1 — 50–70 m; 2 — 90–110 m; 3 — 130–150 m.

Seas [36] have shown that oscillations within wave trains are frequently quasi-sine and are of two to dozens of periods. Train frequencies are close to the maximum Brunt–Väisälä frequency and correspond to 2–5

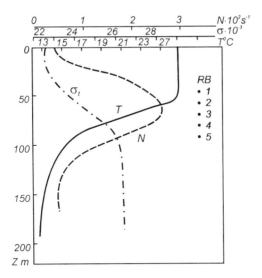

Figure 12.10. The depth distribution of the temperature $(T°C)$ the conventional density $(\sigma \times 10^{-3})$ and the Brunt–Väisälä frequency $(N \times 10^2 \text{ s}^{-1})$ for the Atlantic test region (dots are the positions of radio buoys).

minute periods in seas and 10–30 minute in the ocean [171], [172]. Train oscillations, as a rule, are spatially localized, their period records have several maxima and the number of maxima is variable in time. These characteristic features of short period internal wave trains are obvious in Fig. 12.8, where temperature variations and corresponding periods for the tropical Atlantics [172] are shown.

A detailed study by distributed sensors revealed that short period internal waves contain mainly one mode (see [36], [316]). In Fig. 12.9 the example of temperature fluctuations at three depths are presented. Obviously oscillations are synchronous versus depth, which means the first mode predominates. From theoretical estimations follows that the lowest normal mode will be pronounced in the sharp pycnocline. Coherence and phase shift functions for different horizons, measured by a radio buoy (the Brunt–Väisälä frequency profile and sensors position are given in Fig. 12.10) in the tropical Atlantics [165], are shown in Fig. 12.11. The coherence between the levels 2–3 and 3–4 up to frequencies $\omega \approx 2 \times 10^{-2} \text{ s}^{-1} \approx N$ is rather high and in the range $\omega \leq 2 \times 10^{-2}$ s^{-1} the phase shift is almost zero.

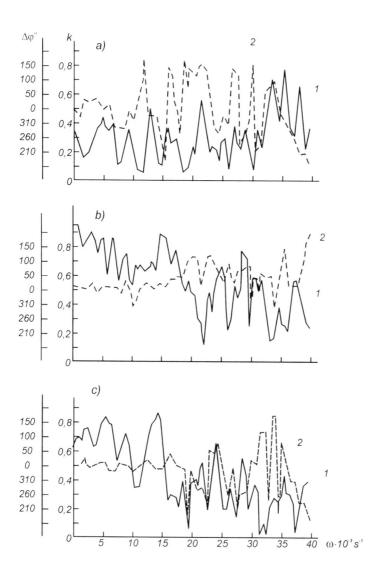

Figure 12.11. Coherence (1) and phase shift (2) variations versus the frequency (from the radio buoy record) [92]. Horizons: a — 2–5 m; b — 3–4 m; c — 2–3 m.

At the same time, less coherence between the levels 2–5 may be explained by the lower position of the fifth horizon, where the density gradient is considerably less and the contribution of lower modes in the total wave field is sufficient. The same conclusion follows from low coherence between the horizons 2–3 and 3–4 at frequencies $\omega > 3 \times 10^{-2}$ s^{-1} and simultaneous coherence increase for 2–5 horizons. It proves that first-mode oscillations are not dominant below the thermocline (the phase shift is non-zero here).

Thus, the mutual analysis of temperature fluctuations show that the first (the lowest) mode may be predominant in a relatively narrow frequency range. And the multi-mode description [92] is not common for the internal waves field.

3.   DISTRIBUTION FUNCTION OF HYDRODYNAMIC PARAMETERS IN INTERNAL WAVE FIELDS

The considerations above dealt with correlation and spectral functions of oscillations of the hydrological field in the scale of internal waves. And the question on empirical laws for the probability distribution of wave parameters was discounted, although their role (even for one dimension) is quite important. There are few works [45], [96], [249] concerning empirical distribution laws for the variations of temperature and the level displacement ζ.

The results of processing evidence have shown that the displacement probability distribution is considerably removed from the normal law. In Fig. 12.12 the histograms of the probability distribution density for the temperature $P(T)$ and the isotherms' displacement $P(\zeta)$ obtained at two stations in the Atlantic Ocean are shown [249].

If the oscillations of the temperature T and displacement ζ observed have the forms of internal wave, and nonlinear effects are negligible, then the probability distribution of the displacement should be normal. The latter statement is based on the fact of ocean waves consisting of a wave component manifold generated in different ocean regions and run to the same point. If the wave generation zone is of moderate size

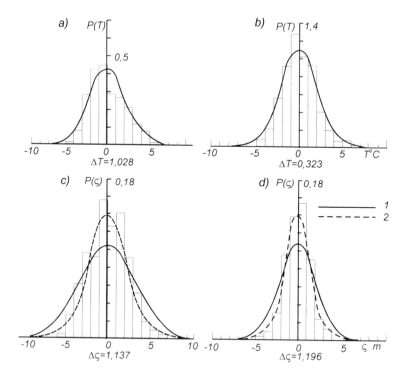

Figure 12.12. The histograms of the probability distribution density [249]. Depth: a — 88 m; b — 108 m. Temperature: c — 18°C; d — 16°C (1 — the normal distribution; 2 — the Gramat–Charlier distribution).

relative to a distance at which wave displacements are correlated, then for the linear approximation (the interaction of waves is neglected) the wave displacement at a given point may be considered as a sum of numerous independent amplitudes with random phases. Then, in general, a quite simple form of the central limit theorem for waves displacements (see, e.g., [218]) is valid, and the probability distribution for ζ will be normal. In the case of the absence of non-adiabatic effects the depth of the isotherm may be identified with the water level, and thus $P(\zeta)$ will have the normal distribution as well. The same conclusion is valid for the temperature probability density at the fixed horizon $P(T)$ because in linear internal waves $T(t) = \zeta(t)\overline{T}/dz$ ($d\overline{T}/dz$ is the mean temperature gradient). In this connection empirical distribution densities for $P(\zeta)$ and $P(T)$ should be compared with the normal distribution law. From the figures it follows that all of the e distribution laws have considerable a discrepancy with the normal distribution, which is also confirmed by χ^2 test and non-zero values of the asymmetry s and the excess B (see all the above works).

The hypothesis of the influence of weak nonlinearity was suggested in [249]. It was based on the results by Longuet–Higgins [218] which showed that taking quadric and higher interaction orders of the statistical variable components into consideration leads to non-zero values of the asymmetry, of the excess, etc., i.e., to the discrepancy from the Gaussian law. The theory developed was applied in [218] to the calculation of the statistical distribution of the free surface displacement. In [29] a similar theory was used for the analysis of the statistical distribution of the internal waves. In particular, it was shown that by analogy with surface waves' the internal waves' displacement, when considering nonlinearity, is described by the Gramat–Charlier series in the Edgeworth form:

$$P\left(\zeta\right) = (2\pi\sigma^2)^{-1/2} e^{-f^2/2} \left[1 + \frac{1}{6}sH_3 + \left(\frac{1}{24}BH_4 + \frac{1}{72}s^2H_6\right) + \dots\right],$$

$$(12.3)$$

where

$$H_3 = f^3 - 3f \; ; \quad H_4 = f^4 - 6f^2 - 3 \; ; \quad H_6 = f^6 - 15f^4 + 45f^2 - 15$$

are the Hermite polynomials and $f = \zeta/\sigma$. For linear waves $s = B = 0$ and Eq. (12.3) relates to the normal law. In Fig. 12.12 the distribution laws (12.3) with empirical s and B values are shown by the dashed line. Eq. (12.3) agrees better with the empirical distribution than with the Gaussian curve. Thus the empirical laws' discrepancy are, probably, caused by the nonlinearity of the internal waves observed, which was indirectly proved by the manifestation of weak nonlinearity in the slopes of waves.

The first-moment distribution formulae were derived in [29] from the theory of internal waves. In particular, it was shown that in the case of the lowest mode's predominance the asymmetry coefficient is positive, which agrees well with the evidence.

4.   INTERNAL WAVES LABORATORY EXPERIMENTS

Along with natural observations there are numerous laboratory experimental data on the investigation of internal waves (see the review by Thorpe [356]).

The modelling of processes in the upper ocean and at its surface needs special tanks with controlled stratification. The dynamics of stratified fluid have long been studied in relatively small and shallow water tanks with salt stratification quite different from that characteristic of the real ocean. In 1943 Hertler modelled an internal wave in a vessel with a stratified fluid and demonstrated a number of peculiarities of the propagation of the internal wave. Information about rather large setups with salt [47], [73] and temperature fluid stratification has appeared recently [27].

Lofquist and Purtell [214], Mowbray and Rarity [267], Sutherland et al. [344], Makarov and Chashechkin [346], Merzkirch and Peters [240] studied short small amplitude internal waves in a continuously stratified fluid by the schlieren method. Maas et al. used the dye technique for visualization of monochromatic internal wave structure in an exponentially stratified fluid [223]. They concluded that the behaviour of internal waves is well described by the linear theory in the ray approximation. The role of viscosity in determining the evolution of the waves was investigated in [344] along with simultaneous comparison of their results with the theoretical ones of Hurley and Keady [134]. A fluorescent dye technique was used by Chomaz et al. [52] to distinguish lee and random waves emitted by the turbulent wake of a sphere in a fluid with linear Brunt–Väisälä frequency gradient. Spedding et al. [337] analysed turbulent structures in a wake past a towed sphere with a digitized particle image velocimetry technique.

Different mechanisms of the generation of internal waves were studied as well. Orlansky [280] studied internal waves generated by a wavemaker in a volume with a stratified fluid. His experiments proved that internal waves may lose their stability under a strong enough influence, which leads to hydrostatically unstable density inversions and, as a result, to a convective mixture. In [163] the generation of internal waves by a pressure variation at the free surface of the stratified fluid was studied. The generation of internal waves by a region of mixed fluid past a moving body was tested by Wu [394]. Similar experiments were carried out by Pao [288] demonstrated the radiation of internal waves by a turbulent jet past a moving cylinder.

Many laboratory experiments have dealt with internal waves in shear flows [176], [321], [354], [355], [358], [360]. Their results showed that internal waves' instability appears at Richardson numbers slightly lower than the theoretical value $Ri = \frac{1}{4}$. And unstable internal waves grow more quickly than was predicted by the Miles–Howard linear theory. Interaction between stratification and turbulence was examined by E and Hopfinger [69], Fernando with co-authors [76], [77], [297]. The influence of rotation on the propagation of nonlinear internal waves was studied by Renouard and Germain [311]. Interaction of internal waves past a thin barrier and a shear flow was investigated by Sutherland and Linden [345]. The tilt of the phase lines of propagating waves lies within a narrow range. Assuming the waves are spanwise uniform, their amplitude in space and time was measured non-intrusively using a recently developed 'synthetic schlieren' technique.

Laboratory experiments used to study the initial response of a stratified fluid (a three-layer initial profile) to the action of a wind stress have been performed by Imberger and Stevens [340]. Consideration of the energy budget showed that little of the input energy (not more than 30%) was stored in the system. Peak mixing efficiency occurred when the surface stress was just sufficient to bring the interfacial region to the surface.

Finite amplitude internal waves in a two-layer fluid were studied by Keulegan and Carpenter [162]. They observed both the periodic running waves and solitary internal waves, the behaviour of which was in the framework of the nonlinear theory of internal waves in a two-layer fluid. Pinettes et al. studied oblique propagation of a solitary wave over a shelf in a two-layer fluid [301]. Running finite-amplitude internal waves in a continuously stratified fluid were investigated in [353]. The wave shapes observed were, to some extent, close to numerical shapes calculated by the perturbation theory. The results of an experimental investigation dealing with finite amplitude internal solitary waves in a two-fluid system were presented in papers by Koop and Butler [175] and Segur and Hammack [322]. Interfacial disturbances were generated by a wavemaker of displacement type. The correlation of the solitary wave's amplitude and length in deep and shallow water conditions was examined.

A study of the internal Korteveg–de Vries (KdV) solitary wave in the pycnocline of a continuously stratified fluid, with the initial disturbance condition as a step-like pool of light water, was completed by Kao et al. [155]. The production of nonlinear dispersive waves by a region of mixed fluid in a continuously stratified liquid with a pycnocline was examined by Kao and Pao [156] and Maxworthy [233]. The observations of an experimental study of internal gravity and, in particular, solitary waves produced by the motion of a self-propelled model in a stratified fluid were presented by Gilreath and Brandt [101]. In this paper, existence conditions of the attached solitary wave versus the ratio of the pycnocline thickness to the body's height were studied. [330] described an experimental investigation of the nonlinearity of long three-dimensional internal waves in a rather thin pycnocline in a fluid of finite depth, studying an intermediate case of the formation of a solitary wave: dispersive wave train propagation and resonant generation of internal waves by a submerged body. When the body's speed and the linear long wave mode-2 phase speed were equal, an internal solitary wave of the 'bulge' type was observed. An experimental investigation of mode-2 solitary waves propagating on a thin interface between two deep layers of different densities was performed in [338]. It was observed that the wavelength of large waves increased with increasing amplitude. Weakly nonlinear theory anticipated the results for amplitudes, but did not provide even a qualitative description of wave amplitudes.

The resonant interaction of internal waves has been intensively studied by laboratory modelling methods in recent years. Davis and Acrivos [59] showed that internal waves are unstable with respect to triple-wave resonant interaction. Similar experiments were performed in [230], [235]. The latter works showed that mixing vortices may be generated as a result of internal waves breaking because of resonant interaction.

The vertical phase configuration of two-dimensional internal waves generated by a cylinder was studied by Stevenson et al. [342] in a pycnocline and by Aksenov et al. [1] in an exponentially stratified fluid placed between two homogeneous layers. The prevalence of each mode was attained by the cylinder driving at a zero level of the vertical displacements profile for the corresponding mode. A study of the generation of internal modes showed that there is a number of critical values of $Fi^* = 1/\pi n$ (n is the mode number) so that for $Fi < Fi^*$ the excitation of modes with $n < n^*$ is impossible. The exception is only for the first mode ($n = 1$).

A series of experiments presented internal waves in the wake of an obstacle moving at the bottom of a stratified fluid (Long [215]) and at the free surface (Hunt and Snyder [133]). Experimental study of three-dimensional linear internal waves for supercritical flow was performed by Ma and Tulin [222]. They found a long narrow V-wake behind a ship (see also [386]), which was in good agreement with the theory of Miloh, Tulin and Zilman [247]. Three-dimensional internal waves past a sphere in a three-layer fluid with a stratified intermediate layer were investigated by Aksenov et al. [2]. Horizontal phase angles of the first IW mode were studied versus Froude number. Excitation of certain parts of the IW spectrum was provided by variation of the sphere's depth in the lower homogeneous layer. A thorough study of laminar-turbulent transformations in the wake of a linearly stratified flow past a sphere is presented by Boyer et al. [32]. A certain difference between lee waves and wake collapse waves was found by Hopfinger et al. [129] during experiments in a fluid with uniform stratification. An amplitude and spectral analysis of the mode structure of the wave field past a horizontally and uniformly driven sphere in the thermocline stratification carried out by Shishkina [328] provided the same conclusion.

There are several experimental works concerning the study of drag on uniformly driven solid bodies caused by the generation of internal waves in a stratified fluid, e.g., [46], [214], [231], [275], [385]. Shishkina [329] studied both the generation of internal waves and drag force in the densimetric Froude number range $Fi = (UN/R) = 0.15$–2.20 (here U is the uniform velocity, N is the Brunt–Väisälä frequency, and R is the characteristic vertical dimension of the body). Results of the comparative analysis of the drag coefficient versus Fi, based on the drag force measurements for models driven uniformly and horizontally in a two-layer liquid [275], an exponentially stratified fluid [214], and a thermocline [328] revealed similarity of Froude number dependences of the drag coefficient's increment (owed to internal waves) for the stated stratification profiles. A linear dependence of Froude numbers for continuously stratified fluids and a square dependence between Fi for the two-layer liquid and an exponential stratification were obtained by basing on the so called 'dividing streamline concept' (see [334]).

5. SOME QUALITATIVE SCHEMES OF INTERNAL WAVES DEGENERATION IN THE OCEAN

Direct damping of internal waves by molecular viscosity in accordance with the law $\exp\{-\nu k^2 t\}$ is negligible. When introducing *a priori* the terms of turbulent viscosity and heat transfer and considering the propagation of internal waves in a turbulent medium we estimate a decay time of about one day (the detailed calculations are given in [200]). But internal waves themselves are one of the sources of turbulence, so the turbulent viscosity should be connected with the parameters of internal waves. For a correct description of internal waves' breaking, dissipation, and probable generation of turbulence we need a thorough study of nonlinear wave effects, wave stability, and of irreversible energy transfer to the dissipative part of the spectrum. Such a sufficiently deep theory is still absent (some examples are presented in Part 3). But there is a number of qualitative phenomenological models of internal waves' breaking and dissipation, which, being far from strict, try to describe these processes on the basis of the evidence. A brief survey of such models is presented below[2].

The breaking of internal waves itself is impossible when the local accelerations $\omega^2 a$ are of the order of g, since the internal waves' frequencies ω are quite small. Phillips [299] assumes that one of the effective mechanisms of the degeneration of internal waves may be their hydrodynamic instability (the generation of turbulence) in the regions where the wave Richardson number $Ri < \frac{1}{4}$. For the lowest wave mode in the pycnocline, when $\omega \ll N$ the limiting steady Richardson number results in the form $\zeta_n^2 = 4\omega^2 (k^2 N^2)^{-1}$. A two-dimensional spectrum of the pycnocline oscillations, bounded by this value, is derived from the relation $E_\zeta (k) \sim \zeta^2 k^{-2}$, which taking into account the frequency dispersion relation yields

$$E_\zeta (k) \sim \delta (1 + \operatorname{cth} kh)^{-1} k^{-3}, \qquad (12.4)$$

where δ is the pycnocline's thickness. For short waves ($kh \gg 1$, $\omega^2 \sim k$) this spectrum is proportional to k^{-3}, for long waves ($kh \ll 1$, $\omega^2 \sim k^2$) it is proportional to k^{-2}. The one-dimensional spectrum results from multiplication by k. The frequency spectrum $E_\zeta (\omega) \sim \zeta^2 \omega^{-1}$ is proportional to ω^{-3} for short waves and to ω^{-1} for long waves. Though experimental data sometimes coincide with the Phillips theory, it is not universal for the ocean and requires large wave slopes ak to meet the condition $Ri < \frac{1}{4}$.

In Fig. 12.13 taken from [256] the experimental dependences of the wave slopes ak, obtained in the ocean thermocline (curves *2, 3, 4*) and its critical value (ak_*) for $Ri < \frac{1}{4}$ are shown. The experimentally observed slopes (typical for intense ocean internal waves) are considerably less than ak_*. And even when meeting the condition $Ri < \frac{1}{4}$ (which means that small disturbances increase) the developed turbulence appearance and the equilibrium internal waves spectrum saturation are not obligatory.

If the horizontal wave velocity component u is greater than the phase velocity c the so called 'advective instability' may occur. For $u \geq c$ a local density gradient is hydrostatically unstable and wave breaking takes place. This mechanism of wave instability was analyzed by Orlansky and Bryan [281]. They made numerical calculations on the development of plane internal waves from a state of rest under a periodic inducing force by the nonlinear equation set in the Boussinesq approximation. The result was that the density inversion appeared at a certain instant. On the basis of this numerical experiment they admitted the opportunity of advective instability of internal waves to be a source of the ocean's fine structure. Though the advective instability is a sufficiently nonlinear process, Orlansky and Bryan obtained its test number $Ri \leq 1 + k^2/l^2$ from the linear dispersion relation, which is more flexible than the shear

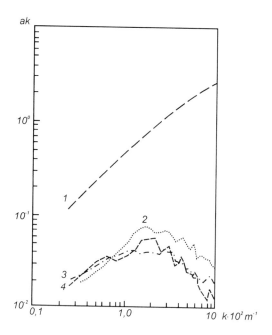

Figure 12.13. Internal wave slopes calculated by the Phillips model (1) and by the experimental data (2,3,4) in three different directions of the Atlantic Ocean bottom [256].

instability test number $Ri \leq \frac{1}{4}$. In his next work [280] Orlansky obtained the internal waves' spectrum, considering it to be saturated for $u \geq c$,

$$S_\zeta \sim \left\{ K_0 \left(x \right) + \left[\int\limits_0^x K_0 \left(x' \right) dx' \right]^2 \right\},$$

where $K_0 \left(x \right)$ is the Kelvin function, x is the dimensionless horizontal wave number, and S_ζ is the vertical displacement spectrum. Such a formula is in line with the presented evidence results.

The comparative analysis of advective and shear internal waves instabilities in a shear flow was performed by Frankignoul [82], who noticed that in the case $N^2 = \text{const}$ and $\Gamma = dV/dz = \text{const}$ (V is the mean current velocity) the solutions $W \sim w \left(t \right) \exp \left\{ i \left[k \left(x - \Gamma z t \right) + l y \right] \right\}$ (analyzed

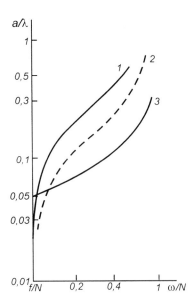

Figure 12.14. Neutral curves for advective and dynamic instability [82]: 1 — viscous dynamic instability; 2 — inviscid dynamic instability; 3 — the neutral line for shear instability $Ri = \frac{1}{4}$.

in Section 4.3) exactly satisfy the nonlinear hydrodynamics equations. Using these solutions and some empirical estimations Frankignoul developed the neutral curves for shear and gravity instabilities presented in Fig. 12.14. From this figure it follows that the advective instability is important for high internal waves and that it may occur before the shear instability.

Natural observations of two mechanisms of internal waves destabilization mentioned were described by Mack and Hebert [224]. High frequency temperature data collected from a vertical array of thermistors towed along the equator in the upper 125 m of the eastern Pacific Ocean was studied. It was noticed that wave-induced shear instability is likely to occur only near critical depths, where the wave phase velocity is equal to the mean flow velocity. In order to view directly the process of turbulent mixing owed to internal waves the observed turbulent mixing within a wave cycle was examined. The observational results suggested that wave-induced shear instability dominates over advective instability in the investigated region as the main mechanism for destabilization of internal waves and turbulent mixing. The same conclusion resulted the theoretical estimations by Thorpe [359].

Garrett and Munk [93] had an opposite viewpoint based on their previous semi-empirical model of the spectrum of internal waves [92] and the test Richardson number

$$Ri = \frac{2}{\pi^2 j^2 E} \frac{N_m}{N},$$

where $j = 20$ is the number of normal wave modes; N_m is the maximum value of N; $E = 2\pi \times 10^{-5}$ is the empirical constant. They stated that the shear instability ($Ri \leq 1/4$) should prevail in the ocean, contrary to the advective one governed mainly by wave slopes (see also [224]). It happens because the most part of ocean energy is contained in the low (inertial) frequencies, when waves have considerably greater shear to slope ratio. The shear instability of low frequency waves produces the fine structure which then breaks weak internal waves. The authors note that such a process is similar to the 'ultraviolet catastrophe' and, seemingly, is the basic mechanism of the degeneration of internal waves. According to [93], the kinetic energy dissipation rate (the turbulence production energy) in this process is equal to $5/2k (2\pi N)^2$, where k is the mixing coefficient and N has the dimension of cycles per second.

Note, in conclusion, the qualitative considerations by Woods [392], [393] on the importance of Kelvin–Helmholtz instability for the breaking of internal waves based on a series of his unique natural observations in the Mediterranean Sea. But more detailed estimations of the mechanism of generation of the unstable regions of static instability in standing or progressive internal waves were not performed.

Notes

1 A survey of oceanographic research of the ocean in Russia during the last decades may be found in [194], [271].
2 Most of the possible mechanisms of the static and dynamic instability of internal waves are described in [339], [360].

References

[1]   Aksenov AV, Mozhaev VV, Skorovarov VE, Sheronov AA. Struc-
        ture of internal waves in a three-layer fluid with the stratified
        intermediate layer. Fluid Dyn 1987; 22:437–441

[2]   Aksenov AV, Mozhaev VV, Skorovarov VE, Sheronov AA. Struc-
        ture of ship internal waves in a three-layer fluid with the stratified
        intermediate layer. J Appl Mech Techn Phys 1989; 30:101–105

[3]   Akylas TR, Grimshaw RH. Solitary internal waves with oscillatory
        tails. J Fluid Mech 1992; 242:279–298

[4]   Amick CJ, Turner REL. A global theory of internal solitary waves
        in two-fluid system. Trans Am Math Soc 1986; 298:431–439

[5]   Anderson DLT, Gill A. Beta dispersion of inertial waves. J Geo-
        phys Res 1979; 84:1836–1842

[6]   Appel JR, Byrro HM, Pronti JR, Charnell RL. Observations of
        oceanic internal and surface waves from the earth resources tech-
        nology satellite. J Geophys Res 1975; 80(6):865–881

[7]   Appel JR, Gasparovich RF, Thompson DR, Gotwols BL. Signa-
        tures of surface waves-internal wave interaction: experiment and
        theory. Dyn Atmos Oceans 1988; 12:89–106

[8]   "A Discussion on Nonlinear Theory of Waves Propagation in Dispersive Systems." In *Collection of Translations*, ed. Gregory Barenblatt, Moscow: Mir, 1970 (in Russian).

[9]   Badulin SI, Shrira VI. On the irreversibility of internal wave dynamics due to wave trapping by mean flow inhomogeneities. J Fluid Mech 1993; 251:21–53

[10]   Badulin SI, Shrira VI, Tsimring LSh. On the trapping and vertical focusing of internal waves in a pycnocline due to the horizontal inhomogeneities of stratification and current. J Fluid Mech 1985; 158:199–218

[11]   Badulin SI, Vasilenko VM, Golenko NN. Transformation of internal waves in the equatorial Lomonosov current. Izv Atmos Ocean Phys 1990; 26: 279–289

[12]   Baines PG. The reflection of internal/inertial waves from bumpy surface. Split reflection and diffraction. J Fluid Mech 1971; 46(2):273–291

[13]   Baines PG. On internal tide generation models. Deep Sea Res II 1982; 29:307–338

[14]   Ball K. Energy transfer between external and internal gravity waves. J Fluid Mech 1964; 19(4):465–478

[15]   Banks WHH, Drazin PG, Zaturska MB. On the normal modes of parallel flow of inviscid stratified fluid. J Fluid Mech 1976; 75:149–171

[16]   Barcilon A, Drasin P. Dust devil formation. Geophys Fluid Dyn 1972; 4(2):147–158

[17]   Barenblatt, Gregory, *Second-Genus Self-Similar Solutions. Classification of Self-Similar Solutions.* Moscow: Institute for Problems in Mechanics RAS, 1975 (in Russian).

[18]   Becker JM, Grimshaw RHJ. Explosive resonant triads in a continuously stratified shear flow. J Fluid Mech 1993; 257:219–228

[19]   Bell TH. Topographically generated internal waves in the open ocean. J Geophys Res 1975; 80(3):320–327

[20]   Benjamin TB. Internal waves of finite amplitude and permanent form. J Fluid Mech 1966; 25(2):241–270

[21]   Benjamin TB. Internal waves of permanent form in fluids of great depth. J Fluid Mech 1967; 29(3):559–592

[22]   Benjamin TB, Lighthill MJ. On cnoidal waves and bores. Proc Roy Soc Lond 1954; A224:448–456

[23]   Benney DJ, Saffman PG. Nonlinear interactions of random waves in a dispersive medium. Proc Roy Soc Lond 1966; A289(1418):301–320

[24]   Berezin YuA, Karpman VI. On nonlinear evolution of perturbations in plasma and other dispersive media. Sov Phys JETP 1966; 51(5):1557–1568

[25]   Bloembergen, Nicolaas, *Nonlinear optics*. Redwood City, California: Addison-Wesley, 1992.

[26]   Bockel M. Travaux océanographigues de l' "Origny" a Gibraltar. Cahier Océanogr 1962; 14(4):325–329

[27]   Bogatyrev SD, Bredikhin VV, Druzhinin OA, Gil'man OA, Kazakov VI, Korotkov DP, Kozyrev OR, Krivatkina NB, Ostrovsky LA, Shishkina OD, Stepanyants YuA, Talanov VI, Zaborskikh DV. *Dynamics of internal waves in rotating or moving fluids: laboratory experiments, numerical modelling and theoretical estimations.* Report INTAS 94-4097. 1996.

[28]   Booker JB, Bretherton FP. The critical layer for internal gravity waves in a shear flow. J Fluid Mech 1967; 27(4):513–539

[29]   Borisenko YuD, Miropol'sky YuZ. On nonlinearity effect on statistical distribution of internal waves in the ocean. Oceanology 1974; 14(5):788–796 (in Russian).

[30] Borisenko YuD, Voronovich AG, Leonov AI, Miropol'sky YuZ. To-
      wards a theory of non-stationary weakly non-linear internal waves
      in a stratified fluid. Izv Atmos Ocean Phys 1976; 12(3):174–178

[31] Born, M., Wolf, E., *Principles of Optics*. Oxford: Pergamon Press,
      1969.

[32] Boyer DL, Fernando HJS, Lin Q, Lindberg WR. Stratified flow
      past a sphere. J Fluid Mech 1992; 240:315–354

[33] Brekhovskikh, Leonid, *Waves in Layered Media*. New York: Aca-
      demic Press, 1960.

[34] Brekhovskikh LM, Goncharov VV, Kurtepov VM, Nau-
      gol'nykh KA. On resonance generation of internal waves under
      non-linear surface waves interaction. Izv Atmos Ocean Phys 1972;
      8(2):192–203

[35] Brekhovskikh LM, Goncharov VV, Naugol'nykh KA, Rybak SA.
      Waves in the ocean. Radiophys and Quant Electron 1976;
      19(6):598–620

[36] Brekhovskikh LM, Konjaev KV, Sabinin KD, Serikov AN. Short-
      period internal waves in the sea. J Geophys Res 1975; 80(6):856–
      864

[37] Bretherton FP. The propagation of groups of internal gravity
      waves in a shear flow. Quart. J Roy Met Soc 1966; 92:466–480

[38] Bretherton FP, Garrett CIP. Wave trains in inhomogeneous mov-
      ing media. Proc Roy Soc Lond 1968; A302:529–554

[39] Briscoe MG. Introduction to collection of papers of oceanic inter-
      nal waves. J Geophys Res 1975; 80 (3):289–290

[40] Briscoe MG. Preliminary results from the trimoored internal wave
      experiment (IWEX). J Geophys Res 1975; 80(27):3872–3884

[41] Brown DJ, Christie DR. Fully nonlinear solitary waves in con-
      tinuously stratified incompressible Boussinesq fluids. Phys Fluids
      1998; 10(10):2569–2586

[42]   Bruhwiler DL, Kaper TJ. Wavenumber transport: scattering of small-scale internal waves by large-scale wavepackets. J Fluid Mech 1995; 289:379–405

[43]   Bunimovich LA, Zhmur VV. Internal waves scattering in a horizontally inhomogeneous ocean. Sov Phys Dokl 1986; 286:197–200

[44]   Cairns RA. The role of negative energy waves in some instabilities of parallel flows. J Fluid Mech 1979; 92:1–14

[45]   Carsola AJ, Callaway EB. Two short-periodic internal wave frequency spectra in the sea of southern California. Limnol and Oceanogr 1962; 7(2):115–120

[46]   Castro IP, Snyder WH, Bains PG. Obstacle drag in stratified flow. Proc Royal Soc London 1990; 429(1876):119–140

[47]   Chabert d'Hieres G, Didelle H, Obaton D. A laboratory study of surface boundary currents: application to the Algerian current. J Geophys Res 1991; 96(C7):12539–12548

[48]   Chen CT, Millero PS. The specific volume of sea water at high pressures. Deep-Sea Res 1976; 23(7):595–612

[49]   Cherkesov, Leonid, *Surface and Internal Waves*. Kiev: Naukova Dumka, 1973 (in Russian).

[50]   Chernikov GP. Rossby solitons: cyclones in rotating shallow waver. Dokl Phys Sect 1994; 39(1):73–75

[51]   Choi W, Camassa R. Fully nonlinear internal waves in a two-fluid system. J Fluid Mech 1999; 396:1–36

[52]   Chomaz JM, Bonneton P, Butet A, Perrier M. Froude number dependence of the flow separation line on a sphere towed in a stratified fluid. Phys Fluids 1992; A4(2):254–258

[53]   Cox CS, Nagata V, Osborn T. Oceanic fine structure and internal waves. Bull Japan Soc Fish Oceanogr 1969; special issue.

[54] Cox CS, Sandström H. Coupling of internal and surface waves in water of variable depth. J. Oceanogr. Soc. Japan, 1962; 20-th anniv. vol:499–513

[55] Craik, Alexander, *Wave Interactions and Fluid Flows.* London: Cambridge University Press, 1988.

[56] Curtin TB, Mooers CN. Observation and interpretation of a high-frequency internal wave packet and surface slick pattern. J Geophys Res 1975; 80(6):882–894

[57] D'Asaro EA. The energy flux from the wind to near-inertial motions in the surface mixed layer. J Phys Oceanogr 1985; 15:1043–1959

[58] Davis RE, Acrivos AW. Solitary internal waves in deep water. J Fluid Mech 1967; 29(3):593–607

[59] Davis RE, Acrivos AW. The stability of oscillatory internal waves. J Fluid Mech 1967; 30(4):723–736

[60] Defant A. Über interne Wellen, besonders solche mit Gezeitencharakter. Deutsch Hydrogr Z 1952; 5: 231–245

[61] De Groot, S., Mazur P. *Non-Equilibrium Thermodynamics.* Amsterdam: North-Holland, 1962.

[62] Demidova TA, Korchagin NN, Maslov VP. On generation of bottom currents by a fluid pulse distortion on the surface of the ocean. Oceanology 1998; 38(4):486–491

[63] Desaubies YJ. Internal waves near the turning point. Geophys Fluid Dyn 1973; 5:143–154

[64] Dikiy, Leonid, *The Theory of the Earth Atmosphere Oscillations.* Leningrad: Gidrometeoizdat, 1969 (in Russian).

[65] Drazin PG, Zaturska MB, Banks WHH. On the normal modes of parallel flow of inviscid stratified fluid. Part 2. Unbounded flow with propagation at infinity. J Fluid Mech 1979; 95:681–705

[66] Dubriel-Jacotin L. Sur les ondes type permanent dans les liquides heterogenes. Atti della Reale Academic Nationale dci Lincei 1932; 15(6):44–52

[67] Dulov VA, Klyushnikov SI, Kudryavtsev VN. Influence of internal waves on intensity of wind wave breaking. Morskoy Gidrofiz Zh 1986; 6:14–21 (in Russian)

[68] Dysthe KB, Das KP. Coupling between a surface-wave spectrum and an internal wave: modulational interaction. J Fluid Mech 1981; 104:483–503

[69] E X, Hopfinger EJ. On mixing across an interface in stably stratified fluid. J Fluid Mech 1986; 166:227–244

[70] Eckart, Carl, *Hydrodynamics of Oceans and Atmospheres*. Oxford: Pergamon Press, 1960.

[71] Eriksen CC. On wind forcing and observed wavenumber spectra. J Geophys Res 1988; 93:4985–4992

[72] Eriksen CC. Equatorial ocean response to rapidly translating wind bursts. J Phys Oceanogr 1993; 23:1208–1230

[73] Etling D, Gelhardt F, Schrader U, Brennnecke F, Kuhn G, Chabert d'Hieres G, Didelle H. Experiments with density currents on a sloping bottom in a rotating fluid. Dyn Atmos Oceans 2000; 31(1-4):139–164

[74] Farmer DM, Smith JD. Tidal interaction of stratified flow with a sill in Knight Inlet. Deep-Sea Res 1980; 27:239–254

[75] Fedorov, Konstantin, *Fine Thermochalinic Structure of the Ocean Water*. Leningrad: Gidrometeoizdat, 1976 (in Russian).

[76] Fernando HJS. Turbulent mixing in stratified fluids. Ann Rev Fluid Mech 1991; 23:455–493

[77] Fernando HJS, Hunt JCR. Some aspects of turbulence and mixing in stably stratified layers. Dyn Atmos Oceans 1996; 23(1-4):35–62

[78] Filyushkin YuB. Influence of the horizontal component of the angular velocity of the Earth's rotation on the propagation of internal waves in the ocean. Oceanology 1982: 22(5):705–710

[79] Fofonoff NP. Physical properties of sea water. The Sea 1962; 1:3–30

[80] Fofonoff NP. Spectral characteristics of internal waves in the ocean. Deep-Sea Res 1969; 16:59–71

[81] Frankignoul CJ. The effect of weak shear and rotation on internal waves. Tellus 1970; 12(2):194–203

[82] Frankignoul CJ. Stability of finite amplitude internal waves in a shear flow. Geophys Fluid Dyn 1972; 4(2):91–100

[83] Fu LL. Observations and models of inertial waves in the deep ocean. Rev Geophys Space Phys 1981; 19:141–170

[84] Funakoshi M, Oikawa M. Long internal waves of large amplitude in a two-layer fluid. J Phys Soc Japan 1986; 55:128–144

[85] Gardner CS, Green JM, Kruskal MD, Miura RM. Method for solving the Korteweg–de Vries equation. Phys Rev Lett 1967; 19:1095–1097

[86] Gardner CS, Su CH. Korteweg–de Vries equation and generalizations. III. Derivation of the Korteweg–de Vries equation and Burgers equation. J Math Phys 1969; 10(3):536–539

[87] Gargett AE. Generation of internal waves in the Strait of Georgia, British Columbia. Deep-Sea Res 1976; 23(3):17–32

[88] Garnich NG, Miropol'sky YuZ. Some properties of the fine thermal structure of the ocean. Oceanology 1974; 14(4):596–601 (in Russian)

[89] Garnich NG, Miropol'sky YuZ, Prokhorov VI, Fedorov KN. Some results from a statistical analysis of the thermochaline microstructure. Izv Atmos Ocean Phys 1973; 9(2):83–90

[90] Garrett CJ. On the interaction between internal gravity waves and a shear flow. J Fluid Mech 1968; 34(4):711–720

[91] Garrett CJ, Munk W. Internal wave spectra in the presence of fine structure. J Phys Oceanogr 1971; 1(3):196–202

[92] Garrett CJ, Munk W. Space–time scales of internal waves. Geophys Fluid Dyn 1972; 2(4):225–264

[93] Garrett CJ, Munk W. Oceanic mixing by breaking internal waves. Deep-Sea Res 1972; 19:823–932

[94] Garrett CJ, Munk W. Space–time scales of internal waves: a progress report. J Geophys Res 1975; 80 (3):291–297

[95] Garrett CJR, Munk WH. Internal waves in the ocean. Ann Rev Fluid Mech 1979; 11:339–369

[96] Gaul RD. Observations of internal waves near Hudson Canyon. J Geophys Res 1961; 66(11):3821–3830

[97] Gear JA, Grimshaw R. A second-order theory for solitary waves in shallow fluids. Phys Fluids 1983; 26:14–29

[98] Gerkema T. A unified model for the generation and fission of internal tides in a rotating ocean. J Marine Res 1996; 54:421–450

[99] Gerkema T, Zimmerman JTF. Generation of nonlinear internal tides and solitary waves. J Phys Oceanogr 1995; 25:1081–1094

[100] Gill, Adrian, *Atmosphere–Ocean Dynamics*. New York etc.: Academic Press, 1983.

[101] Gilreath HE, Brandt A. Experiments on the generation of internal waves in a stratified fluid. AIAA J 1985; 23:693–700

[102] Ginzburg, Vitaliy, *Propagation of Electromagnetic Waves in Plasma*. New York: Gordon and Breach, 1961.

[103] Goncharov, Valeriy. "On Some Peculiarities of Internal Waves in the Ocean." In *Tsunami and Internal Waves*, ed. Boris Nelepo, Sevastopol': Marine Hydrophysical Institute, 1976 (in Russian).

[104] Goncharov VP, Krasil'nikov VA, Pavlov VI. A contribution to the theory of wave interactions in stratified media. Izv Atmos Ocean Phys 1976; 12(11):704–709

[105] Goncharov V.P., Pavlov V.I., *Problems of Hydrodynamics in the Hamiltonian Description.* Moscow: Moscow University Press, 1993 (in Russian).

[106] Gould WJ, Schmitz WJ, Winsch C. Preliminary field results for a Mid-Ocean Dynamic Experiment (MODE-O). Deep-Sea Res 1974; 21(11):911–932

[107] Grimshaw R. Nonlinear internal gravity waves in a slowly varying medium. J Fluid Mech 1972; 54:193–207

[108] Grimshaw R. Evolution equation for weakly nonlinear long internal waves in a rotating fluid. Stud Appl Math 1985; 73:1–33

[109] Grimshaw R, Ostrovsky LA, Shrira VI, Stepanyants YuA. Nonlinear surface and internal gravity waves in rotating ocean. Surveys Geophys 1998; 19(4):289–338

[110] Grimshaw R, Pelinovsky E, Talipova T. The modified Korteweg-de Vries equation in the theory of large amplitude internal waves. Nonlinear Processes in Geophys 1997; 4:237–350

[111] Grimshaw R, Pelinovsky E, Talipova T. Solitary wave transformation in a medium with sign variable quadratic and cubic nonlinearity. Physica D 1999; 132:40–62

[112] Grimshaw R, Pudjaprasetya SR. Hamiltonian formulation for the description of interfacial solitary waves. Nonlinear Processes Geophys 1998; 5:3–12

[113] Grimshaw RHJ, Smyth N. Resonant flow of a stratified fluid over topography. J Fluid Mech 1986; 169:429-464

[114] Grimshaw RHJ, Yi Z. Resonant generation of finite amplitude waves by the flow of a uniformly stratified fluid over topography. J Fluid Mech 1991; 229:603–628

[115] Grue J, Friis A, Palm E, Rusås P-O. A method for computing unsteady fully nonlinear interfacial waves. J Fluid Mech 1997; 351:223–252

[116] Grue J, Jensen A, Rusås P-O, Sveen JK. Properties of large amplitude internal waves. J Fluid Mech 1999; 380:257–278

[117] Guizien K, Barthélemy E, Inall ME. Internal tide generation at a shelf break by an oblique barotropic tide: observations and analytical modeling. J Geophys Res 1999; 104(C7):15655–151978

[118] Hanazaki H. On the three-dimensional internal waves excited by topography in the flow of a stratified fluid. J Fluid Mech 1994; 263:293–318

[119] Hasselmann K. On the non-linear energy transfer in a gravity wave spectra. J Fluid Mech 1962; 12(1):481–500

[120] Hasselmann K. A criterion for non-linear wave stability. J Fluid Mech 1967; 30(3):737–741

[121] Haurwitz B. The occurrence of internal tides in the ocean. Arch Met Geophys Bioclimatol 1954; A7:406–424

[122] Hecht A, Hyghes P. Observations of temperature fluctuations in the upper layer of the Bay of Biscay. Deep-Sea Res 1971; 18(7):663–684

[123] Hecht A, White R. Temperature fluctuations in the upper layer of the ocean. Deep-Sea Res 1968; 15(3):339–354

[124] Hector D, Cohen J, Bleistein N. Ray method expansions for surface and internal waves in inhomogeneous oceans of variable depth. Studies in Appl Math 1972; 51(2):121–137

[125] Helfrich, K., Melville, W.K. "Review of Dispersive and Resonant Effects in Internal Wave Propagation." In *The Physical Oceanography of Sea Straits*, ed. Lawrence J. Pratt, Dordrecht: Kluwer Academic Publishers, 1990.

[126] Henyey F. Hamiltonian description of stratified fluid dynamics. Phys Fluids 1983; 26:40–47

[127] Holloway PE. Internal hydraulic jumps and solitons at a shelf break region on the Australian north west shelf. J Geophys Res 1987; 92:5405–5416

[128] Holloway PE, Pelinovsky E, Talipova T. A generalized Korteweg–de Vries model of internal tide transformation in the coastal zone. J Geophys Res 1999; 104(C8):18333–18350

[129] Hopfinger EJ, Flor JB, Chomas JM, Bonneton P. Internal waves generated by a moving sphere and its wake in a stratified fluid. Exp Fluids 1991; 11:255–261

[130] Howard LN. Note on a paper of John W. Miles. J Fluid Mech 1961; 10(4):509–514

[131] Hughes BA, Grant HL. The effect of internal waves on surface wind waves. Part 1. Experimental Measurements. J Geophys Res 1978; 83:443–454

[132] Hughes BA, Grant HL. The effect of internal waves on surface wind waves. Part 2. Theoretical Analysis. J Geophys Res 1978; 83:455–465

[133] Hunt JCR, Snyder WH. Experiments on stably and neutrally stratified flow over a model three-dimensional hill. J Fluid Mech 1980; 96:671–704

[134] Hurley DG, Keady G. The generation of internal wave by vibrating elliptic cylinders. Part 2. Approximate viscous solution. J Fluid Mech 1997; 351:119–138

[135] *Investigations of Oceanic Turbulence*, ed. Rostislav Ozmidov, Moscow: Nauka, 1973 (in Russian).

[136] Ivanov AV. Internal waves generation by an oscillating source. Izv Atmos Ocean Phys 1989; 25:84–89 (in Russian)

[137] Ivanov VA, Konyaev KV, Serebryany AN. Groups of intense internal waves in the shelf zone of the sea. Izv Atmos Ocean Phys 1981; 17(12):1302–1309

[138] Ivanov VA, Leont'eva EA, Serebryany AN. Space structure of internal waves in the Eygenian Sea. Morskoy Gidrifiz Zh 1986; 2(6):61–64 (in Russian)

[139] Ivanov VA, Serebryany AN. Frequency spectra of short–period internal waves in non-tidal sea. Izv Atmos Ocean Phys 1982; 18(6):52–529

[140] Ivanov VA, Serebryany AN. Internal waves on the shallow water shelf of a non-tidal sea. Izv Atmos Ocean Phys 1983; 19(6):495–498

[141] Ivanov VA, Serebryany AN. Short–period internal waves in the coastal zone of a non-tidal sea. Izv Atmos Ocean Phys 1985; 21(6):496–501

[142] Ivanov YuA, Morozov EG. Internal gravity waves deformation by a flow with horizontal velocity shear. Oceanology, 1974; 14(3):457–461 (in Russian)

[143] Ivanov YuA, Smirnov BA, Tareyev BA, Filyushkin BN. The experimental investigation of temperature oscillations in the frequency interval of internal gravity waves. Izv Atmos Ocean Phys 1969; 5(4):230–235

[144] Jahnke, E., Emde, F. and Lösch, F., *Tafeln Höherer Funktionen (Tables of Higher Functions)*. Stuttgart: Teubner, 1966.

[145] Jones WL. Propagation of internal gravity waves in fluids with shear flow and rotation. J Fluid Mech 1967; 30 (3):439–448

[146] Jones WL. Ray tracing for internal gravity waves. J Geophys Res 1969; 74(8):2028–2033

[147] Jonsson IG. Wave-current interactions. The Sea 1998; 9:65–120

[148] Kadomcev, Boris, *Plasma Turbulence*. London: Academic Press, 1965.

[149] Kadomcev BB, Karpman VI. Nonlinear waves. Sov Phys Uspehi 1971; 103(2):193–211

[150] Kadomcev BB, Petviashvili VI. Stability of solitary waves in weakly dispersing media. Sov Phys Dokl 1970; 15:539–542

[151] Kakutani T, Yamasaki N. Solitary waves on a two-layer fluid. J Phys Soc Japan 1978; 45:674–679

[152] Kamenkovič, Vladimir, *Fundamentals of Ocean Dynamics*. Amsterdam: Elsevier Science, 1977.

[153] Kamenkovich VM, Kulakov AV. Influence of rotation on waves in a stratified ocean. Oceanology 1977; 17(3):260–266

[154] Kamenkovich VM, Odulo AB. Free oscillations in a stratified compressible ocean of constant depth. Izv Atmos Ocean Phys 1972; 8(11):693–700

[155] Kao TW, Pan F-S, Renouard DP. Internal solitons on the pycnocline. J Fluid Mech 1985; 159:19–53

[156] Kao TW, Pao H-P. Wake collapse in the thermocline and internal solitary waves. J Fluid Mech 1980; 97:115–127

[157] Karabasheva EI, Kozhelupova NG, Miropol'sky YuZ. Some evidence of spatial structure of internal wave field in the ocean. Oceanology 1974; 14(3):462–467 (in Russian)

[158] Karpman, Vladimir, *Nonlinear Waves in Dispersive Media*. Oxford: Pergamon, 1975.

[159] Katz E. Tow spectra from MODE. J Geophys Res 1975; 80(9): 1163–1167

[160] Keller JB, Mow VC. Internal wave propagation in an inhomogeneous fluid of non-uniform depth. J Fluid Mech 1969; 38(2):365–374

[161] Kenyon KE. Wave refraction in ocean currents. Deep-Sea Res 1971; 18:1023–1034

[162] Keulegan GH, Carpenter LH. An experimental study of internal progressive oscillatory waves. Nat Bur Stan Rep 7319, 1961.

[163] Keunecke KH. Stehende intern wellen in rechteckigen becken. Dtsch Hydrogr Z 1970; 23:61–79

[164] Kielman J, Krauss W, Keunecke K. Currents and stratification in the Balt Sea and the Arkona Basin during 1962–1968. Kiel Meeresforsebungen, Institut für Meereskunde und der Universität Kiel 1973; 29(2):1–47

[165] Kitaygorodskiy SA, Miropol'sky YuZ, Filyushkin BN. Use of ocean temperature fluctuation data to distinguish internal waves from turbulence. Izv Atmos Ocean Phys 1973; 9(3):149–159

[166] Klaassen GP, Peltier WR. The influence of stratification on secondary instability in free shear layers. J Fluid Mech 1991; 227:71–106

[167] Klostermeyer J. On parametric instabilities of finite amplitude internal gravity waves. J Fluid Mech 1982; 119:367–377

[168] Kluwick A, Cox EA. Propagation of weakly nonlinear waves in stratified media having mixed nonlinearity. J Fluid Mech 1992; 244:171–185

[169] Kneser, Adolf, *Die Integralgleichungen und ihre Anwendung in der mathematische Physik.* Berlin: Akademie-Verlag, 1922.

[170] Koniaev KV. Measurement of the spatial structure of internal waves in the ocean by means of a gradient system of temperature sensors. Izv Atmos Ocean Phys 1975; 11(7):458–463

[171] Koniaev KV, Sabinin KD. New data concerning internal waves in the sea obtained using distributed temperature sensors. Sov Phys Dokl 1973; 209(1):86–89

[172] Koniaev KV, Sabinin KD. Resonator hypothesis of the generation of internal waves in sea. Sov Phys Dokl 1973; 210(6):1342–1345

[173] Konyaev KV, Sabinin KD, Serebryany AN. Large-amplitude internal waves at the Mascarene Ridge in the Indian Ocean. Deep-Sea Res 1995; 42(11/12):2075–2091

[174] Konyaev KV, Sabinin KD. Intensive internal waves near the Pacific coast of Kamchatka. Oceanology 1998; 38(1):27–32

[175] Koop CG, Butler G. An investigation of internal solitary waves in a two-fluid system. J Fluid Mech 1981; 112:225–251

[176] Koop CG, McGee B. Measurements of internal gravity waves in a continuously stratified shear flow. J Fluid Mech 1986; 172:453–480

[177] Korteweg DJ, de Vries G. On the change of form of long waves advancing in a rectangular channel, and on a new type of long stationary waves. Phil Mag 1895; 17(39):422–443

[178] Kozhelupova NG, Miropol'sky YuZ, Filyushkin BN. Vertical variations of a spatial structure of the internal waves field in the ocean. Oceanology 1975; 15(6):962–965 (in Russian)

[179] Krauss, Wolfgang, *Interne Wellen*. Berlin: Gebrüder Bornträger, 1966.

[180] Krauss W. Wind generated internal waves and inertial-period motions. Deutsch Hydrogr Z 1972; 25(2):241–250

[181] Krauss W. Internal tides resulting from the passage of surface tides through an eddy field. J Geophys Res 1999; 104(C8):18323–18331

[182] Kubota T, Ko DRS, Dobbs L. Propagation of weakly nonlinear internal waves in a stratified fluid of finite depth. AIAA J 1978; 12:157–169

[183] Kudryavtsev VN. On the growth and decay of internal waves in their interaction with wind waves. Morskoy Gidrofiz Zh 1988; 4:3–13 (in Russian)

[184] Kudryavtsev VN. The coupling of wind and internal waves: modulation and friction mechanisms. J Fluid Mech 1994; 278:33–62

[185] Kundu PK. On internal waves generated by travelling wind. J Fluid Mech 1993; 254:529–559

[186] Kundu PK, Allen JS, Smith RL. Modal decomposition of the velocity field near the Oregon coast. J Phys Oceanogr 1975; 5(4):683–704

[187] Kundu PK, Thomson RE. Inertial oscillations due to a moving front. J Phys Oceanogr 1985; 15:1076–1084

[188] Kunze E. Near-inertial wave propagation in geostrophic shear. J Phys Oceanogr 1985; 15:544–565

[189] Lamb, Horace, *Hydrodynamics*. London: Cambridge University Press, 1932.

[190] Lamb KG, Pierrehumbert RT. Steady-state nonlinear internal gravity wave critical layers satisfying an upper radiation condition. J Fluid Mech 1992; 238:371–404

[191] Lamb KG, Yan L. The evolution of internal wave undular bores: comparisons of a fully nonlinear numerical model with weakly nonlinear theory. J Phys Oceanogr 1996; 26:2712–2734

[192] Landau, L.D., Lifshits, E.M., *Mechanics of Continuous Media*. Oxford: Pergamon Press, 1960.

[193] Landau, L.D., Lifshits, E.M., *Statische Physik*. Berlin: Akademie-Verlag, 1992.

[194] Lappo SS. Oceanologic Research of the World Ocean in Russia. Oceanology 1999; 39(5):645–653

[195] Larsen H. Internal waves incident upon knife edge barrier. Deep-Sea Res 1969; 16(5):411–416

[196] Lax P. Integrals of nonlinear equations of evolution and solitary waves. Comm Pure Appl Math 1968; 21:467–490

[197] Laykhtman DL, Leonov AI, Miropol'sky YuZ. Interpretation of measurements of statistical parameters of scalar fields in the ocean

in the presence of internal gravity waves. Izv Atmos Ocean Phys 1971; 7(4):291–295

[198] Laykhtman DL, Leonov AI, Miropol'sky YuZ. Determination of the two-dimensional statistical characteristics of a scalar field in the ocean from measurements in the presence of internal gravity waves. Izv Atmos Ocean Phys 1971; 7(6):427–431

[199] Leaman K, Stanford T. Vertical energy propagation of internal waves. a vector spectral analysis of velocity profiles. J Geophys Res 1975; 80(15):1975–1978

[200] Le Blond PH. On the damping of internal gravity waves in a continuous stratified ocean. J Fluid Mech 1966; 25(1):121–142

[201] Le Blond, P., Mysak, L., *Waves in the Ocean.* Oceanography Series 20, Amsterdam: Else Science, 1978.

[202] Lee C-V, Beardsley RC. The generation of long nonlinear internal waves in a weakly stratified shear flow. J Geophys Res 1974; 79:453–462

[203] Leonov AI. On the Korteweg–de Vries equations in the nonlinear theory of surface and internal waves. Sov Phys Dokl 1976; 229(4):820–824

[204] Leonov AI. The effect of the Earth's rotation on the propagation of weakly nonlinear surface and internal long oceanic waves. Ann NY Acad Sci 1981; 373:150–159

[205] Leonov AI, Miropol'sky YuZ. Resonant excitation of internal gravity waves in the ocean by atmospheric pressure fluctuations. Izv Atmos Ocean Phys 1973; 9(8):480–485

[206] Leonov AI, Miropol'sky YuZ. On steady internal gravitational waves of finite amplitude. Sov Phys Dokl 1974; 218(6):1287–1290

[207] Leonov AI, Miropol'sky YuZ. Toward a theory of stationary nonlinear internal gravity waves. Izv Atmos Ocean Phys 1975; 11(5):298–304

[208] Leonov AI, Miropol'sky YuZ. Short wave approximation in the theory of non-linear steady state internal gravity waves. Izv Atmos Ocean Phys 1975; 11(11):732–736

[209] Leonov AI, Miropol'sky YuZ, Tamsalu RE. On calculation of the fine structure of density and velocity fields (example for the Balt sea). Oceanology 1977; 17(3):250–256

[210] Levine MD. Internal waves in the ocean: a review. Rev Geophys Space Phys 1983; 21:1206–1216

[211] Levikov SV. Nonstationary and slightly nonlinear internal waves in a deep ocean. Oceanology 1976; 16(6):551–554

[212] Lighthill, James, *Waves in Fluids*. London: Cambridge University Press, 1978.

[213] Lighthill J. Internal waves and related initial value problems. Dyn Atmos Oceans 1996; 23(1-4):3–17

[214] Lofquist K, Purtell L. Drag on a sphere moving horizontally through a stratified liquid. J Fluid Mech 1984; 148:271–284

[215] Long RR. Some aspects of the flow of stratified fluids. Pt. 1. Tellus 1953; 5(1):42–57

[216] Long RR. Solitary waves in one and two fluid systems. Tellus 1956; 8:460–472

[217] Long RR. On the Boussinesq approximation and its role in the theory of internal waves. Tellus 1965; 17:46–52

[218] Longuet-Higgins MS. The effect of non-linearities on statistical distributions in the theory of sea waves. J Fluid Mech 1963; 17(3):459–480

[219] Luke JG. A perturbation method for nonlinear dispersive wave problems. Proc Roy Soc Lond 1966; A292:403–412

[220] Lumley, J., Panofsky, H., *The Structure of Atmospheric Turbulence*. New York: Interscience, 1964.

[221] Ma H. Trapped internal gravity waves in a geostrophic boundary current. J Fluid Mech 1993; 247:205–229

[222] Ma H, Tulin MP. Experimental study of ship internal waves: the supersonic case. J Offshore Mech Arctic Edging 1993; 115:16–22

[223] Maas LRM, Benielli D, Sommeria J, Lam F-PA. Observation of an internal wave attractor in a confined, stable stratified fluid. Nature 1997; 388(7):557–561

[224] Mack AP, Hebert D. Mixing structure of high-frequency internal waves in the upper Eastern equatorial Pacific. J Phys Oceanogr 1999; 29:3090–3100

[225] Magaard L. Zur Berechnung interner Wellen in Meeresräumen mit nichtebenen Böden bei einer speziellen Dichteverteilung. Kieler Meeresforsch 1962; 18:161–183

[226] Magaard L. Zur Theorie zweidimensionaler nicht-linearer interner Wellen in stetig geschicheten Medien. Kieler Meeresforsch 1965; 21:22–31

[227] Magaard L. On the generation of internal gravity waves by a fluctuating buoyancy flux at the sea surface. Geophys Fluid Dyn 1973; 5:101–111

[228] Makarov SA, Neklyudov VI, Chashechkin YuD. Space structure of two-dimensional monochromatic internal wave packets in exponentially stratified fluid. Izv Atmos Ocean Phys 1990; 26:744–754 (in Russian)

[229] Marchuk GI, Kagan BA. Internal gravity waves in a real stratified ocean. Izv Atmos Ocean Phys 1970; 6(4):236–241

[230] Martin S, Simmons W, Wunch C. The excitation of resonant triads by single internal waves. J Fluid Mech 1972; 53:17–44

[231] Mason PJ. Forces on spheres moving horizontally in rotating stratified fluid. Geophys Astrophys Fluid Dyn 1977; 8:137–154

[232] Matygin AS, Sabinin KD, Filonov AYe. Mean spatial spectra of internal tides at the hydrophysical Polygon-1970 in the Atlantic Ocean. Izv Atmos Ocean Phys 1982; 18(2):172–177

[233] Maxworthy T. On the formation of nonlinear internal waves from the gravitational collapse of mixed regions in two and three dimensions. J Fluid Mech 1980; 96:47–64

[234] McCorman RE, Mysak LA. Internal waves in a randomly stratified fluid. Geophys Fluid Dyn 1973; 4(3):243–266

[235] McEwan A. Degeneration of resonantly excited standing internal gravity waves. J Fluid Mech 1971; 50:431–448

[236] McEwan A, Mander D, Smith R. Forced resonant second-order interaction between damped internal waves. J Fluid Mech 1972; 55:589–608

[237] McGoldrick LF. Resonant interaction among capillary-gravity waves. J Fluid Mech 1965; 21:305–332

[238] McIntyre ME. Mean motions and impulse of a guided internal gravity wave packet. J Fluid Mech 1975; 60:801–812

[239] McLean JW. Instabilities of finite amplitude water waves. J Fluid Mech 1982; 114:315–341

[240] Merzkirch W, Peters F. Optical visualization of internal gravity waves in stratified fluid. Opt Lasers Eng 1992; 16:411–425

[241] Milder DM. Hamiltonian dynamics of internal waves. J Fluid Mech 1982; 119:269–282

[242] Miles JW. On the stability of heterogeneous shear flows. J Fluid Mech 1961; 10:496–508

[243] Miles JW. On the stability of heterogeneous shear flows. Pt. 2. J Fluid Mech 1961; 10:496–508

[244] Miles JW. On internal solitary waves. Tellus 1979; 31:456–462

[245] Miles JW. On internal solitary waves. Tellus 1981; 33:397–401

[246] Millero FJ. The physical chemistry of sea water. Annual Rev Earth and Planet Sci 1974; 2:101–150

[247] Miloh T, Tulin MP, Zilman G. Dead-water effects of a ship moving in stratified seas. J Offshore Mech Arctic Enging 1993; 115:105–110

[248] Miropol'sky YuZ. The effect of the microstructure of the density field in the sea on the propagation of internal gravity waves. Izv Atmos Ocean Phys 1972; 8(8):515–517

[249] Miropol'sky YuZ. Probability distribution of certain characteristics of internal waves in the ocean. Izv Atmos Ocean Phys 1973; 9(4):226–230

[250] Miropol'sky YuZ. Propagation of internal waves in an ocean with horizontal density field non-uniformities. Izv Atmos Ocean Phys 1974; 10(5):312–318

[251] Miropol'sky YuZ. On internal waves generation in the ocean by a wind field. Oceanology 1975; 10(3):389–396 (in Russian)

[252] Miropol'sky YuZ. Influence of shear flow on the generation of short-period internal waves in the ocean. Izv Atmos Ocean Phys 1975; 11(9):585–589

[253] Miropol'sky YuZ. Pulse propagation in a stratified rotating fluid. Izv Atmos Ocean Phys 1975; 11(12):819–823

[254] Miropol'sky YuZ. On the instability of weakly nonlinear waves in anisotropic dispersive media applied to internal gravitational waves and Rossby waves. Sov Phys Dokl 1975; 223(4):848–851

[255] Miropol'sky YuZ. Self-similar solutions of the Cauchy problem for internal waves in an unbounded fluid. Izv Atmos Ocean Phys 1978; 14(9):673–679

[256] Miropol'sky YuZ, Filyushkin BN. Temperature fluctuations in the upper ocean comparable in scale to internal gravity waves. Izv Atmos Ocean Phys 1971; 7(7):523–535

[257] Miropol'sky, Yu.Z., Filyushkin, B.N. "Some Statistical Charac-teristics of Isotherm Oscillations in the Presence of Sea Inter-nal Waves". In *Internal Waves in the Ocean*, Proceedings of the Russian-French Meeting on Internal Waves in the Ocean; 1971 June 8–11; Novosibirsk. Novosibirsk: Computing Center of the Siberian Brunch RAS, 1972 (in Russian).

[258] Miropol'sky YuZ, Neyman VG. Internal waves and temperature microstructure in the Timor Sea. Izv Atmos Ocean Phys 1974; 10:730–737

[259] Miropol'sky YuZ, Solntseva NI, Filyushkin BN. On horizontal vari-ability of the Brunt– Väisälä frequency in the ocean. Oceanology 1975; 15(1):25–32 (in Russian)

[260] Moiseev, Nikita, *Asymptotic Methods of Nonlinear Mechanics*. Moscow: Nauka, 1981 (in Russian).

[261] Monin AS. The hydrothermodynamics of the ocean. Izv Atmos Ocean Phys 1973; 9(10):602–604

[262] Monin, A., Kamenkovich, V. and Kort, V., *Variability of the Oceans*. New York: Interscience, 1997.

[263] Monin AS, Piterbarg LI. On the statistical description of internal waves. Sov Phys Dokl 1977; 234(3):564–567

[264] Monin AS, Fedorov KN, Shevtsov VP. On the vertical meso- and microstructure of ocean currents. Sov Phys Dokl 1973; 208(4):833–836

[265] Montgomery, RB. "Oceanographic Data." In: *American Institute of Physics Handbook*, New York et al.: McGraw-Hill, 1957.

[266] Moore D. Empirical orthogonal function — a non-statistical view. Mode News, 1974; 67:1–6

[267] Mowbray D, Rarity R. A theoretical and experimental investiga-tion of the phase configuration of internal waves of small amplitude in a density stratified liquid. J Fluid Mech 1967; 28(1):1–18

[268] Munk, Walter. "Internal Waves." In *Evolution of Physical Ocea-nography*, eds. Bruce A. Warren and Carl Wunsch, Cambridge-London: MIT Press, 1980.

[269] Nagovitsyn A, Pelinovsky E, Stepaniants Yu. Observation and analysis of solitary internal waves at the coastal zone of the Sea of Okhotsk. Sov J Phys Oceanogr 1991; 2(1):65–70

[270] Nayfeh, Ali Hasan, *Perturbation Methods*. New York: Interscience, 1973.

[271] Neiman VG. Pages of history of experimental research in the ocean physics in the Russian Academy of Sciences. Oceanology 1999; 39(5):654–660

[272] Neshyby S, Neal V, Denner W. Spectra of internal waves: *in situ* measurements in a multilayered structure. J Phys Oceanogr 1972; 2:91–95

[273] New AL, Pingree RD. Large amplitude internal soliton packets in the central Bay of Biscay. Deep-Sea Res 1990; 37:513–524

[274] Newberger PA, Allen JS. On the use of the Boussinesq equations, the reduced system, and the primitive equations for the computation of geophysical flows. Dyn Atmos Oceans 1996; 25(1):1–24

[275] Nikitina YeA. Drag of the ships in a "dead water". Izvestiya Akademii Nauk. Mechanika i Machinostroyeniye 1959; 1:188–192 (in Russian)

[276] *Ocean Acoustics*, ed. Leonid Brekhovskikh, Moscow: Nauka, 1974 (in Russian).

[277] Olbers, Dirk. 'Internal Gravity Waves." In *Oceanography Series*, ed. Jürgen Sündermann, Berlin: Springer, 1983.

[278] Olbers DJ, Herterich K. The spectral energy transfer from surface waves to internal waves. J Fluid Mech 1979; 92:349–380

[279] Ono H. Algebraic solitary waves in stratified fluids. J Phys Soc Japan 1975; 39:1082–1093

[280] Orlansky I. Energy spectrum of small scale internal gravity waves. J Geophys Res 1971; 76(24):5829–5835

[281] Orlansky I, Brian K. Formation of the thermoclinic step structure by large amplitude internal gravity waves. J Geophys Res 1969; 74(28):6975–6983

[282] Ostrovskii LA. Approximate methods in the theory of nonlinear waves. Radiophys and Quant Electron 1974; 17(4):344–360

[283] Ostrovskiy LA. On the cluster nature of internal waves dispersion in an ocean with periodic vertical structure. Izv Atmos Ocean Phys 1977; 13(7):529–530

[284] Ostrovsky LA. Nonlinear internal waves in a rotating ocean. Oceanology 1978; 18:119–125

[285] Ostrovskii LA, Rybak SA, Tsimring LSh. Negative energy waves in hydrodynamics. Sov Phys Usp 1986; 29:1040–1052

[286] Ostrovsky LA, Stepanyants YuA. Do internal solitons exist in the ocean? Rev Geophys 1989; 27:293–310

[287] Ostrovsky, L.A., Stepanyants, Yu.A. "Nonlinear Surface and Internal Waves in Rotating Fluids." In *Nonlinear Waves. 3. Physics and Astrophysics.* New York: Springer Verlag, 1990.

[288] Pao Y. Spectra of internal waves and turbulence in stratified fluids. Radio Sci 1969; 4(12):1315–1320

[289] Pekeris CL. Propagation of sound in the ocean. Geol Soc Amer Memor 1948; 27:1–117

[290] Pelinovskiy EN, Raevskiy MA. Weak turbulence of internal waves in the ocean. Izv Atmos Ocean Phys 1977; 13(2):130–134

[291] Pelinovskiy EN, Raevskiy MA, Shavratskiy SKh. The Korteweg-de Vries equation for nonstationary internal waves in an inhomogeneous ocean. Izv Atmos Ocean Phys 1977; 13(3):226–228

[292] Pelinovskiy EN, Romanova NI. Nonlinear stationary waves in the atmosphere. Izv Atmos Ocean Phys 1977; 13(11):804–807

[293] Pelinovskiy EN, Shavratskiy SKh. Propagation of nonlinear internal waves in the inhomogeneous ocean. Izv Atmos Ocean Phys 1976; 12(1):41–44

[294] Pelinovsky E, Stepanyants Yu, Talipova T. Modelling of the propagation of nonlinear internal waves in horizontally inhomogeneous ocean. Izv Atmos Oceanic Phys 1994; 30:77–83

[295] Peregrine DH. Interaction of water waves and currents. Adv Appl Mech 1976; 16:9–117

[296] Perelman TL, Fridman AKh, Elyashevich MN. On the relationship between the N soliton solution of the modified Korteweg–de Vries equation and the KdV equation solution. Phys Lett 1974; A47:321–323

[297] Perera JAM, Fernando HJS, Boyer D. Turbulent mixing at an inversion layer. J Fluid Mech 1994; 267:275–298

[298] Phillips OM. On the dynamics of unsteady gravity waves of finite amplitude. Part I. J Fluid Mech 1960; 9(2):193–217

[299] Phillips, Owen, *The Dynamics of the Upper Ocean*. Cambridge: Cambridge University Press, 1977.

[300] Phillips OM. Spectral and statistical properties of the equilibrium range in wind generated gravity waves. J Fluid Mech 1985; 156:505–531

[301] Pinettes M-J, Renouard D, Germain J-P. Analytical and experimental study of the oblique passing of a solitary wave over a shelf in a two-layer fluid. Fluid Dyn Res 1995; 16:217–235

[302] Pochapsky H. Internal waves and turbulence in the deep ocean. J Phys Oceanogr 1972; 2:96–103

[303] Pollard R. On the generation by wind of inertial waves in the ocean. Deep-Sea Res 1970; 17:795–812

[304] Pollard R, Millard R. Comparison between observed and simulated wind-generated inertial oscillations. Deep-Sea Res 1970; 17:813–821

[305] Prinsenberg SJ, Rattray M. Effect of continental slope and variable Brunt–Väisälä frequency on the coastal generation of internal tides. Deep Sea Res I 1975; 22:251–263

[306] Pullin DI, Grimshaw RHJ. Large amplitude solitary waves at the interface between two homogeneous fluids. Phys Fluids 1988; 31:3550–3559

[307] Raevsky MA. Gravity wave propagation at randomly inhomogeneous nonstationary currents. Izv Atmos Ocean Phys 1983; 19(6):639–645 (in Russian)

[308] Rarity BS. A theory of the propagation of internal gravity waves of finite amplitude. J Fluid Mech 1969; 39(3):497–510

[309] Rattray M. On the coastal generation of internal tides. Tellus 1960; 12(1):54–62

[310] Rattray M, Dworski JG, Kovala PE. Generation of long internal waves at the continental slope. Deep Sea Res I 1969; 16:179–195

[311] Renouard D, Germain J-P. Experimental study of long nonlinear internal waves in rotating fluid. Ann Geophysicae 1994; 12:254–264

[312] Romanova NN, Shrira VI. Explosive generation of surface waves by the wind. Izv Atmos Ocean Phys 1988; 24:528–535

[313] Sabinin KD. Determination of the parameters of internal waves using data from a towed chain of thermistors. Izv Atmos Ocean Phys 1969; 5(2):113–115

[314] Sabinin KD. Certain features of short period internal waves in the ocean. Izv Atmos Ocean Phys 1973; 9(1):32–36

[315] Sabinin KD, Nazarov AA, Filonov AYe. Internal wave trains above the Mascarene Ridge. Izv Atmos Ocean Phys 1992; 28(6):473–480

[316] Sabinin KD, Nazarov AA, Serikov AN. Association of short period internal wave trains with the relief of thermocline in the ocean. Izv Atmos Ocean Phys 1982; 18(4):416–425

[317] Sabinin KD, Shulepov VA. Short period internal waves in the Norway Sea. Oceanology 1965; 5(2):264–275 (in Russian)

[318] Samodurov AS. Internal waves in a medium with a horizontally varying Väisäla–Brunt frequency. Izv Atmos Ocean Phys 1974; 10(3):185–187

[319] Sandström H. Effect of topography on propagation of waves in stratified fluids. Deep-Sea Res 1969; 16(5):405–410

[320] Sandström H, Oakey NS. Dissipation in internal tides and solitary waves. J Phys Oceanogr 1995; 25:604–614

[321] Scotti RS, Corcos GM. Measurements of the growth of small disturbances in a stratified shear layer. Radio Sci 1969; 4(12):1309–1313

[322] Segur H, Hammack JL. Soliton models of long internal waves. J Fluid Mech 1982; 118:285–304

[323] Seliger RL, Whitham GB. Variational principles in continuum mechanics. Proc Roy Soc Lond 1968; A305(1480):1–25

[324] Serebryany AN. Internal waves in the coastal zone of a tidal sea. Oceanology 1985; 25(5):578–582

[325] Serebryany AN. Internal waves on a shelf and near the continental slope from data for a towed distributed temperature sensor. Oceanology 1987; 27(2):225–226 (in Russian)

[326] Serebryany AN. Nonlinearity effects in internal waves on a shelf. Izv Atmos Ocean Phys 1990; 26(3):206-212

[327] Serebryany AN. Steepening of the leading and back faces of solitary internal wave depressions and its connection with tidal currents. Dyn Atmos Oceans 1996; 23(1-4):393–402

[328] Shishkina OD. Wake regimes influence on hydrodynamic characteristics of a submerged sphere. Preprints of the Fourth International Symposium on Stratified Flows; 1994 29 June – 2 July; Grenoble. Grenoble: Institut de Mécanique de Grenoble, 1994.

[329] Shishkina OD. Comparison of the drag coefficients of bodies moving in liquids with various stratification profiles. Fluid Dyn 1996; 31:484–489

[330] Shishkina OD. Resonant generation of solitary wave in thermocline. Exp Fluids 1996; 21:374–379

[331] Smirnov, Vladimir, *Lehrgang der höheren Mathematik*. Berlin: Deutcher Verlag d Wiss, 1991.

[332] Smirnov VN. Oscillations of the ice cover caused by internal waves in the Arctic Ocean. Sov Phys Dokl 1972; 206(5):1106–1108

[333] Snyder, A., Love, J., *Optical Waveguide Theory*. London– New York: Chapman and Hall, 1983.

[334] Snyder WH, Thompson AS, Eskridge RE, Lawson RE, Castro IP, Lee JT, Hunt JCR, Ogawa Y. The structure of strongly stratified flow over hills: dividing streamline concept. J Fluid Mech 1985; 152:249–288

[335] Sobolev SL. On one new problem of mathematical physics. Izv Mat Sect 1954; 18(1):3–50

[336] Sonmor LJ, Klaasen GP. Higher order resonant instabilities of internal gravity waves. J Fluid Mech 1996; 324:1–23

[337] Spedding GR, Browand FK, Fincham AM. Turbulence, similarity scaling and vortex geòmetry in the wake of a towed sphere in a stably stratified fluid. J Fluid Mech 1996; 314:53–103

[338] Stamp AP, Jacka M. Deep water internal solitary waves. J Fluid Mech 1995; 305:347–371

[339] Staquet C, Sommeria J. Internal waves, turbulence and mixing in stratified flows: a report on Euromech Colloquium 339. J Fluid Mech 1996; 314:349–371

[340] Stevens C, Imberger J. The initial response of a stratified lake to a surface shear stress. J Fluid Mech 1996; 312:39–66

[341] Stevens C, Lawrence G, Hamblin P, Carmack E. Wind forcing of internal waves in a long narrow stratified lake. Dyn Atmos Oceans 1996; 24(1-4):41–50

[342] Stevenson TN, Kanellopulos D, Constantinides T. Two-dimensional internal waves systems generated by a cylinder. The phase configuration. Appl Sci Res 1986; 43:91-105

[343] Sutherland BR. Internal gravity wave radiation into weakly stratified fluid. Phys Fluids 1996; 8:430–441

[344] Sutherland BR, Dalziel SB, Hughes GO, Linden PF. Visualization and measurement of internal waves by 'synthetic schlieren'. Part 1. Vertically oscillating cylinder. J Fluid Mech 1999; 390:93–126

[345] Sutherland DR, Linden PF. Internal wave excitation from stratified flow over a thin barrier. J Fluid Mech 1998; 377:223–252

[346] Sysoeva EYa, Chashechkin YuD. Vortex systems in the stratified wake of a sphere. Fluid Dyn 1991; 26:544–551.

[347] Talipova T, Pelinovsky E, Lamb K, Grimshaw R, Holloway P. Cubic nonlinearity effects in the propagation of intense internal waves. Dokl Earth Sci Sect 1999; 365(2):241–244

[348] Tareev BA. Internal baroclinic waves in the flow past a bottom topography and their influence on processes of sediments production in the ocean. Oceanology 1965; 5(1):45–52 (in Russian)

[349] Tareev BA. On the dynamics of internal gravity waves in a continuously stratified ocean. Izv Atmos Ocean Phys 1966; 2(10):640–646

[350] Ter-Krikorov AM. On internal waves in a inhomogeneous fluid. Prikl Mat Mech 1962; 26(6):1067–1976 (in Russian)

[351] Ter-Krikorov AM. On the theory of steady waves in an inhomogeneous fluid. Prikl Mat Mech 1965; 29(3):440–452 (in Russian)

[352] Thorpe SA. On wave interaction in a stratified fluid. J Fluid Mech 1966; 24(4):737–751

[353] Thorpe SA. On the shape of progressive internal waves. Phil Trans Roy Soc London 1968; A263:563–614

[354] Thorpe SA. Experiments on the instability of stratified shear flows: immiscible fluids. J Fluid Mech 1969; 39(1):25–48

[355] Thorpe SA. Experiments on the instability of stratified shear flows: miscible fluids. J Fluid Mech 1971; 46(2):299–319

[356] Thorpe SA. Turbulence in stratified fluids: A review of laboratory experiments. Boundary Layer Met 1973; 5:95–119

[357] Thorpe SA. The excitation, dissipation and interaction of internal waves in deep ocean. J Geophys Res 1975; 80(3):329–338

[358] Thorpe SA. An experimental study of critical layers. J Fluid Mech 1981; 103:321–344

[359] Thorpe SA. The stability of statically unstable layers. J Fluid Mech 1994; 260:315–331

[360] Thorpe SA. Statically unstable layers produces by overturning internal gravity waves. J Fluid Mech 1994; 260:333–350

[361] Tolstoy, Ivan, *Wave Propagation*. New York et al.: McGraw Hill, 1973.

[362] Tolstoy, I., Clay, C., *Ocean Acoustics*. New York et al.: McGraw-Hill, 1966.

[363] Townsend AA. Excitation of internal waves in stably stratified atmosphere with considerable wind shear. J Fluid Mech 1968; 32(1):145–171

[364] Trofimov MYu. Calculations of normal modes of internal waves on a weakly sheared flow. Izv Atmos Ocean Phys 2000; 36(2):294–301 (in Russian)

[365] Tsutahara M, Hashimoto K. Three-wave resonant interactions and multiple resonances in two-layer and three-layer flows. Phys Fluids 1986; 29:2812–2818

[366] Tung K-K, Chan TF, Kubota T. Large amplitude internal waves of permanent form. Stud Appl Math 1982; 66:1–44

[367] Turner, John, *Buoyancy Effects in Fluids*. Cambridge et al.: Cambridge University Press, 1973.

[368] Turner REL, Vanden-Broeck J-M. The limiting configuration of interfacial gravity waves. Phys Fluids 1986; 29:372–385

[369] Turner REL, Vanden-Broeck J-M. Long periodic internal waves. Phys Fluids 1992; A4(9):1929–1935

[370] Unna PJH. Waves and tidal streams. Nature 1942; 149:219–220

[371] Väisälä V. Über die Wirkung der Windschwankungen auf die Pilotbeobachtungen. Soc Sci Finnica, Comm Phys Math 1925; 2(19):1–46

[372] Vlasenko VI. Multimodal soliton of internal waves. Izv Atmos Ocean Phys 1994; 30(2):161–169

[373] Voisin B. Internal wave generation in uniformly stratified fluids. Part 1. Green's function and point sources. J Fluid Mech 1991; 231:439–480

[374] Voisin B. Internal wave generation in uniformly stratified fluids. Part 2. Moving point sources. J Fluid Mech 1994; 261:333–347

[375] Volkov AP, Fedorov KN, Shevtsov VP. Ocean current velocity sounding by a crossed beam method. Izv Atmos Ocean Phys 1975; 11(2):109–117

[376] Volosov, Vladimir, *Asymptotic Methods of Nonlinear Waves Study in a Stratified Medium with Application to the Ocean Internal Waves Theory.* Moscow: MGU, 1972 (in Russian).

[377] Voorhis AD. Measurements of vertical motion and the partition of energy in the New England slope water. Deep-Sea Res 1968; 15(4):599–608

[378] Voorhis AD, Perkins HT. the spatial spectrum of shortwave temperature fluctuations in the near surface thermocline. Deep-Sea Res 1966; 13(4):641–654

[379] Voronovich AG. Resonance three-wave interaction of the internal waves. Oceanology 1975; 10(5):773–780 (in Russian)

[380] Voronovich AG. Propagation of internal and surface gravity waves in the approximation of geometrical optics. Izv Atmos Ocean Phys 1976; 12(8):519–523

[381] Voronovich AG. Hamiltonian formalism for internal waves in the ocean. Izv Atmos Ocean Phys 1979; 15(1):58–62

[382] Voronovich AG, Leonov AI, Miropol'sky YuZ. Contribution to the theory of formation of the fine structure of hydrophysical fields in the ocean. Oceanology 1976; 16(5):431–436

[383] Voronovich AG, Rybak SA. Explosive instability of stratified flows. Sov Phys Dokl 1978; 239(6):1457–1460

[384] Voronovich VV, Pelinovsky DE, Shrira VI. On internal wave – shear flow resonance in shallow water. J Fluid Mech 1998; 354:209–237

[385] Vosper SB, Castro IP, Snyder WH, Mobbs SD. Experimental studies of strongly stratified flow past three-dimensional orography. J Fluid Mech 1999; 390:223–249

[386] Watson G. Measurements of the internal wave wake of a ship in a highly stratified sea loch. J Geophys Res 1992; 97(C6):9689–9703

[387] Watson KM. The coupling of surface and internal gravity waves: revisited. J Phys Oceanogr 1990; 20:1233–1248

[388] Watson KM, West BJ, Cohen BI. Coupling of surface and internal gravity waves: a mode coupling model. J Fluid Mech 1976; 77:185–208

[389] Webster F. Turbulence spectra in the ocean. Deep-Sea Res 1969; 16(2):357–368

[390] Whitham GB. Two-timing variational principles and waves. J Fluid Mech 1970; 44(2):373–396

[391] Whitham, Gerald Beresford, *Linear and Nonlinear Waves*. Chichester–Sussex: John Wiley, 1974.

[392] Woods JD. Wave induced instability in the summer thermocline. J Fluid Mech 1968; 32(4):791–800

[393] Woods JD. An investigation of some physical processes associated with the vertical flow of heat through the upper ocean. Met Mag 1968; 97:65–72

[394] Wu J. Mixed region collapse with internal wave generation in a density stratified medium. J Fluid Mech 1969; 35(4):531–544

[395] Wunsch C. Progressive internal waves on slopes. J Fluid Mech 1969; 35(1):131–144

[396] Wunsch C. On oceanic boundary mixing. Deep-Sea Res 1970; 17(2):293–301

[397] Wunsch C. Deep ocean internal waves: what do we really know. J Geophys Res 1975; 80(3):339–343

[398] Wunsch C, Dahlen J. Preliminary results of internal wave measurements in the main thermocline at Bermuda. J Geophys Res 1970; 75(30):5899–5908

[399] Wunsch C, Hendry R. Array measurements of the bottom boundary layer and the internal wave field on the continental slope. Geophys Fluid Dyn 1972; 4(2):101–144

[400] Xing J, Davies AM. A three-dimensional model of internal tides on the Malin-Hebrides Shelf and shelf edge. J Geophys Res 1998; 103(C12):27821–27847

[401] Yampol'sky AD. On internal waves in the North-East part of the Atlantic Ocean. Transactions of the Shirshov Oceanology Institute RAS 1962; 56:229–240 (in Russian)

[402] Yanovitch M. Gravity waves in a heterogeneous incompressible fluid. Comm Pure Appl Mathem 1962; 15(1):45–52

[403] Yih, Chia-Shun, *Stratified Flows*. New York etc.: Academic Press, 1980.

[404] Zabusky NJ, Kruskal MD. Interaction of "solitons" in a collisionless plasma and the reccurence of initial states. Phys Rev Lett 1965; 15(6):240–243

[405] Zakharov VE. Weak turbulence in media with a decay spectrum. J Appl Mech Tech Phys 1965; 4:22–24

[406] Zakharov VE. Stability of periodic waves of finite amplitude on the surface of a deep fluid. J Appl Mech Tech Phys 1968; 9(2):190–194

[407] Zakharov VE. The Hamiltonian formalism for waves in nonlinear media having dispersion. Radiophys and Quant Electron 1974; 17(4):326–343

[408] Zakharov, Vladimir. "Method of Inverse Problem of Scattering Theory." In *Elastic Media with Microstructure*, ed. Isaak Kunin, Berlin: Springer Verlag, 1982.

[409] Zakharov VE, Faddeev LD. The Korteweg–de Vries equation — the quite integrable Hamiltonian system. Functional Analysis and Its Applications 1971; 5(4):18–27 (in Russian).

[410] Zakharov VE, Filonenko NN. Weak turbulence of capillary waves. J Appl Mech Tech Phys 1967; 8(5):37–40

[411]   Zakharov, V.E., L'vov, V.S., *Statistical Description of Nonlinear Wave Fields.* Moscow: Landau Institute of Theoretical Physics, Chernogolovka, 1975 (in Russian).

[412]   Zakharov VE, Shabat AB. An exact theory of two-dimensional self-focusing and of one-dimensional self-modulation of waves in nonlinear media. Sov Phys JETP 1970; 61(1):118–134

[413]   Zakharov VE, Shabat AB. On solitons interaction in a stationary medium. Sov Phys JETP 1973; 64(5):1627–1639

[414]   Zakharov VE, Kusnetsov EA. Hamiltonian formalism for nonlinear waves. Physics – Uspekhi 1997; 40(11):1087–1116

[415]   Zalkan R. High frequency internal waves in the Pacific Ocean. Deep-Sea Res 1970; 17(1):91–108

[416]   Zeilon N. Experiments on boundary tides. Medd Göteborgs Högskolas Oceanogr Inst 1934; 8.

[417]   Zhou X, Grimshaw R. The effect of variable currents on internal solitary waves. Dyn Atmos Oceans 1989; 14:17–39

Author Index

393

Topic Index